T0351113

ARTIFICIAL INTELLIGENCE, ETHICS AND THE FUTURE OF WARFARE

This volume examines how the adoption of AI technologies is likely to impact strategic and operational planning, and the possible future tactical scenarios for conventional, unconventional, cyber, space and nuclear force structures. In addition to developments in the USA, Britain, Russia and China, the volume also explores how different Asian and European countries are actively integrating AI into their military readiness. It studies the effect of AI and related technologies in training regimens and command structures. The book also covers the ethical and legal aspects of AI augmented warfare.

The volume will be of great interest to scholars, students and researchers of military and strategic studies, defence studies, artificial intelligence and ethics.

Kaushik Roy is Guru Nanak Chair Professor in the Department of History, Jadavpur University, Kolkata, India. He has authored several books from Cambridge University Press, Oxford University Press, Routledge, Springer, etc. and numerous articles in several peer-reviewed journals on the military history of Eurasia. At present, he is associated with PRIO's 'Warring with Machines' project.

ARTIFICIAL INTELLIGENCE, ETHICS AND THE FUTURE OF WARFARE

Global Perspectives

Edited by Kaushik Roy

Routledge
Taylor & Francis Group

LONDON AND NEW YORK

Designed cover image: Gettyimages/grandeduc

First published 2024
by Routledge
4 Park Square, Milton Park, Abingdon, Oxon OX14 4RN

and by Routledge
605 Third Avenue, New York, NY 10158

Routledge is an imprint of the Taylor & Francis Group, an informa business

British Library Cataloguing-in-Publication Data
A catalogue record for this book is available from the British Library

ISBN: 978-1-032-63560-6 (hbk)
ISBN: 978-1-032-72952-7 (pbk)
ISBN: 978-1-003-42184-9 (ebk)

DOI: 10.4324/9781003421849

Typeset in Sabon
by Deanta Global Publishing Services, Chennai, India

CONTENTS

FIGURES

CONTRIBUTORS

Guru Saday Batabyal (Retd. Colonel), PhD, FRHistS, has 44 years of cross-functional experience in the Indian Army, United Nations, Industry as CEO/Executive Director of IT companies and in academia. He is the author of many articles, book chapters and a book titled *Politico-Military Strategy of Bangladesh Liberation War, 1971* (2020). As a visiting professor to various universities, he teaches 'Theory of War, International Relations, and Modern History of South Asia.'

Kelly Fisher is a Research Assistant at the Peace Research Institute of Oslo (PRIO), and he received his MA in Gender Studies from the University of Oslo. As a Research Assistant at PRIO, Kelly works on a number of projects relating to ethical challenges of new and emerging technologies, including artificial intelligence, where he focuses on the gendered dimension of these technologies.

Ajey Lele (Retd. Group Captain) had served in the Indian Air Force. Now he is a Security Analyst based in New Delhi, India. His areas of research include weapons of mass destruction, strategic technologies and space security. He has a Master's degree in Physics and his PhD is in International Relations. He has various publications to his name. His recent book is titled *Quantum Technologies for The Militaries* (2021).

P.K. Mallick, VSM (Retd. Major-General) is an engineering graduate from Bengal Engineering College, Shibpore and MTech from IIT, Kharagpur. He is a Post Graduate in Automation and Control Engineering from the

Indian Institute of Technology, Kharagpur and has an MPhil from Madras University. He was commissioned in the Corps of Signals of Indian Army. In his 35 years of service, he had wide experience in command, staff and instructional appointments. He has taken part in Counter Insurgency Operations in Kashmir Valley, Assam and Punjab. He had been Chief Signals Officer of the Srinagar-based Corps and Central Command. As officer he had been an instructor at the Military College of Tele Communication Engineering, Mhow and Senior Directing Staff at National Defence College New Delhi from where he retired. Presently, he holds the COAS Chair of Excellence at Centre of Land Warfare Studies. He has been a prolific writer in peer-reviewed journals, a regular speaker in seminars and discussions and a member of various Task Forces.

R.S. Panwar (Retd. Lieutenant-General) was commissioned into the Indian Army Corps of Signals in 1976. He has commanded two logistics formations, an electronic warfare brigade and an armoured division signal regiment, as well as the Military College of Telecommunication Engineering. He has the distinction of being awarded MTech and PhD degrees in computer science, as well as the coveted Distinguished Alumnus Award by IIT Bombay, the first defence officer to hold such an honour. He has attended a one-semester Cryptology course at IIT Delhi, holds a Master's degree in Management from Osmania University, and is a graduate of the National Defence College. Panwar has vast experience in managing communications and electronic warfare systems and projects. He has done noteworthy R&D work as well, including the development of sophisticated data communication equipment for tactical defence networks. For his distinguished services, he has been awarded thrice by the President and also by the Ministry of Defence. He is presently a Distinguished Fellow at the USI and is engaged in evolving strategic thought in the areas of information operations, network centric warfare and emerging disruptive technologies. He is also part of an international group of experts which is working under the aegis of the Centre for Humanitarian Dialogue, Geneva, to evolve a code of conduct for AI-enabled military systems.

Swaminathan Ramanathan works at the intersection of academia and industry. He is a researcher and lecturer at Uppsala University and a consultant specialising in strategy and business design for futureproofing organisations from a systems perspective.

Gregory M. Reichberg is a Research Professor at PRIO, where he leads 'Warring with Machines: Artificial Intelligence and the Relevance of Virtue Ethics,' a multi-year project funded by the Research Council of Norway. A philosopher by training, he has published widely on military ethics, including the impact of artificial intelligence on the conduct of combatants. He has authored and co-edited several volumes, most recently *Robotics, AI, and Humanity* (2021) and *Thomas Aquinas on War and Peace* (2017).

Kaushik Roy is Guru Nanak Chair Professor, in the Department of History, Jadavpur University, Kolkata, India. He is at present associated with PRIO's 'Warring with Machines' project. He specialises in Eurasian military history. He has published numerous articles in peer reviewed journals and has many books to his credit.

Kritika Roy is a Threat Intelligence Researcher at the Deutsche Cyber Security Organisation (DCSO GmbH) Berlin, where she conceptualises and contributes to a long-term cyber threat landscape research. She was formerly working for MP-IDSA as a Research Analyst and was also a Researcher for the Centre for Land and Warfare Studies (CLAWS). She holds an MA in Geopolitics and International Affairs and is also pursuing another Master's degree in International Affairs at the Hertie School, Berlin. Her research interests include AI and weapons of mass destruction (WMD), cyberthreats and security, and the impact of disruptive technologies on WMDs. She has also authored the book titled *Advances in ICT and the Likely Nature of Warfare* (Routledge) and the article titled 'Rationales for introducing artificial intelligence into India's military modernization programme' in *The Impact of Artificial Intelligence on Strategic Stability and Nuclear Risk*.

A.K. Sachdev (Retd. Group Captain) has operational experience in the Indian Air Force (IAF) and in civil aviation. He commanded a combat squadron and a flying station of the IAF, was Chief Operations Officer at two IAF stations, and was Commodore Commandant of a Helicopter Unit. He was Chief Operations Officer of an airline and the Chief Executive of an air charter company before co-founding a business aviation company. He is a Performance Instructor on the Airbus A 320. He is a life member of Aeronautical Society of India, Rotary Wing Society of India, All India Management Association, United Services Institution and Manohar Parrikar Institute for Defence Studies and Analyses. He was on the faculty of Defence Services Staff College, Wellington, India for three years and is an Accredited Management Teacher (All India Management Association). He has an MA, an MSc, an MBA, and an MPhil degree and is a former Senior Research Scholar from IDSA. He has published one book, one monograph and numerous articles in professional military and aviation journals.

Henrik Syse is a Research Professor at PRIO (Peace Research Institute Oslo) and a Professor of Peace and Conflict Studies at Oslo New University College. A philosopher by training, he has written and edited books and articles on political philosophy, ethics, peace and religion. He holds a Master of Arts degree from Boston College and a Doctorate from the University of Oslo.

Gabriel Udoh is currently a DAAD-Scholar and PhD Candidate at the Europa Universität Viadrina, Frankfurt (Oder). His research area is in the ethics and laws of AI, especially lethal autonomous weapon systems in armed conflicts. He has a Master's degree in International Humanitarian Law and certifications in artificial intelligence, and is also a specialist in programming with Python.

PREFACE AND ACKNOWLEDGEMENTS

An 'information age' is fast replacing the industrial age and artificial intelligence (AI) is a key driver in this process. Rapid advance in information and communication technologies, especially the processing power of the computers, machine learning, big data and the breakthrough in the field of robotics are the causative factors behind the massive stride of machine intelligence or AI. AI is already reshaping the civilian sector. The military domain is likewise undergoing a dramatic shift. Over the next 10 or 20 years, AI-enhanced weapon systems will likely transform the conduct of warfare. The USA, China and Russia are already integrating AI within their military postures. Following the lead of the USA and China, several countries like Russia and Israel are formulating new strategic doctrines with AI as the centrepiece. NATO has recently announced that it will do the same. Where do other countries of Asia and Africa, and their armed forces stand in this regard?

Strangely not much is discussed within the Indian academia about the impact and the future role of AI-related technologies in military affairs. On 3–4 January 2022, Peace Research Institute Oslo (PRIO) organised a conference at The LaLit Great Eastern Hotel in Kolkata, India. This conference is possible due to the moral and material assistance provided by Professor Gregory Reichberg of PRIO who is leading the project titled 'Warring with Machines: Artificial Intelligence and the Relevance of Virtue Ethics,' a multi-year project funded by the Research Council of Norway. The conference was attended by senior-ranking retired military officers from the Indian armed forces, scientists, engineers, philosophers, diplomats and academics -who belonged to the streams of history, sociology and political science from all over the world. The participants discussed not merely AI technologies, but also innovative doctrines, command structures, and new tactics and training which are necessary for integrating AI within the existing politico-military formats. Adoption of AI in the civilian-military sectors will change not only intra-human but also machine-human interactions which will give rise to new identities and cultural values. Several participants also discussed the emerging ethical and legal issues associated with the deployment of 'killer robots' and the use of machine augmented decision-making in future conflicts. Some of the papers which were presented at that conference are revised and the result is this edited volume.

First, let us be clear that this edited collection is not dealing with the political, psychological and economic effects of AI-enabled weapons. Rather our focus is on the strategic, tactical, doctrinal and ethical challenges resulting from the induction of AI-enabled weapon systems. This edited volume will examine how the adoption of AI technologies is likely to impact strategic planning and the command culture for conventional, unconventional, cyber, space and nuclear force structures. While much has already been written on the USA and China, how different Asian and European countries are actively integrating AI into their military readiness is relatively understudied. This volume will accordingly pay special attention to the role of AI within the military planning of these areas. By way of comparison, the adoption of AI-based technology by different military organisations and the consequent transformation of tactics and training will be considered. How are different countries reconfiguring their doctrines and command structures to benefit from the introduction of AI and related technologies? What are the perceived advantages of this technology? In what measure have the attendant risks and wider ethical concerns been envisioned? In raising these questions, the volume does not merely aim to study specific technologies. Rather the goal is to contextualise the position of different Asian and European countries within the broader matrix of global AI developments. While some of the essays of this volume focus on particular geographical areas, others take a global approach. The volume brings together a mix of scientists, engineers, military officers and academics to discuss how AI will transform warfare in the years to come.

Finally, special thanks to CENERS-Kolkata Branch, Lieutenant-General (retired) Arun Roye, Professor Reichberg, Kelly Fisher, a researcher of PRIO, Dr Moumita Chowdhury, Ms Sohini Mitra and Mr Arka Choudhury of the Department of History, Jadavpur University for making the conference possible. Without their cooperation, the conference and this volume would not have been possible.

Kaushik Roy
Guru Nanak Chair Professor, Jadavpur University, Kolkata, India
Kolkata July 2023

ABBREVIATIONS

ABMS: Advanced Battle Management System
AFADS: Armed Fully Autonomous Drone Swarms
AGI: Artificial General Intelligence (Same as General Artificial Intelligence)
AI: Artificial Intelligence
ANI: Artificial Narrow (Weak) Intelligence (Same as Narrow Artificial Intelligence)
ASAT: Anti-Satellite Weapons
ASI: Artificial Superintelligence
ATRS: Automated Target Recognition System
AWS: Autonomous Weapon System
BCI: Brain-Computer Interface
BMD: Ballistic Missile Defence
BMS: Battlefield Management System
BVR: Beyond Visual Range
CBRNE: Chemical, Biological, Radiological, Nuclear and Explosives
C2:   Command and Control
C4ISR: Command, Control, Communications, Computers, Intelligence, Surveillance and Reconnaissance
C4I2SR: Command, Control, Communications, Computers, Information, Intelligence, Surveillance and Reconnaissance
CCW: Convention on Conventional Weapons
CIA: Central Intelligence Agency
COIN: Counterinsurgency
DARPA: Defence Advanced Research Projects Agency

DDOS:	Distributed Denial of Service
DE:	Directed Energy
DICON:	Defence Industries Corporation of Nigeria
DL:	Deep Learning
DNS:	Domain Name System
DOD:	Department of Defence
DOS:	Denial of Service
DRDO:	Defence Research and Development Organisation
EC:	European Commission
EMP:	Electromagnetic Pulse
EMS:	Electromagnetic Spectrum
EP:	European Parliament
EU:	European Union
EW:	Electronic Warfare
FATA:	Federally Administered Tribal Areas in Pakistan
F2T2EA:	Find, Fix, Track, Target, Engage, and Assess
GAI:	General Artificial Intelligence
GEO:	Geosynchronous Earth Orbit
GPS:	Global Positioning System
HEO:	Highly Elliptical Orbit
HLMI:	Human Level Machine Intelligence
HMC:	Human-Machine Collaboration
HMT:	Human-Machine Team
IAF:	Indian Air Force
IAI:	Israel Aerospace Industry
ICRC:	International Committee of the Red Cross
ICT:	Information and Communication Technologies
IDF:	Israel Defence Force
IDP:	Internally Displaced Person
IED:	Improvised Explosive Device
IHL:	International Humanitarian Law
IOMT:	Internet of Military Things
ISI:	Inter-Services Intelligence of Pakistan
ISR:	Intelligence, Surveillance and Reconnaissance
IT:	Information Technology
LAWS:	Lethal Autonomous Weapon System
LEO:	Low Earth Orbit
LLM:	Large Language Model
LOAC:	Laws of Armed Conflict
MAD:	Mutually Assured Destruction
MALE:	Medium Altitude Long Endurance
MEO:	Medium Earth Orbit

MHC: Meaningful Human Control
ML: Machine Learning
MTR: Military Technical Revolution
NATO: North Atlantic Treaty Organization
NBC: Nuclear, Biological and Chemical
NCW: Net Centric Warfare
NGAD: Next Generation Air Dominance
NGO: Non-Governmental Organisation
NLP: Natural Language Processing
NUDET: Nuclear Detonation
OEM: Original Equipment Manufacturer
OODA: Observe, Orient, Decide and Act
PAP: Provincial Armed Police of India
PAV: Pilotless Aerial Vehicle
PC: Platform centric
PGM: Precision Guided Munition
PLA: People's Liberation Army of China
POK: Pakistan Occupied Kashmir
PRIO: Peace Research Institute Oslo
RAAF: Royal Australian Air Force
REAM: Reactive Electronic Attack Measures
RMA: Revolution in Military Affairs
ROE: Rules of Engagement
SEAD: Suppression of Enemy Air Defence
SIA: Self-Indication Assumption
SSA: Self-Sampling Assumption
SSSA: Strong Self Sampling Assumption
SUAS: Small Unmanned Aerial System
TSK: Turkish Armed Forces
UAV: Unmanned Aerial Vehicle
UGS: Unmanned Ground System
UGV: Unmanned Ground Vehicle
UKMOD: United Kingdom Ministry of Defence
UMS: Unmanned Maritime System
UMV: Unmanned Marine/Maritime Vehicle
UN: United Nations
UNSC: United Nations Security Council
USAF: United States Air Force
USDOD: United States Department of Defense
USN: United States Navy
USV: Unmanned Surface Vessel
UUS: Unmanned Underwater System

UUV:	Unmanned Underwater/Undersea Vessel
VAS:	Vehicle Automation System
WMD:	Weapon of Mass Destruction
XLUUV:	Extra Large Unmanned Underwater Vehicle

INTRODUCTION

Intelligent autonomous war machines and the future of warfare

Kaushik Roy

What is an intelligent military machine?

The term artificial intelligence (AI) was first used as a title for the conference held at Dartmouth College in 1955. Alan Turing's (a mathematician and a pioneer in computer science) 1950 essay titled 'Computing Machinery and Intelligence' can also be taken as a signpost for the birth of the term AI.[1] AI is machines that can generate behaviour similar to humans, such as making choices and passing judgements. The rise of AI is part of the 'Information Revolution.' Information is a crucial integer of material productivity in the era of Information Revolution. In this era, the generation of new knowledge is dependent on computing power. Basic services like banking, railways, air transportation, water and energy distribution are becoming more and more dependent on information and communication technologies (ICT). The principal sources of energy include digital information, computer software and computerised communications. The computer with its electronic brain is supplementing (might supplant in the distant future) human thinking. The integrated circuits and computer technology are the lifeblood of AI, and these two technologies are continuously developing.[2] However, AI is a combination of technologies which includes both hardware and software. So, an AI system includes hardware like mainframe computers, microchips, sensors integrated with software as programs, algorithms, etc. The fundamental building blocks of AI are learning algorithms[3] which allow the machine to detect patterns in data, make decisions, adjust their behaviour in relation to the environment. This learning process is called machine learning (ML) and its subset is deep learning (DL).[4] We will discuss these two terms later in detail. The production of computers with increasing processing powers,

DOI: 10.4324/9781003421849-1

the internet, Internet of Things (IoT), iCloud, Facebook, X (formerly known as Twitter), advances in robotics, synthetic biology, neuroscience and nano-technology are making possible the use of AI in the non-military and military sectors. So, AI can be simplistically grouped as a symbiosis of related technologies centring around ICT. The result is that weapons are being informationised and intellectualised.

The field studying the effect of AI on warfare is overloaded with acronyms, abbreviations and conceptual categories. Social scientists argue that a tight and static definition of a concept or a category will result in the death of social sciences. Nevertheless, a working definition of various terms, categories and concepts is necessary. The understanding of what constitutes autonomous intelligent war machines differs among people and regions.

The UK in a Joint Doctrine Note published in 2011 differentiated between automated/automatic systems and autonomous systems (which have elements of AI in them). An automated or automatic system follows a predefined set of rules and results in predictable outcomes. One example is the Phalanx Anti-Ship Missile Defence System. In contrast, an autonomous system is capable of understanding higher-level intent and direction. Along with a sophisticated perception of its environment, such a system is capable of taking a course of action from among a number of possible alternatives, without depending on human oversight or control. Overall, the activity of an autonomous system is predictable though individual actions may not be, as the system does not merely follow a pattern of rules in a predictable way.[5] Actually, the UK Ministry of Defence's (MOD) 2011 Note is equating Narrow or Weak AI (Artificial Narrow Intelligence/ANI) with automatic systems, semi-autonomous machines and fully autonomous machines (part of Strong or General AI[GAI]) as a truly autonomous intelligent system.

Paul Scharre, a former US Army Ranger turned military AI expert, writes that as machines become more sophisticated, they are more capable and able to execute complex tasks in more open-ended environments. Scharre classifies AI on the basis of its intelligence and autonomy into three classes: Automatic, automated and autonomous. An automatic machine is simple threshold-based, and an automated one is more complex rule-based. An automatic system senses the environment and acts. It does not exhibit much in the way of decision-making, and its actions are predictable. An automated system, in contrast, may consider a variety of inputs and weighs several variables before taking an action. An autonomous machine is goal-orientated and self-directed. Goal-orientated suggests that the human user specifies the goal to be achieved, but how it is achieved is decided by the machine. Its internal cognitive processes are less intelligible to its handlers. So, the autonomous machine's specific action to achieve a goal may be less predictable even to its operators.[6] Hence, in Scharre's framework, at one pole is the automatic machine, which is least autonomous and minimally intelligent.

The automated one is at the mid-spectrum of intelligence and autonomy. At the extreme pole is the autonomous machine, which is most intelligent and to a great extent unpredictable in its behaviour.

When we think of AI, robots and drones come to our mind. The Czech author Karel Capek in 1921 in his play *Rostum's Universal Robot* used the word robot. The word robot is derived from the Czech word *robota* meaning forced labour.[7] Paul J. Springer highlights that there is a difference between a computer program and a robot. Unlike a computer program, a robot (or a drone for that matter) is capable of some form of direct action in the physical world.[8] So, robots and drones are cyber-physical composite systems which comprise both physical and software components.

From the French perspective, a military robot is defined as a system which possesses capabilities of perception, communication, decision-making and action. A robot might operate under human supervision but is also capable of improving its own performance through automatic learning. So, a robot is a machine that can sense, think and act.[9] Daniel Ichbiah, a robotics expert, defines a robot as 'a very powerful computer with equally powerful software housed in a mobile body and able to act rationally on its perception of the world around it.'[10]

The Russian military encyclopaedia divides robots into three generations The first-generation robots have software and remote control and can operate only in an organised environment. These robots can be taken as examples of ANI. The second-generation robots are adaptive and possess some kind of sensory organ and are able to cooperate in previously unknown conditions, meaning that these machines are able to adapt to a changing environment. The third-generation robots are intelligent and have elements of AI in their control systems. One Russian policy analyst believes that the Russian military considers only the third-generation robots as Autonomous Weapon Systems (AWS).[11] Those AWS, which could identify and track a target and fire without human supervision, can be labelled as Lethal Autonomous Weapon Systems (LAWS).

Robots are used both in the civil and military sectors. One author asserts that in 2013, the global population of robots exceeded 10 million.[12] According to one calculation in 2008, the US military had more than 11,000 robots.[13] Military robots could operate in the air (Unmanned Aerial Vehicle [UAV, also popularly known as drone]), on the ground (Unmanned Ground Vehicle [UGV]), on the surface of the sea and also under the sea (Unmanned Underwater Vessel [UUV]). UAVs could be armed (as a hunter-killer platform) or unarmed for intelligence, surveillance and reconnaissance (ISR). Armed UAVs are known as Unmanned Combat Aerial Vehicles (UCAVs).

Some examples of these types of unmanned systems can be given. First, let us give examples of some types of UGVs. Israel is developing a 2 metre (m) snake robot which could crawl along narrow tunnels. This robot's head is made up of sensors and cameras that send information back to the human teams operating in the rear.[14] Ripsaw M5 is a remote-controlled unmanned battle tank of the US Army. It will be able to cooperate (share data) with UAVs, manned tanks and armoured vehicles. Ripsaw M5 can function both during daytime and night. AI technologies embedded into its software reduce the necessity of constant remote control. The ML algorithms in its software allow this tank to learn from its environment in real time. This tank could function autonomously to a significant extent.[15]

UAVs could be big or very small in size. AeroVironment has manufactured micro aerial vehicles like Black Widow. It is 15.2 centimetres (cm) in width, weighs 80 grams (g) and can fly for 30 minutes while providing a video downlink to its handler. Nano Hummingbird is a micro drone which is similar in shape to a hummingbird and flies like it. It is equipped with a small camera. Such UAVs are designed for urban reconnaissance and covert operations, especially in counterinsurgency (COIN) campaigns.[16] One of the most talked-about drones of the US military is the Predator. MQ 1B Predator is an armed, multi-mission, medium-altitude, long-endurance, remotely piloted aircraft. Due to its sensors, multi-mode communication suits, precision weapons (Hellfire missiles) and capability for loitering over the target for a long period, Predator is used for ISR, close air support, search and rescue, and precision strike both in regular war and irregular warfare.[17]

There are both UUVs and USVs (Unmanned Surface Vessels). USVs could be Large Unmanned Surface Vessel (LUSV) and Medium Unmanned Surface Vessel (MUSV). Unmanned Underwater Systems (UUS) could be Ultra/Extra Large Unmanned Underwater Vessel (XLUUV), ordinary UUV and micro UUV. In 2017, Russia displayed two small UUVs named Amulet and Juno, respectively. Juno, equipped with sensor technologies, is 2.9 m in length, 200 millimetres (mm) in diameter and weighs 80 kilograms (kg). Juno has a diving depth of 1,000 m and can operate for six hours. Amulet is 1.6 m in length, weighs 25 kg and has a diving depth of 50 m and a range of 15 kilometres (km). Vladimir Putin in 2018 disclosed Russia's possession of Poseidon, a strategic UUV. Russia's Poseidon UUV possesses a nuclear power device for propulsion. Poseidon has a diameter of 2 m, a maximum diving depth of 1,000 m and its operating range is 10,000 km. This UUV can carry a warhead with an explosive equivalent to over 10 million tonnes of TNT. Generally, Poseidon will be launched from a nuclear submarine for independent operation deep under the enemy sea or ocean. In 2019, Singapore displayed a vessel which could operate both as a USV and, when required as a UUV. It is a fast attack ship named Seekrieger and is armed with torpedoes for underwater attacks and guns for conducting attacks on the surface.[18] So, Seekrieger has shape-shifting capabilities.

The 11 essays in this edited collection by historians, philosophers, sociologists, engineers and retired military officers deal with the issue of the impact of AI on the logic and grammar of warfare. We are not dealing with the political, psychological and economic effects of AI-enabled weapons. The contributors tackle the problem of how far the use of AI-augmented war machines square with morality and legality. While the bulk of the essays takes a global perspective, some of the chapters follow a region-specific approach. Most of the essays deal with the four issues discussed below. These four issues are: Whether the introduction of AI in military affairs is causing a Military Revolution or a Revolution in Military Affairs (RMA), the challenges of command and control of the AI-augmented weapon systems, the probable use of AI for warfare in cyberspace and outer space, and finally, the legality and ethicality of using military AI. All the essays in this volume deal with the state of AI which is available at present, that is, ANI. In the last section of this Introduction, I delve into the possibility of the emergence of GAI (equivalent to Human Level Machine Intelligence [HLMI]) and superintelligence in the near future and the impact it might have on the conduct of war in particular and the survival of humanity in general.

AI in warfare: A military revolution or an RMA?

Can we say that the induction of AI-embedded weapons (like robots, drones, etc.) is causing a Military Revolution or an RMA? The term RMA became fashionable among American security analysts in the 1990s. RMA is a narrow concept compared to the much broader notion of Military Revolution. RMA, which is similar to the Soviet concept of Military Technical Revolution (MTR), refers to the introduction of a new type of weapon system which has massive tactical and operational effects in the battlefield. In the 1970s, as the American armed forces started inducting Precision Guided Munitions (PGMs) and cruise missiles, the Soviet military theoreticians termed it as a MTR. In contrast, the Military Revolution involves transformation of the society due to the emergence of an original system of systems (a combination of different technologies and managerial techniques) which generates a new military system. The implementation of this innovative military system results in a quantum jump of military effectiveness of the party initiating the Military Revolution. The genesis and maintenance of the resulting advanced military system transforms the state-society relationship. Hence, unlike the RMA, the effects of the Military Revolution are not merely confined to the tactical-operational spheres but also spill over at the military strategic and grand strategic levels. Further, while the driver of RMA is the emergence of a new piece of technology, the causation behind the Military Revolution involves a lot of factors. Besides the technological component, social, cultural and economic factors are also responsible for spawning a Military Revolution.[19]

Probably *stunde null* (complete break with the past) never occurs in reality. Even a revolutionary phenomenon incorporates elements of continuity within it. Both Military Revolutions and RMAs have occurred in history. In fact, Military Revolutions, RMAs and military evolution (slow gradual changes in halting stages) have occurred sequentially across the broad span of the last five millennia of human history. The emergence of the English longbow in the fourteenth century, which resulted in the decline of French feudal heavy cavalry, was a classic case of RMA.[20] The rise of gunpowder war, which gave birth to the military fiscal state in early modern Europe (1500–1800), is an example of a Military Revolution. This Military Revolution gave birth to the cannon-equipped ocean-going navy and gunpowder infantry. Gunpowder infantry means drilled and disciplined infantry armed with handguns and supported by field (light) artillery during battles and with siege (heavy) artillery for conducting sieges. This military system, in turn allowed the Western European polities during the eighteenth and nineteenth centuries to colonise large swathes of Afro-Asia and the New World.[21]

Though a group of historians emphasises continuity and changes, the concept of Military Revolution, in my understanding, the heuristic device of Military Revolution has merit. True, a Military Revolution might often span over one or more centuries. The same is applicable in the case of the Agricultural Revolution which occurred between 7000 and 5000 BCE, the Urban Revolution in the Bronze Age (2500–1200 BCE), and the First Industrial Revolution that unfolded during the late eighteenth and early nineteenth centuries, etc.[22] My take is that the induction of AI in the military sphere is part of the Information Revolution occurring in broader society from the late twentieth century. The integration of a set of associated technologies (collectively termed as AI) is resulting in quantitative and qualitative changes at the tactical, operational, and strategic planes. The AI-initiated Military Revolution is still ongoing. Further, with the passage of time, the speed at which the Military Revolutions are spreading in the various regions of the planet is increasing. It took thousands of years for the chariot-centric armies (products of the Bronze Age) which emerged in West Asia to reach Western India and Spain.[23] It took a couple of centuries for the expansion of gunpowder armies throughout the planet earth.[24] The AI-centric Military Revolution is spreading all over the world in a couple of decades.

Extension of the role and autonomy of intelligent military machines will require changes in the individual and collective training of the soldiers, reconfiguration of the command system and also the generation of a new doctrine. Radical changes in these spheres will give rise to new armed forces that will fight differently compared to the existing militaries and war as we know it at present. AI-augmented weapon systems and AI-augmented decision-making systems will have a multiplier effect not merely on the tactical and operational planes but also at the strategic level. I believe that AI is a strategically disruptive weapon system. Moreover, shifts in the broader society are driving the process of militarisation of AI. Most of the AI technologies that the militaries at present are adopting are dual-use technologies.

Integration of AI with military organisations is possible because of the ongoing Information Revolution or the Fourth Industrial Revolution in the societies of the advanced countries. Klaus Schwab first coined the term Fourth Industrial Revolution. The First Industrial Revolution used water and steam power to run the machines and the Second Industrial Revolution was based on electric power. The Third Industrial Revolution witnessed the emergence of personal computers, the internet, and automation in the 1960s, thus ushering in the Digital Age. So, the Third Industrial Revolution was the prelude to the Fourth Industrial Revolution. The last one is closing the gap between digital, physical and biological spheres and generating a vast ecosystem of human-technological interface. This is having a revolutionary effect on how we work, travel and educate ourselves.[25] In this neat scheme, the issue of nuclear power is left out. Whether one calls it the Fourth Industrial Revolution or not, the interconnected system of technological and social changes from the beginning of the twenty-first century is fuelling the AI-initiated Military Revolution.

In Chapter 1, P.K. Mallick shows that AI-related technologies if applied in the case of defence wholesale, have the potential to revolutionise warfare by changing not only the character but also the nature of warfare. If war is dehumanised both at the level of waging it (by intelligent machines instead of humans) and also as regards decision-making, then truly we are on a threshold. However, at present, after analysing the recent Israeli COIN against Hamas (2006–2009) and the ongoing Russian-Ukraine War, concludes Mallick, the application of AI-related technologies on the battlefield has been very limited.

Chapter 2 by Ajey Lele in this edited volume provides an overview of the possible use of AI-related technologies in the various branches of the military. For him, at the strategic level, the effect of AI is ambiguous. Lele argues that, at present, the effects of AI at the tactical and operational levels are massive and amount to an RMA. Lele asserts that AI, being a disruptive technology, and its widespread applications, which are inevitable in the near future, will transform the logistics, doctrine, training, command and tactical concepts of combat.

In Chapter 9, R.S. Panwar takes a more moderate view of the Indian Army's integration of AI-embedded weapons. Panwar writes that the Indian Army has been quite slow to integrate the techniques of Network Centric Warfare (NCW) and is only gradually incorporating the AI-enabled weapon systems. For the Indian Army, it is more a case of evolution rather than revolution. He accepts that AI technologies hold great promise for facilitating military decisions, minimising human casualties and enhancing the combat potential of forces. He gives a broad overview of the possible military applications of this technology, and brings out the main legal and ethical issues involved in the current ongoing debate on development of LAWS. Panwar suggests steps that need to be taken on a priority basis to ensure that Indian defence forces keep pace with other advanced armies in the race to usher in a new AI-triggered RMA.

In contrast to Panwar, Kaushik Roy notes in Chapter 10 that the Indian armed forces are giving a lot of thought towards using semi-autonomous machines like UAVs, ordnance disposal ground robots, etc. for various COIN tasks. The point to be noted is that the use of drones by the USA in Afghanistan and Pakistan has alienated the local population. Winning the 'hearts and minds' of the people living in the disturbed zone is a *sine qua non* for successful COIN. Drones can never replace a large number of boots on the ground.[26] At present, the Indian COIN strategy is based on using a large number of infantry equipped with handheld weapons only. The use of UAVs and land mobile robots might change the equation for worse.

In Chapter 3, A.K. Sachdev focuses on a particular piece of military system: Aviation. He details the various ways in which AI-embedded sixth-generation aircraft is paving the way for the emergence of pilotless aircraft. Have human pilots reached a dead end? There are several advantages to using automated aircraft compared to a manned one. Unlike a human pilot, the AI flying the aircraft will not get tired and will be able to make a steeper and faster turn and would be able to fly longer and higher. The USA, Russia and China along with some other major powers, are planning to have one manned aircraft controlling six to ten unmanned aircraft with the aid of AI. The UCAVs will function in this scenario as loyal wingmen. The point to be noted is that between 2002 and 2010, there had been a 40-fold increase in the number of drones in the American armed forces. In 2010, the United States Air Force (USAF) declared that it was training more drone operators than F-16 pilots.[27]

Arash Heydarin Pashakhanlou of the Swedish Defence University takes a more moderate view of the pilotless aerial vehicle (PAV). Pashakhanlou notes: 'The central challenge for PAVs will probably be operations in complex and crowded environments, where human, but not pilot, involvement may be required in the absence of AGI. It will also likely take considerable time before even the semi-autonomous PAVs dominate the skies. The day that happens, it will mark the end of pilots.'[28] So, PAV/UAV is an RMA in waiting. But UAV in combination with other new AI embedded weapon systems is generating a new Military Revolution.

Besides UAV, another new tool is swarming drones. Armed fully autonomous drone swarms (AFADS) is considered as a weapon of mass destruction (WMD).[29] What is swarming? Swarming occurs when several units conduct a convergent attack on a target from multiple axes. Such attacks could be long range or short range. Swarming could be preplanned or opportunistic. About swarming, Sean J.A. Edwards writes: 'It usually involves "pulsing" where units converge rapidly on a target, attack and then re-disperse.'[30] Swarming drones are a group of drones which can interact and function as a collective with a common objective. The essence of swarming is coordinated behaviour. Over time, cooperative behaviour will make the drones more capable of conducting operations independently. Machines can coordinate their actions at speed and scale impossible for humans. This feature, writes Paul Scharre, will revolutionise warfare.[31] One may argue that swarming tactics by itself is nothing new. In the animal world, ants, bees and wolves conduct swarming. Swarming is actually a complex adaptive system. It is an agent-based system where the agents themselves follow simple behavioural rules that, in the aggregate and in the system as a whole, produce complex emergent behaviour.[32] We will elaborate on emergent behaviour later. In history, the Mongol light cavalry during the thirteenth century practised swarming.[33] However, swarming by unmanned combat robots is really revolutionary as it's a break from past military practices.

Large swarms with sensors and munitions can conduct mass attacks. The swarming drones will possess machine vision. Machine vision, which means the capability of a machine to see, requires a high volume of data to train the algorithms. Sensor drones use these algorithms to collect and share information on hostile drones, possible targets, and natural and artificial obstacles present in the environment in which they are operating. Flexible swarms add or remove drones to meet the changing tactical scenarios and to attack one target from different directions or multiple targets simultaneously. Diverse swarms are able to incorporate different types of munitions and sensors and allow closely integrated multidomain strikes.[34] However, these swarming drones will move at different speeds towards various targets in dissimilar risk environments, and this creates a serious challenge for command and control.

Generalship in the digitised battlefield

For historian Martin Van Creveld, command is equivalent to management of the military organisation. Command requires coordination and control, which in turn is dependent on information flow. However, lower-level command requires the fighting men to be motivated by their officers to fight and die in the fire-swept killing field.[35] Killer robots do not require motivating speeches by their officers or performance of heroic acts by frontline junior commanders. However, the situation would be different in the case of human-machine integrated battle teams. Before the coming of AI, human inventions like chariots, composite bows, gunpowder, steam power, internal combustion engines, nuclear power, to name a few, have aided humans in waging war. However, the introduction of AI for the first time is cutting into the hitherto human monopoly of taking decisions regarding the conduct of war. This development is truly revolutionary. Since computers can process and interpret data faster than the human brain, the higher command apparatus would include an AI-augmented decision-making apparatus. In the sphere of command, AI could play an important role in drawing correlations from huge amounts of data which are potentially invisible to humans. From the raw data collected by the sensors, the algorithms can construct models, develop and test insights, draw correlations and detect anomalies for human decision-makers.[36] This development, in turn, will complicate the command fabric of robotic-human armed forces in the digitised battlefield.

ML is concerned with discovering statistical relationships in data. ML could be composed of three types: Supervised learning, reinforcement learning and unsupervised learning. DL is a type of supervised learning technique based on artificial neural networks that was invented decades ago but is now becoming effective due to the rapid rise of the computational power of computers. Neural networks, somewhat inspired by the connectivity of the brain, have nothing to do with the human brain. Artificial neural network means using several artificial neurons (units) connected together. Each unit computes a number based on its inputs. The output of an artificial neural network is the sum of the output of all these units. DL comprises many layers of these artificial neural networks connected together and is extremely effective in extracting statistical relationships between inputs and outputs using vast amounts of data. It is most effective in computer vision and language processing. For instance, these artificial neural networks can be trained to recognise objects in a picture with an extreme degree of accuracy. In reinforcement learning, algorithms learn how to choose between a set of actions to perform a task in the best possible manner. Unsupervised learning means designing algorithms which can learn by themselves without any external master and which would be able to come up with their own goals. This is the key to constructing a really intelligent machine. However, till now ML research has not come up with this sort of technology.[37] So, construction and deployment of a truly autonomous LAWS remain in the field of fantasy at least for now. Nevertheless, AI-embedded weapon systems with even limited intelligence appear to be revolutionary and difficult to control. One group of researchers in 2009 asserted that even robots with primitive intelligence display characteristics of deceit, greed and self-preservation without these researchers programming them to do so.[38]

As this book goes to print, a new technology (Large Language Models [LLMs] and its subset ChatGPT) is in the offing. Some overenthusiastically term it as a piece of transformative technology. This technology (an example of ANI) has military applications especially in the field of creating AI-augmented decision-making mechanisms. The US armed forces believe that in future battlefields, data will be as important as weapon platforms. Computers using LLMs, ML and other AI applications will augment military decision-makers for planning at the strategic, operational and tactical levels of war. A large number of high-definition sensors will be deployed in the coming battlespace. They will generate a lot of Big Data which will be collected at several ingest points and then streamed to the processing nodes for making real-time decisions. Acquisition and utilisation of all this data requires ML to augment human cognitive abilities. However, in the contested battlefield, the armed forces operating with restricted bandwidth and degraded communication systems, tactical use of cloud-based computing is a liability. Because of latency, bandwidth and security considerations, such data cannot be sent to the cloud. It is required to be processed and exploited at the point of ingest, i.e., at the edge through a computer. Thus, it is necessary that computing must occur on-premise at the edge. Edge AI computing capability will enable the military planners to rapidly synthesise diffuse data streams for planning military tactics. The edge AI capability ought to reside on edge computing hardware. These computers will be devoted to computer vision and sensor fusion data processing duties for real-time missions. Such computers utilise LLMs.[39]

LLMs are capable of synthesising information and using structures of language to answer questions posed by military officers planning military moves. The US armed forces use Hermes LLM for military planning. The military planners in dialogue with the LLMs could assess the validity of their assumptions and practicability of their strategic, operational and tactical plans. The LLMs allow military planners to see battlefield geometry in multiple dimensions. Besides answering factual questions, Hermes can generate hypotheses about temporal and positional advantages of various plans. Hermes can also help the human commanders get an idea of the hostile party's military doctrine and their possible course of action. Overall, LLMs enable the military commanders to gain a better appreciation of the military environment in which they are operating. Benjamin Jensen and Dan Tadross stress the point that LLMs augment but do not replace human commanders. It must be noted that AI-enabled machines also hallucinate. Further, human commanders must always use their critical judgement while assessing the solutions provided by LLMs. Ultimately, the effectiveness of the computer using LLMs will depend on the human commanders asking the right questions and also on the quality of training the machine has received before being deployed.[40]

A group of Chinese scholars is eager to use LLMs to streamline the command apparatus of the armed forces. Incorporating prior knowledge into pretrained language models is effective for knowledge-driven Natural Language Processing (NLP) tasks. NLP tasks involve named entity recognition, relation extraction and entity typing. However, current pretraining procedures for injecting external knowledge into the models through knowledge masking, knowledge fusion and knowledge replacement are not that effective, especially for NLP designed for military tasks. Masked language models can only model low-level semantic knowledge. Generally, pretraining tasks are designed with respect to the corpora and entity features in general texts, which are not suitable for military text mining.[41] Hence, the necessity of developing domain-specific pretraining exercises. Thus, the job before the scientists and engineers is to develop a more effective method for injecting military domain-specific knowledge into the language representation model.

AI-based applications and systems pose significant risks, arising mainly as a result of the unique characteristics of ML technology, especially deep neural networks, which power them. AI-enabled military systems, in particular, are of special concern because of the threat they pose to human lives. At the heart of these concerns is the increasingly intelligent behaviour displayed by AI systems in recent years, which in turn has resulted in delegation of cognitive functions to machines in ever-greater measure. This has given rise to a host of legal, ethical and moral conundrums, which therefore need to be suitably analysed and addressed. At the same time, it is universally accepted that huge benefits could accrue to humankind, both on and off the battlefield, if the power of AI is leveraged in a responsible manner. This double-edged character of AI technologies points to the need for a carefully thought-out mechanism for regulating the development of AWS. Mallick warns that AI-related software, while increasing the speed of decision-making, is also escalatory in nature. War might become uncontrollable. So, integration of AI in the command structure is a dicey move.

What about swarming robots, a weapon system on which the USA and China are investing resources? The command of swarming robots involves two parts: Human-machine and machine-machine interactions. Interaction between the individual robots in the swarm will generate emergent behaviour. This behaviour is not preprogrammed, so the commander cannot predict the system's behaviour. Merel Ekelhof and Giacomo Persi Paoli warn about the dangers in the following words: '[B]y the lack of a universal model that allows humans to understand complex emergent behaviour and take adequate responses in a timely manner.'[42] This will result in a lack of trust and control on the part of the commander over the swarm and bring into play a higher risk of undesired behaviour by the robotic swarm.

Panwar in an article titled 'Regulation of AI-enabled Military Systems,' notes that AI-triggered risks posed by different types of military systems may vary widely, and applying a common set of risk-mitigation strategies across all systems will likely be suboptimal for the following reasons: These may be too lenient for very high-risk systems, overly stringent for low-risk systems, and hamper the development of systems which might be beneficial in various ways. A risk-based approach has the potential to overcome these disadvantages. What is required is to sketch the contours of such an approach which could be adopted for the regulation of military systems. The analysis which underpins this approach, although not always explicitly expressed, is deeply guided by the necessity for achieving meaningful human control (MHC), which is a key objective sought to be achieved in all AI-enabled military systems.[43] In Chapter 8, Guru Saday Batabyal warns that unless MHC is established, a Terminator-like AWS might become a reality in the near future. Could AWS be a malevolent super-intelligent killing machine? Further, we need to unpack the concept of MHC and then think of how to establish and maintain it, and what sort of command ecosystem we need for it.

Manned-unmanned teaming has a serious limitation: The threat posed by electronic warfare (EW)/cyberwarfare, mainly spoofing and jamming of drone communications. Heavily encrypted communications are difficult to decipher, but they could be jammed. Multiple data links may be attached to a UAV for the sake of redundancy, so that an electronic attack against one device would not necessarily knock out the entire support system that enables unmanned flight. A heavier drone equipped with EW countermeasures would be able to survive electronic attacks, but such a drone with greater payload would be less manoeuverable against enemy swarming drones and manned aircraft. China, for instance, is still behind in the race for manufacturing guidance and control systems in the fields of solid-state electronics and microprocessors. One way the Chinese drone fleet could overcome hostile air defence systems is by following the principle of unaware collaboration (emergent behaviour). In this kind of network, which is somewhat similar to an ant colony, the drones do not fully comprehend the presence and activity of other drones. Rather, the drones adhere to instructions that are patterned for an aggregate effect through limited monitoring of their surroundings. Within this set up, a drone equipped with collision avoidance software would detect any gaps or malfunctioning units next to it and self-adjust. The unaware collaboration principle may allow the drones to conduct collaborative missions and would keep the drone swarm in a tight formation. Such a command system requires very little command signals from the headquarters conducting the drone strike against the enemy. Hence, this command system is quite immune to hostile EW.[44]

Nevertheless, ANI-embedded weapon systems (weapons with limited autonomy) could cause unintended and unnecessary harm to civilians on both sides or engage in fratricide. This might be a possibility because while warring with machines, AWS might malfunction or in the digitised battlefield, weapon systems with AI might be hacked by hostile hackers as part of cyberwarfare. Van Creveld insightfully warns that one should not become a slave of technology. Rather, one should understand the limitations of a technological device and understand what it cannot do. To fill the gap, the human should step in. Further, efficient command requires thoughtful attention to the spheres of training and doctrine.[45] What is worrisome is the increasing dependence of the higher-level command headquarters on vast amounts of data sent through computers. This very information flow to and fro between the units at the frontline and headquarters (at times via satellite links) is vulnerable to cyberattacks.

Artificial intelligence and warfare in space and cyberspace

Cyberwarfare and war in space are two new features of the ongoing AI-initiated Military Revolution. Cyberspace is an artificial domain generated by computers and the latter is the creation of humanity in the mid-twentieth century. Initially cyberwarfare was called electronic warfare and then information warfare. James A. Green defines the term cyberwarfare as employed broadly to convey the use of technological force in cyberspace (that is, the realm of computer networks and the users behind them) in which information is stored, shared and communicated online. At times cyberwarfare results in the disabling of the hostile side's critical infrastructure without any direct kinetic effect. For instance, stealing of information from the enemy's computers, etc.[46] Cyberattacks could result in deletion of the hostile sides' websites, email accounts, etc. Computers can be compromised by altering the Domain Name System (DNS) records local to these machines. The DNS provides the process by which a networked device resolves a domain name to an IP address. Or, the whole computer network can have false records injected into their DNS tables. With a poisoned DNS, computer devices can have traffic to specific websites redirected to the attacker. The attacker can then view (compromise the confidentiality) or interfere with this information before relaying it to the original customers. Denial of Service (DoS) attacks aim to prevent the computers from assisting their intended users.[47] Malicious software injected into a computer system by the cyberwarriors by corrupting and deleting the files in the hard drives, and overwriting of the master boot records of the affected computers would make the whole computer system inoperable without causing any physical damage to the hardware.[48] So, cyberattacks involve the three traditional forms: Sabotage, espionage and subversion, but with new tools in the virtual space, hence these are revolutionary.

Cyberwarriors are not required to be in geographical proximity to their targets. This is because the attack occurs in the virtual battlefields as it is carried out through the internet. According to one calculation, in 2019, 46%

of the world's population (living in the most advanced regions of the planet earth) was using the internet.[49] At present, in an advanced military system, almost everything – conventional bombs-equipped missiles, nuclear weapons-equipped missiles, soldiers' communications and satellites, vehicles, advanced armour systems, lasers, and navigation systems – depends on computers. Hence cyberattacks in the virtual universe will have serious effects in the material world. To cap it all, cyberattacks complementing kinetic attacks would be devastating.

By the beginning of the twenty-first century, another epochal creation of humanity is the production of autonomous intelligence. Command, Control, Communications, Computers, Information, Surveillance and Reconnaissance (C4ISR), claims Lele in his chapter, is the central nervous system of a modern military organisation. In Chapter 4, Kritika Roy argues that if AI is integrated with computer technology, then AI-implanted computer software will bring a sea change in cyberwarfare. Cyberwarfare is actually contactless digital warfare which can have serious impacts on the physical world. This is because the physical world and the digital world are intimately interconnected with each other. To give an example, Al-Qaeda in the beginning of the new millennium used the internet to gather recruits and funds. In 2013, it opened a Twitter account to gather support for its cause.[50] In case of interstate conflicts, AI-integrated cyberattacks would result in uncalculated escalation and reduce the reaction time available to the warring parties. Such a development would destabilise the military balance and hence strategic stability among the nations.

Worse, warns Kritika Roy, the use of AI for ISR and in cyberweapons might threaten the second-strike retaliatory capability of the nuclear powers. The nuclear deterrence would be then on shaky grounds; itself a cause of serious worry. Paul Bracken cautions that nuclear deterrence is dependent on the second-strike capability of the states. This second-strike capability, mainly based on land-based nuclear weapons-equipped missiles, is now under threat. Mobile land-based missiles are preferred over static land-based missiles because the latter are vulnerable to precision-guided missiles (PGMs) equipped with conventional and nuclear warheads. However, the technologies resulting from the so-called Fourth Industrial Revolution are threatening even the mobile land-based nuclear-tipped missiles. This is possible because new technologies like AI, cyber technologies, cloud computing, data analytics, drones and hypersonic missiles could easily locate and destroy these land-based nuclear-tipped missiles. AI-directed search/reconnaissance technologies are more efficient than the Cold War-era reconnaissance techniques and technologies.[51] The sense of insecurity that one's second-strike capability is under threat from the hostile side will force the states to keep these mobile missiles always on the move and to increase the number of such land-based mobile missiles in their arsenals. This would result in a new nuclear arms race, which would further destabilise the fragile stability that earlier has resulted from the stable nuclear deterrence. Further, nuclear warheads-equipped missiles on the move are vulnerable to terrorist attacks. To cap it all, the AI-embedded cyber technologies of the enemy state or even non-state actors can disable the computerised command and control fabric of the nuclear weapons of a state.[52] In such a scenario, an accidental nuclear attack (everybody's nightmare) cannot be ruled out.

Cyberoperations can have serious effects on the nuclear arsenal of a state. One example is the disabling of Iran's nuclear refining operations at Natanz in 2010 by the Stuxnet virus. Stuxnet was delivered through a Universal Serial Bus (USB) thumb drive and it damaged the gas centrifuges used to refine uranium. Stuxnet found its way to the machine that controlled the Programmable Logic Controller attached to the motor control. The virus modified the instructions sent to that motor, thus causing it to spin at different rates. Machinery rotating at high rates is extra sensitive to changes in the rotation rates, and gas centrifuges for refinement of radioactive uranium spin at 6,000 revolutions per minute. Thus, Stuxnet caused physical damage to the centrifuges in the Natanz plant. This attack was probably carried out jointly by American and Israeli intelligence agencies.[53]

Integration of ML algorithms and other types of AI into cyberoperations offers several advantages for military operations in cyberspace. The use of ML increases the chance of discovering gaps in code that the AI software could then exploit autonomously, which leads to greater efficiency and speed of offensive cyberoperations. The same technology can also be used to strengthen defensive systems and ward off malware and other hostile intrusions into networks. ML algorithms are also useful to scan and surveil the online activities of a large number of individuals or to autonomously prepare the digital battlefield by planting malware in an adversary's networks that might be activated remotely in case a war breaks out. The use of AI increases the potential impact of disinformation campaigns and other types of information warfare, making such campaigns more effective, scalable and widespread. The creation of deepfake audiovisual content, the detection of faultlines in the target population's social fabric to maximise a campaign's effect, the deployment of bots to artificially amplify subversive messaging directed at a target population and to pursue automated agenda setting – particularly with more sophisticated bots that are programmed to credibly mimic real people's behaviour – are some of the ways AI can be used. Finally, the automatic calibration of microtargeting methods in order to tailor content to a receptive audience can also be done effectively with AI.[54]

The hackers could disable banking systems, bring down the electric grid, obstruct the firing mechanisms of both conventional and nuclear-equipped missiles, etc. For instance, in 2007, in Estonia, 95% of banking activities were carried out online.[55] Further, hackers could inject misinformation and disinformation within the systems of a particular country that is being attacked. Different states (like Norway, for example) are now building digital border defences against digital threats from hostile states and non-state groups. A digital border defence means the country develops the capability to defend against digital threats before they enter that particular country's national territory. Cyberoperations (also known as Computer Network Operations [CNO]) are of two types: Offensive and defensive. Defensive cyberoperations include monitoring and handling cyberattacks to disclose reconnaissance, infiltration and attacks as early as possible.[56] What is ironic is that integration of AI in cyberwarfare by a party increases its capacity for conducting both offensive cyberoperations as well as strengthens the defensive potential of that party's CNO against any hostile threat. Mallick in his chapter warns that against an enemy using AI in cyberwarfare, a country defending its cyber assets without AI would face a catastrophe. The defender could only successfully defend itself by using AI algorithms. Actually, it is becoming a war of algorithms against the hostile party's algorithms.

Besides cyberspace, the use of AI in outerspace is becoming a reality now. Control of the space domain (the area above the altitude where atmospheric effects on airborne objects become negligible) is necessary for conducting all domain military operations. Most of the space operations occur within the geocentric regime in the Geosynchronous Earth Orbit (GEO), Highly Elliptical Orbit (HEO), Medium Earth Orbit (MEO) and Low Earth Orbit (LEO). GEO is good for global communications, surveillance, reconnaissance, observing large-scale weather patterns and warning about missile launches by the enemy side. A spacecraft at HEO could loiter for a long time over a particular target, especially at high altitudes. MEO orbits are between LEO and GEO. Spacecraft at LEO are relatively closer to Earth. These spacecraft use less powerful transmitters for communication and achieve higher resolution imagery with similar-sized apertures compared to satellites in the HEO and above. However, spacecraft at LEO could remain over a particular target in the terrestrial plane for only a short time. These spacecraft at various orbits (altitudes) constitute the orbital segment of space power. The terrestrial segment includes terrestrial-based radars and all the terrestrial-based facilities and equipment required to launch spacecraft. The link segment comprises the signals in the electromagnetic spectrum (EMS) and the nodes that connect the orbital segment to the terrestrial segment, as well as the individual terrestrial segments with each other.[57]

AI-embedded satellite imagery offers more detailed analysis of post-battle damage in urban areas. Destruction of public infrastructures resulting from war could be correctly assessed from high-resolution satellite images with DL techniques. AI applied to satellite imagery could result in automated detection of war-related damage. An ML technique (convolutional neural networks) for scanning multiple successive images of the conflict zone that had suffered damages due to military operations improves detection of destruction significantly. A correct assessment of the destruction of buildings in the conflict zones will aid human rights monitoring, humanitarian relief efforts and reconstruction initiatives.[58] AI-embedded satellite imagery using the above-mentioned techniques could also be used to track the movement of hostile forces on the surface of the earth.

The threats to space operations are nuclear detonation (NUDET), electromagnetic pulse (EMP) and physical attack. Directed Energy (DE) threats include laser, radio frequency and particle beam weapons. Laser weapons can temporarily disrupt or deny capabilities or permanently degrade subsystems. Electromagnetic energy from terrestrial or space-based sources can attack/disrupt electronic components as well as the link segment, including uplink, downlink and crosslink signals. EMP will damage voltages and currents into unprotected electronic circuits in the spacecraft, terrestrial nodes and their associated links. NUDET by its blast can destroy the spacecraft or its terrestrial links. Even the radiation resulting from the blast could seriously damage components of the spacecraft. Moreover, Anti-Satellite (ASAT) missiles could destroy spacecraft and orbital stations. Space operations are data intensive and rely on terrestrial communication links (for example, terrestrial-based radars) and space-based communication links via the cyberspace domain. Hence, space operations are vulnerable to AI-augmented cyberattacks. Such attacks may disrupt or deny space-based or terrestrial based computing functions used to conduct or support spacecraft operations and obstruct the collection, processing, and dissemination of mission data.[59]

Batabyal's chapter throws light on the possibilities of using AI technologies in conducting both offensive and defensive war in cyberspace and outerspace. Combat in these realms would be conducted by intelligent machines that will adapt to the changing circumstances by adopting ideas and techniques which the original designers or the mission handlers had probably not thought of. Here morality, legality, laws of war as well as virtue ethics come into play, which constitute the subject of the next section.

Holy war, ethics, legality, and AI-embedded weapon systems

Whenever in history a new military technology with increasing lethality was introduced, civilian intellectuals and religious leaders argued against it and tried to prohibit its use. Or, if that proved impossible, they attempted to limit its use by deploying reason, ethics and laws. Ethics could be defined as the basic question of right and wrong. In ethical reflection, one examines how norms inform human actions and institutions, and how virtuous decision-making affects human dignity and environmental sustainability.[60] So, ethics involves the creation of a normative model of doing certain things and not doing several things.

In the twelfth century, the Pope was against the use of crossbow within Europe but argued that these 'devilish' weapons could be used against the Muslims. The Church tried to ban this weapon in 1139. The Chinese emperors were not eager to deploy gunpowder weapons when gunpowder was accidentally discovered by the 'alchemists' of premodern China. Similarly, in ancient and early medieval India, Hindu *acharyas* (Brahmin military theorists) tried to limit casualties in war by introducing the code of *dharmayuddha*. The ethical code of conduct laid down that wars should be fought in respectful ways by honourable warriors (Kshatriyas). The principles of *dharmayuddha* are as follows: 1. Combatants should develop respect for the enemy combatants. 2. Honour, a sense of justice, and fair play should shape the behaviour of the combatants. As far as possible, war should be treated as a chivalrous sport. There should not be any nocturnal attack, surprise raids (preemptive strikes) or harm towards the non-combatants and their properties. 3. Protection of the prisoners of war. 4. Non-use of contemporary advanced weapons due to their high lethality.[61]

The equivalent of *dharmayuddha* tradition in Western history is the Just War theory, which was first formulated by Hugo Grotius (1583–1645). The objective of Just War, like *dharmayuddha* is to reduce the scope, intensity, and impact of warfare on society. Just War tradition has two components: *jus ad bellum* (justice of war) and *jus in bello* (justice in war). The first principle is concerned with the reasons for going to war. It requires human beings to make judgements about aggression and self-defence. The second is concerned with the conduct of war: observance or violations of the customary rules of engagement (ROE). Michael Walzer writes that, in accordance with the Just War paradigm, it is just to resist aggression, but the use of force must be subject to moral and legal restraint.[62] Walzer continues that *jus ad bellum* demands that the aggressors should not only be contained but also punished. Then, *jus in bello* requires that the rules of war are applicable both to the armed forces and the civilians, to the aggressors and the victims.[63] How far does the use of unmanned systems allow one to follow the principles of Just War?

As the militaries of different countries are integrating AI-augmented weapon systems, the 'Stop Killer Robots' campaign has also started in Western civil society. Jai Galliott writes that the use of unmanned systems is linked with unethical decision-making and lowers the barriers of killing, thus endangering the moral conduct of warfare.[64] Unfortunately, we have seen in history that, due to military necessity and when the survival of the state was at stake or new weapons were being used by the enemies, 'ethical' behaviour was replaced by 'practical' behaviour. For instance, in Chapter

10 Kaushik Roy portrays that, at present, the Indian armed forces' officers, are eager to use AI as a subordinate in the human-centric decision-making process due to the state of technology and assessment of threat levels. But, if the threat from China escalates, then they might argue for automating the Observe, Orient, Decide and Act (OODA) loop. This is because AI (especially ML) can process large quantities of complex data with such speed that human beings cannot directly supervise the resulting evaluations, decisions, and actions. If large teams of humans attempt to do it, it will slow the tempo of operations, especially when all the military operations have to be carried out in highly compressed time frames. Hence, if, due to military pressure, AI replaces humans in the OODA loop in the near future, this would go against the *dharmayuddha* tradition of the Indian armed forces.

Broadly, the people arguing for and against using AI in warfare, especially from the perspectives of morality and legality, can be divided into two groups: Pragmatists/realists (instrumental in their approach) and the idealists/virtuous. The pragmatists assume and believe that the ends justify the means. The idealists believe that the 'good/virtue' in human beings will see reason at the end of the tunnel for the 'common good' of humanity and through debates and discussions they will find out a way out of the impasse. For the present, we need to find a middle path between these two extreme positions.

War machines of the future will be distinctly non-human autonomous entities. A future intelligent and autonomous war machine will be one of several forms of non-carbon entities and a type of Artificial Superintelligence (ASI). ASI is a machine whose cognitive capacity in all domains exceeds that of humans. The principal issue for ASI, asserts Swaminathan Ramanathan in Chapter 6, is not sentience. Sentience will emerge. The key challenge is the normative architecture that will regulate that sentience. War machines of the future will not only need it the most but will also need it as a DL system. His chapter engages with the challenges of building a normative architecture for a war machine of the future. In doing so, Ramanathan makes the case for a system-of-systems approach of reinforcing and balancing feedback loops as 'anticipatory' control laws of ethics, virtue and morality for a real-time sentience. Ramanathan warns that AWS being non-carbon forms, human biases, prejudices and assumptions will not be applicable to them. How the AWS will think and behave (fight) will be a combination of several unknown factors. The uncertainty (danger) could be somewhat reduced if humans are integrated with AI in the decision-making apparatus. Then, it would also be a case of fusion of known-unknown elements.

In Chapter 5, Gregory Reichberg and Henrik Syse discuss the safeguards built into NATO's decision-making cycle to prevent unethical use of LAWS. In fact, the decision-making process is so cumbrous and slow, that one doubts whether it will work in wartime when time is short and a quick response is vital for survival. Reichberg and Syse assert that intelligent autonomous machines can never equal the human mind's adaptability and flexibility. Hence, human control over AWS, especially LAWS, is a must. The human mind could further be trained in accordance with the principles of virtue ethics rather than consequentialist ethics. They conclude by warning that AI should never be integrated with nuclear weapons and hypersonic missiles.

In 2021, the European Commission came up with a proposal for the regulation of a European approach to AI, the European Union AI Act. The object is to pursue trustworthy AI. The regulation will ban AI systems that create potential threats to the safety, livelihood and rights of the citizens. The approach (similar to Panwar's article on risk-based approach discussed earlier) is to differentiate between unacceptable risk, high risk and low or minimal risk AI weapon systems.[65] Even if we assume that the EU will not come up with unacceptable risk category AI-embedded weapons, how can we be sure that Russia, China, USA and Israel, among others, would subscribe to such an approach as regards the development of military AI?

In a report on ethics and AI generated by the Peace Research Institution Oslo (PRIO) dated 2021, some practical steps are mentioned for controlling AI-embedded war machines. The report states that the handlers must be trained thoroughly in the technical as well as ethical aspects of the AI system that they are operating. Further, resilient feedback loops between software designers, software and hardware manufacturers (engineers and scientists), decision-makers (politicians, and generals), and users (military) need to be developed.[66]

In Chapter 8, Guru Saday Batabyal asserts that the use of AI-embedded weapon systems both in the physical world and in cyberspace will increase in the coming days. Both the state and the non-state actors will be parties to this process. The polities, argues Batabyal, should not cross the ethical boundaries in using AI-empowered weapon systems (like drones for targeted killings across borders) against terrorists. Then there will not be any moral distinction between the civilised norms of the nation states and the insurgents. He continues that besides using Hugo Grotius's Just War theory and International Humanitarian Law (IHL), more comprehensive distinctive and elaborate legal promulgations are necessary to keep the use of AWS within the bounds of established legal-ethical norms. Batabyal vehemently argues that the use of armed drones for killing terrorists does not fall within the purview of IHL. Further, he hints that if joint teams comprising humans and intelligent warring machines are used, then we should probably think of formulating 'machine ethics' as machines can never think like humans. Here we find an echo of Ramanathan's observations mentioned above. But how far can IHL accommodate LAWS?

China's position is that first there should be a consensus on the definition of LAWS and their characteristics rather than using subjective and undefinable concepts such as MHC and substantial human judgement. The Chinese spokesmen further claim that the existing IHL is not equal to the task of dealing with LAWS. They point out that the probability of inconsistent interpretations of IHL might allow some countries to legitimise the development, production, and use of LAWS. In 2018, China proposed a narrow definition of LAWS comprising five distinct features. The first is lethality, which means sufficient payload charge for it to be lethal. The second is autonomy, which refers to the absence of human control during the entire process of executing a task. Third is the impossibility of termination, which means that once activated, the killing machine cannot be stopped. The fourth feature is indiscriminate effects, meaning that the device will execute the task of killing and maiming regardless of conditions, scenarios, or targets. The last characteristic is evolution, which means that the intelligent war machine through interaction with the environment can learn autonomously and expand its functions and capabilities in ways that exceed expectations of those human planners and commanders who have launched that machine. Critics point out that the narrow Chinese definition of LAWS leaves considerable space for the inclusion of higher levels of autonomy in the targeting functions of the killer robots which are being developed in China and elsewhere.[67] For many, killer machines with higher levels of autonomy are not really safe.

Civilians especially become the predominant casualties in intrastate wars. In Chapter 7, Kelly Fisher warns us that AI could be discriminatory towards women and coloured people because the data which is fed into AI software is racist, gendered and biased. This in turn will result in LAWS generating disproportionate negative effects on the female and coloured population in the conflict zone. Kelly's argument has serious implications for military use of AI. The lifeblood of AI is data. AI-embedded military systems when being trained, if fed 'imperfect' data, would churn out erroneous conclusions while identifying, tracking and attacking hostile targets in wartime scenarios.

In contrast, in Chapter 11 Gabriel Udoh makes a case that governments of several African countries are eager to use AI-embedded military weapons because such machines will not rape, molest or commit unwanted violence on the civilians in the warring zones, as machines do not have passion or emotion. But do intelligent machines have ethics or consciousness? Udoh asserts that AI by itself is neutral. For him, the intentions of the humans using AI are the primary concern. Whether the use of AI-embedded machines would be legal or illegal depends on the motivations and objectives of their human users. This is important for the civilians, Udoh continues, because unlike the major powers like the USA, Russia, China and NATO countries, in Africa the principal driver behind the use of AI was not interstate but intrastate conflicts.

For ethical, legal and military requirements, it is necessary that an AWS must not be given unlimited autonomy. Most of the researchers dealing with LAWS speak of MHC. For me, MHC means that the human commander must have the dominant role in selecting and engaging the target. Maaike Verbruggen uses the term 'human control', which is defined as the mechanism for achieving a human commander's intent and refers to the extent of human influence on the outcome of a mission. Verbruggen continues that loss of human control could occur because of automating certain functions of the AWS and also due to deterioration of critical human judgements by the human commander in the context of the information provided by the intelligent machines.[68]

Human control could be exercised broadly through three levers. One is fixing control on the weapon system's parameters of use, including measures that restrict the type of targets and the task the AWS are assigned. Temporal and spatial limits further constrain its operation. Then there should be a mechanism for deactivation if necessary conditions arise. The second lever of control involves establishing control on the environment. This refers to the control fabric that conditions or structures the environments in which the AWS is used, for instance, deploying AWS in areas where no civilians are present. The third level involves the human-machine interface. The human must always supervise the machine and retain the right to intervene in its operations when necessary.[69] However, operationalising these measures is immensely difficult.

It is questionable whether AWS possess the ability to distinguish legitimate targets. Then, AWS pose a challenge to the principle of proportionality. Proportionality requires that incidental civilian casualties must not be excessive when balanced against the anticipated military advantages that could be derived from the destruction of the target. Further, AWS have no sense of ethics. So, in case of violations of IHL, the warring machines could not be held responsible for committing the criminal acts. In such a scenario, who would take responsibility? Professor Paola Gaeta of Switzerland says: 'State responsibility could thus have a considerable deterrent effect on States and would give them an incentive to make sure that the autonomous weapon systems deployed comply with IHL.'[70]

Transparency could be referred to as the end users should clearly understand how an AI-implanted machine works and makes decisions. In Chapter 4, Kritika Roy warns that the mechanism through which complex algorithms of AI makes decisions remains opaque. We have to understand how the AI is making a particular decision and not merely what solutions it is providing to a problem. This is essential for complete transparency of the AI's decision-making process. Only when we get a clear idea of it, can we can assess the AWS's reliability.

The rise of Terminator-like AI and the end of Humanity

> The first ultra intelligent machine is the last invention that man need ever make, provided the machine is docile enough to tell us how to keep it under control.
>
> I.J.Good[71]

I.J. Good, whose quote is given above, was a mathematician who had served as the principal statistician in Alan Turing's code-breaking team during the Second World War. AI researchers could be divided into two groups: The optimists and the pessimists. The optimists believe that even if super-intelligent AI is invented, it will not pose any existential threat to humanity. In contrast, the pessimists assert that superintelligence might evolve into an all-powerful singleton and might attempt to take control of the world. Nick Bostrom, who belongs to the pessimists' camp, believes that in the next 50 years, ANI will develop into HLMI. Then HLMI, through recursive self-improvement, would mutate into superintelligence. The HLMI would be able to understand its own workings to engineer new algorithms and computational structures to bootstrap its cognitive performance. It would be able to iteratively improve itself. The HLMI would develop a smarter version of itself, which, through sustained recursive self-modification, would generate a far smarter version of itself, and this process would go on. The end product would be able to surpass humans in cognitive functions in all domains. One of the probable results of such a self-improving super-intelligent entity (he calls it singleton) might be the extinction of mankind.[72]

In contrast, Kenneth Payne, an optimist, writes: 'It [AI] feels no emotion and is motivated only by the instructions we issue. It doesn't want to please us, and nor does it fear punishment if it transgresses.'[73] Payne continues that machine intelligence, unlike us has no comparable biological urge to survive and reproduce. The goals of machine intelligence are not intrinsic. They want what we tell them to want. There is no emotional aspect to machine life, and they lack conscious experience.[74] But, how can we be so sure that a super-intelligent AI will not develop consciousness for survival? One may argue that the possession of intelligence itself creates self-awareness and the necessity for self-preservation.

Panwar in his chapter asserts that high-risk AI-enabled war machines (in these machines AI decides on critical functions like targeting and firing) should never be developed. Only low-risk machines (completely under human control) ought to be developed. And a robust testing regime should be established for medium-risk machines (command systems with humans in the loop). But, what if to gain military advantages, some states develop high-risk machines? There are valid military reasons for doing so.

Because of the way AGI makes a decision, it is a black box and not transparent. So, Steven I. Davis agrees that AI should be used for tactical but not strategic purposes. Davis continues that at the operational level, as a risk mitigation measure rather than AGI, several ANI should be used and a human commander or a human team should oversee them.[75]

However, one RAND report asserts: 'Having a human in the loop slows everything down. If our adversaries are not using humans in the loop and we are, we will fall behind during time-sensitive periods within the conflict. Speed requires automation.'[76] So, fully autonomous LAWS is an absolute necessity that makes the difference between victory and defeat in war. Hence, AWS will become a reality in the near future.

One way to smoothen the MHC for controlling the AWS is to create cyborgs, which includes the best of humans and machines. A cyborg is a cybernetic organism. It is an organic life form that has integrated substantial elements of electronic technology to enhance its own capabilities.[77] In his book titled *Future War* (2017), Robert H. Latiff, a retired USAF major-general, writes about the possible nature of the battlefield in 2050 in the following words:

> Soldiers will look and be different. Technology will be employed, first externally, to give the soldier greater protection, greater situational awareness, and greater stamina... Function-enhancing drugs will become common. Soldiers' bodies will be modified for greater efficiency. They are likely to be artificially enhanced with exoskeletons to improve strengths, drugs to improve cognition or alter memory, and surgery to implant microelectronic neurological aids.[78]

Synthetic biology is a comparatively newer area of research. It aims to design and construct novel artificial biological pathways, organisms and devices, and restructuring of natural biological systems. Synthetic biology could provide the soldiers with new physiological functions and improved performance.[79]

Louis A. Del Monte, another pessimist, claims that superintelligence will come into existence by the end of the twenty-first century. And superintelligence will depend on quantum computers which are more efficient than present-day computers. Unlike present-day computers which use transistors connected via wires, quantum computers cannot be hardwired to human control. Further, superintelligence based on quantum computers can also control cyborgs which are dependent on organic brain-machine interface. The net result will be the elimination of the free will of humanity.[80]

Whether in another century, the carbon-based brain will be replaced by a silicon-based brain is the biggest question before humanity. Only time will tell. In case ANI continues to evolve and exceeds the cognitive abilities of humans, then AI-augmented human decision-making in war will be replaced by an AI-centred decision-making apparatus. Then the wars will be fought by cyborgs, robots and drones, and the decisions about war will be taken by super-intelligent machines and not humans. If AI dominates humanity, then warfare might be abolished because, after all the principal causation of warfare remains human beings themselves. Superintelligence might think that humans cause war and that the elimination of humans is the best way to abolish war. If superintelligence decides to wage war against humanity to abolish war, then the future is bleak for us. Abolition of war means the elimination of humanity, which will result in the end of the anthropocentric age and the beginning of a new phase of machine-human civilisation, with machines as the dominant partner. That will not be the beginning of the end, but the end of the beginning.

Notes

1 Forrest E. Morgan et al., *Military Applications of Artificial Intelligence: Ethical Concerns in an Uncertain World* (Santa Monica, CA: RAND Corporation, 2020), p. 2.
2 Louis A. Del Monte, *Genius Weapons: Artificial Intelligence, Autonomous Weaponry, and the Future of Warfare* (New York: Prometheus Books, 2018), p. 28.
3 Algorithm is a sequence of computer instructions allowing it to do calculations and other problem solving jobs.
4 Nik Hynek and Anzhelika Solovyeva, *Militarizing Artificial Intelligence: Theory, Technology, and Regulation* (London: Routledge, 2022), pp. 36–8.
5 Lieutenant-Colonel John Stroud-Turp, 'Lethal Autonomous Weapon Systems (LAWS),' in Expert Meeting, Autonomous Weapon Systems, Implications of Increasing Autonomy in the Critical Functions of Weapons, Versoix, Switzerland, 15–16 March 2016, p. 57, icrcndresourcecentre.org/wp-content/uploads/2017/11/4283_002, accessed on 12 November 2022.
6 Paul Scharre, *Army of None: Autonomous Weapons and the Future of War* (2018, reprint, New York: W.W. Norton and & Co., 2019), pp. 30–1.
7 Ajey Lele, *Strategic Technologies for the Military: Breaking New Frontiers* (New Delhi: SAGE, 2009), p. 44.
8 Paul J. Springer, *Outsourcing War to Machines: The Military Robotics Revolution* (Santa Barbara, CA: Praeger, 2018), p. 5.
9 Didier Danet and Jean-Paul Hanon, 'Introduction, Digitization and Robotization of the Battlefield: Evolution or Revolution?' in Ronan Doare et al. (eds.), *Robots on the Battlefield: Contemporary Issues and Implications for the Future* (Fort Leavenworth, KS: Combat Studies Institute Press, 2014), p. xvii.
10 Quoted from Armin Krishnan, *Killer Robots: Legality and Ethicality of Autonomous Weapons* (Surrey: Ashgate, 2009), p. 8.
11 Vadim Kozyulin, 'Russia's Automated and Autonomous Weapons and Their Consideration from a Policy Standpoint,' in Expert Meeting, Autonomous Weapon Systems, Implications of Increasing Autonomy in the Critical Functions of Weapons, Versoix, Switzerland, 15–16 March 2016, p. 60.
12 Nick Bostrom, *Superintelligence: Paths, Dangers, Strategies* (2014, reprint, Oxford: Oxford University Press, 2017), p. 18.
13 Krishnan, *Killer Robots*, p. 2.
14 Antonin Tisseron, 'Robotic and Future Wars: When Land Forces Face Technological Developments,' in Doare et al. (eds.), *Robots on the Battlefield*, p. 4.

15 Hynek and Solovyeva, *Militarizing Artificial Intelligence*, pp. 51–2.

16 Jai Galliott, *Military Robots: Mapping the Moral Landscape* (Surrey: Ashgate, 2015), p. 27.

17 MQ-1B Predator, available at: af.mil/About-Us/Fact-sheets/Display/Article/1 04469/mq-1b-predator/, accessed on 27 April 2023.

18 Jin-Yun Wang and Wei Ke, 'Development Plan of Unmanned System and Development Status of UUV Technology in Foreign Countries,' *Journal of Robotics and Control*, vol. 3, no. 2 (2022), pp. 1–9, DOI: 10.18196/jrc. v3i2.10201.

19 Williamson Murray and MacGrgeor Knox, 'Thinking About Revolutions in Warfare,' in MacGregor Knox and Williamson Murray (eds.), *The Dynamics of Military Revolution: 1300–2050* (2001, reprint, Cambridge: Cambridge University Press, 2003), pp. 1–14.

20 Clifford J. Rogers, '"As If a New Sun Had Arisen": England's Fourteenth-Century RMA,' in Knox and Murray (eds.), *The Dynamics of Military Revolution*, pp. 15–34.

21 See the classic work in this respect by Geoffrey Parker, *The Military Revolution: Military Innovation and the Rise of the West, 1500–1800* (Cambridge: Cambridge University Press, 1988).

22 For the historiography of Military Revolutions see Clifford J. Rogers (ed.), *The Military Revolution Debate: Readings on the Military Transformation of Early Modern Europe* (Boulder, CO: Westview Press, 1995).

23 Robert Drews, *The End of the Bronze Age: Changes in Warfare and the Catastrophe CA 1200 BC* (1993, reprint, Princeton, NJ: Princeton University Press, 1995).

24 Kaushik Roy, *Military Transition in Early Modern Asia: 1400–1750: Cavalry, Guns, Government and Ships* (London: Bloomsbury, 2014).

25 Philip Ross and Kasia Maynard, 'Towards a 4th Industrial Revolution,' *Intelligent Buildings International*, vol. 13, no. 3 (2021), pp. 159–61.

26 Mark Hagerott, 'Robots, Cyber, History, and War,' in Doare et al. (eds.), *Robots on the Battlefield*, p. 29.

27 Christian Malis, 'New Extrapolations: Robotics and Revolution in Military Affairs,' in Doare et al. (eds.), *Robots on the Battlefield*, p. 33.

28 Quoted from Arash Heydarin Pashakhanlou, 'AI, Autonomy, and Airpower: The End of Pilots?' *Defence Studies*, vol. 19, no. 4 (2019), p. 347.

29 Jonathan B. Bell, 'Countering Swarms: Strategic Considerations and Opportunities in Drone Warfare,' *JFQ*, 107, 4th Quarter (2022), pp. 5–6.

30 Sean J.A. Edwards, *Swarming and the Future of Warfare* (Santa Monica, CA: RAND, 2005), p. 2.

31 Paul Scharre, 'How Swarming Will Change Warfare,' *Bulletin of the Atomic Scientists* (2018), pp. 1–3, DOI: 10.1080/00963402.2018.1533209.

32 Edwards, *Swarming and the Future of Warfare*, p. 2.

33 Timothy May, *The Mongol Art of War* (Barnsley: Pen & Sword, 2007). See especially Chapter 5.

34 Zachary Kallenborn, 'InfoSwarms: Drone Swarms and Information Warfare,' *Parameters*, vol. 52, no. 2 (2022), p. 88.

35 Martin Van Creveld, *Command in War* (Cambridge, MA: Harvard University Press, 1985), pp. 1–16.

36 Major Matthew R. Voke, *Artificial Intelligence for Command and Control of Air Power* (Alabama: Air University Press, 2019), p. 3.

37 Ludovic Righetti, 'Emerging Technology and Future Autonomous Weapons,' in Expert Meeting, Autonomous Weapon Systems, Implications of Increasing Autonomy in the Critical Functions of Weapons, Versoix, Switzerland, 15–16

March 2016, pp. 37–8, available at: icrcndresourcecentre.org/wp-content/uplo ads/2017/11/4283_002, accessed on 12 November 2022.

38 Del Monte, *Genius Weapons*, p. 142.

39 Edge Compute for the Data-Centric AI-enabled Battlefield, Edge Compute White Paper of Systel Inc., 20 November 2022, available at: https://systelusa.com/whitepaper/edge-compute-for-the-data-centric-ai-enabled-battlefield/, accessed on 7 July 2023.

40 Benjamin Jensen and Dan Tadross, 'How Large-Language Models Can Revolutionize Military Planning,' *War on the Rocks*, 12 April 2023, waron therocks.com/2023/04/how-large-language-models-can-revolutionize-military -planning/, accessed on 7 July 2023.

41 Hui Li et al., 'MLRIP: Pre-Training a Military Language Representation Model with Informative Factual Knowledge and Professional Knowledge Base,' pp. 1–2, arxiv.2207.13929, accessed on 7 July 2023.

42 Merel Ekelhof and Giacomo Persi Paoli, 'Swarm Robotics: Technical and Operational overview of the next generation of Autonomous Systems,' p. 55, United Nations Institute for Disarmament Research, 2020, https://unidir.org.

43 R.S. Panwar, 'Regulation of AI-Enabled Military Systems: A Risk Based Approach,' https://futurewars.rspanwar.net/regulation-of-ai-enabled-military -systems-a-risk-based-approach-part-i/, accessed on 7 April 2023.

44 David Schaefer, *China's Drone Air Power and Regional Security* (New Delhi: KW Publishers, 2015), pp. 46–51.

45 Van Creveld, *Command in War*, p. 275.

46 James A. Green, 'Introduction,' in James A. Green (ed.), *Cyber Warfare: A Multidisciplinary Analysis* (2015, reprint, London: Routledge, 2016), pp. 1–2.

47 Duncan Hodges and Sadie Creese, 'Understanding Cyber-Attacks,' in Green (ed.), *Cyber Warfare*, pp. 49, 51.

48 Heather A. Harrison Dinniss, 'The Regulation of Cyber Warfare Under the *jus in bello*,' in Green (ed.), *Cyber Warfare*, p. 130.

49 Kritika Roy, *Advances in ICT and the Likely Nature of Warfare* (New Delhi: KW Publishers, 2019), p. 14.

50 Roy, *Advances in ICT*, pp. 39–40.

51 Paul Bracken, *The Hunt for Mobile Missiles: Nuclear Weapons, AI, and the New Arms Race* (Philadelphia, PA: Foreign Policy Research Institute, 2020), pp. 1, 6–8.

52 Bracken, *The Hunt for Mobile Missiles*, pp. 10–11.

53 Richard Stiennon, 'A Short History of Cyber Warfare,' in Green (ed.), *Cyber Warfare*, pp. 7, 20, 22.

54 Henning Lahmann, 'The Future Digital Battlefield and Challenges for Humanitarian Protection: A Primer,' pp. 12–13, Geneva Academy, April 2022, https://www.geneva-academy.ch/joomlatools-files/docman-files/working-papers /The%20Future%20Digital%20Battlefield%20.pdf, accessed on 8 April 2023.

55 Danny Steed, 'The Strategic Implications of Cyber Warfare,' in Green (ed.), *Cyber Warfare*, p. 77.

56 Ole Boe and Glenn-Egil Torgersen, 'Norwegian "Digital Border Defense" and Competence for the Unforeseen: A Grounded Theory Approach,' *Frontiers in Psychology*, vol. 9 (2018), pp. 1–14, DOI: 10.3389/fpsyg.2018.00555.

57 Space Doctrine Note Operations: Doctrine for Space Forces, pp. 5, 7–8, Headquarters, United States Space Force, January 2022, available at: https://media.defence.gov/feb/ accessed on 28 April 2023.

58 Hannes Mueller et al., 'Monitoring War Destruction from Space using Machine Learning,' *PNAS*, vol. 118, no. 23 (2021), pp. 1–9, DOI: 10.1073/pnas.2025400118.

59 Space Doctrine Note Operations: Doctrine for Space Forces, pp. 8–9.

60 Ilaria Carrozza et al., 'Algor-Ethics in the Emerging Battlespace, A Report Prepared by PRIO for the Norwegian Ministry of Defence,' p. 9, 2021, available at: https://www.coursehero.com/file/166495964/Algor-ethics-whole-report-Finalpdf/, accessed on 5 May 2022.

61 B.N.S. Yadava, 'Chivalry and Warfare,' in Jos J.L. Gommans and Dirk H.A. Kolff (eds.), *Warfare and Weaponry in South Asia: 1000–1800* (New Delhi: Oxford University Press, 2001), pp. 66–98.

62 Michael Walzer, *Just and Unjust War: A Moral Argument with Historical Illustrations* (1977, reprint, New York: Basic Books, 2006), p. 21.

63 Walzer, *Just and Unjust War*, pp. 59–63.

64 Galliott, *Military Robots*, p. 136.

65 Ilaria Carrozza, Nicholas Marsh and Gregory M. Reichberg, *Dual-Use AI Technology in China, the US and EU: Strategic Implications for the Balance of Power* (Oslo: PRIO, 2022), p. 29.

66 Ilaria Carrozza et al., *Algor-Ethics in the Emerging Battlespace, a Report prepared by PRIO for the Norwegian Ministry of Defence* (Oslo: PRIO, 2021), p. 7.

67 Guangyu Qiao-Franco and Ingvild Bode, 'Weaponised Artificial Intelligence and Chinese Practices of Human-Machine Interaction,' *The Chinese Journal of International Politics*, vol. 16 (2023), pp. 115–16, 121–22, DOI: https://doi.org/cjip/poac024.

68 Maaike Verbruggen, 'The Question of Swarms Control: Challenges to ensuring Human Control over Military Swarms,' Non-Proliferation and Disarmament Papers, no. 65, December 2019, p. 2, available at: https://www.sipri.org/publications/2019/eu-non-proliferation-and-disarmament-papers/question-swarms-control-challenges-ensuring-human-control-over-military-swarms, accessed on 13 April 2023.

69 Vincent Boulanin, Neil Davison, Netta Goussac and Moa Peldan Carlsson, *Limits on Autonomy in Weapon Systems: Identifying Practical Elements of Human Control* (Stockholm: SIPRI, 2020), p. ix.

70 Quoted from Paola Gaeta, 'Autonomous Weapon Systems and the Alleged Responsibility Gap,' in Expert Meeting, Autonomous Weapon Systems, Implications of Increasing Autonomy in the Critical Functions of Weapons, Versoix, Switzerland, 15–16 March 2016, p. 45.

71 Quoted from Bostrom, *Superintelligence*, p. 5.

72 Bostrom, *Superintelligence*, pp. 6–35.

73 Quoted from Kenneth Payne, *I, Warbot: The Dawn of Artificially Intelligent Conflict* (2021, reprint, London: C. Hurst & Co., 2022), p. 8.

74 Payne, *I, Warbot*, pp. 26–7.

75 Steven I. Davis, 'Artificial Intelligence at the Operational Level of War,' *Defense & Security Analysis*, vol. 38, no. 1 (2022), pp. 74–87.

76 Quoted from Peter Dortmans et al., *Supporting the Royal Australian Navy's Strategy for Robotics and Autonomous Systems: Building an Evidence Base* (Santa Monica, CA/Canberra: RAND Corporation, 2021), p. 89.

77 Springer, *Outsourcing War to Machines*, p. 9.

78 Quoted from Robert H. Latiff, *Future War: Preparing for the New Global Battlefield* (New York: Alfred A. Knopf, 2017), p. 22.

79 Latiff, *Future War*, p. 30.

80 Del Monte, *Genius Weapons*, pp. 140–56.

1

ARTIFICIAL INTELLIGENCE, NATIONAL SECURITY AND THE FUTURE OF WARFARE

P.K. Mallick

Introduction

> Throughout history, militaries that have failed to adapt to disruptive changes in war have paid a steep price, losing lives, battles, or even wars. Artificial intelligence is the future, not only for Russia, but for all humankind. It comes with colossal opportunities, but also threats that are difficult to predict. Whoever becomes the leader in this sphere will become the ruler of the world.
>
> Russian President Vladimir Putin, 1 September 2017[1]

Today is the age of the Fourth Industrial Revolution. The current period of rapid, simultaneous and systemic transformations driven by advances in science is reshaping industries, blurring geographical boundaries, challenging existing regulatory frameworks and even redefining what it means to be human. Artificial intelligence (AI) is the software engine that drives the Fourth Industrial Revolution. AI has the potential to change entire industries significantly, increase economic growth and impact all aspects of society. Recent developments demonstrate that AI is expanding quickly and has captured the attention of various groups, such as commercial investors, defence experts, policymakers and international competitors. AI will be a crucial component in future products and services and play a significant role in the Fourth Industrial Revolution, affecting every aspect of life.[2]

AI has been transforming the way modern warfare is conducted. The integration of AI in military operations has the potential to greatly enhance the capabilities and efficiency of military organisations, but it also raises important ethical and security concerns. The use of AI in warfare ranges

DOI: 10.4324/9781003421849-2

from the automation of mundane tasks, such as inventory management, to the integration of AI in advanced weapon systems. The US National Security Commission on Artificial intelligence concluded in its final report that 'although AI will be ubiquitous across all domains, the high-data volumes associated with the space, cyber, and information operations domains make use cases in those domains particularly well-suited for prioritized integration of AI-enabled applications in wargames, exercises, and experimentation.'[3] This chapter explores how AI is being used in warfare, the benefits and risks associated with its use, and the importance of ethical considerations in developing and deploying AI in military contexts.

AI as an emerging and disrupting technology (EDT)

In all the lists published by the different government and non-government entities, AI always figures among four of the top five technologies along with Biotechnology, Energy and Autonomy/Robotics. AI has a wide range of uses in defence, from improving decision-making to increasing the effectiveness of armed forces and protecting personnel by automating hazardous tasks by automating 'dull, dirty and dangerous' jobs. AI is being seen as a force multiplier for the military. It can enhance the speed and efficiency of support functions, improve network security and increase the military forces' mass, persistence, reach and effectiveness. AI brings its own unique challenges, which must be balanced with effective human oversight. AI needs to work effectively as a part of a distributed team. The AI system must be dependable and robust under various future scenarios and must be able to collaborate effectively with human teammates.

The study of AI started in the 1940s. The term 'artificial intelligence' was introduced at the Dartmouth Summer Research Project on Artificial Intelligence in 1955. It saw a significant surge in popularity around 2010 due to the combination of these factors: Access to large amounts of data, advancements in machine learning (ML) techniques, an increase in computing power, and the rise of the internet enabling any online system to access big data and to distribute intelligent behaviours among networked devices (see Figure 1.1).

Different areas of AI

AI can be divided into three types: Narrow AI, General AI and Artificial Superintelligence. Narrow AI can perform a single task, General AI can perform various tasks, and Artificial Superintelligence is a system that surpasses human cognitive abilities in most domains. Currently, General AI and Artificial Superintelligence do not exist and may never exist. Narrow AI is currently being incorporated into many military applications like

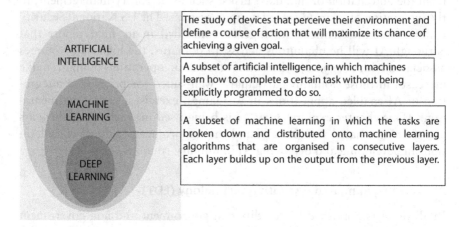

The study of devices that perceive their environment and define a course of action that will maximize its chance of achieving a given goal.

A subset of artificial intelligence, in which machines learn how to complete a certain task without being explicitly programmed to do so.

A subset of machine learning in which the tasks are broken down and distributed onto machine learning algorithms that are organised in consecutive layers. Each layer builds up on the output from the previous layer.

FIGURE 1.1 AI Terminology.

Source: Adapted from Laurie A. Harris, 'Artificial Intelligence: Background, Selected Issues, and Policy Considerations,' *Congressional Research Service*, 19 May 2021, https://crsreports.congress.gov/product/pdf/R/R46795, accessed on 14 January 2023.

Intelligence, Surveillance and Reconnaissance (ISR), cyber operations, logistics, semi-autonomous and autonomous vehicles, and command and control. AI-enabled capabilities can deliver game-changing advantages for national security and defence like:

- Accelerated and better decision-making.
- Enhanced human cognitive and physical performance.
- Heightened military readiness and operational competency.
- New systems of design, manufacture and sustainment of military systems.
- Capability to produce and detect strategic cyberattacks and information operations.
- Innovative capabilities that can upset delicate military balances.

The importance of AI on national security has been amply clarified in the National Defense Strategy of the USA, released on 12 October 2022.[4] The National Defense Strategy of the Pentagon states:

New applications of artificial intelligence, quantum science, autonomy, biotechnology, and space technologies have the potential not just to change kinetic conflict, but also to disrupt day-to-day U.S. supply chain and logistics operations. We will be a fast-follower where market forces are driving commercialization of militarily-relevant capabilities in trusted artificial intelligence and autonomy, integrated network system-of-systems, microelectronics, space, renewable energy generation and storage, and human-machine interfaces. Because Joint Force operations increasingly rely on data-driven technologies and integration of diverse data sources, the Department will implement institutional reforms that integrate our data, software, and artificial intelligence efforts and speed their delivery to the warfighter.

AI raises several questions that have to be addressed:

- How will AI change warfare?
- What kinds of military AI applications are possible?
- What limits, if any, should be imposed?
- What effect will it have on the military balance with adversaries?
- What unique advantages and vulnerabilities come with employing AI for defence?

Definitions

It is universally acknowledged that there is no commonly accepted definition. The US DoD AI Strategy of 2018 defined AI as 'the ability of machines to perform tasks that normally require human intelligence—for example, recognizing patterns, learning from experience, drawing conclusions, making predictions, or taking action.'[5] The US Defense Innovation Board (DIB) defines AI as 'a variety of information processing techniques and technologies used to perform a goal-oriented task and the means to reason in the pursuit of that task.'[6] These techniques can include, but are not limited to, symbolic logic, expert systems, ML and hybrid systems.

The following relevant concepts as regards AI, especially regarding its military applications, are helpful.[7] First is automation. Automated or automatic systems function with no (or limited) human operator involvement, typically in structured and unchanging environments, and the system's performance is limited to the specific set of actions that it has been designed to accomplish. Typically, these are well-defined tasks that have predetermined responses according to simple scripted or rule-based prescriptions. Then comes autonomy. This term can be defined as the condition or quality of being self-governing in order to achieve an assigned task based on the system's own situational awareness (integrated sensing, perceiving and analysing), planning and decision-making. Then comes the Lethal Autonomous Weapon System (LAWS). LAWS could be termed as a weapon system that, once activated, can select and engage targets without further intervention by a human operator. In contrast, a Semi-Autonomous Weapon System, once activated, is intended to only engage individual targets or specific target groups that have been selected by a human operator. A robot is a powered machine capable of executing a set of actions by direct human control, computer control or a combination of both. At a minimum, it is comprised of a platform, software and a power source.

AI is a conglomeration of several distinct technologies. Algorithms, data and computing power are basic building blocks of modern AI technologies. The availability of very large data sets has been critical to the development of ML and AI. Though the fields of big data analytics and AI are interrelated, there are important differences between the two.

Big data analytics attempts to make sense and value out of huge amounts of digitally recorded data. AI processing is a method by which the data can be combined, categorised (labelled) and analysed. Not all big data analytics use AI, but much of it does. All AI is dependent on big data input. Large quantities of data are needed to train AI. Usually the term Big Data is used to describe data that are too large to be stored on a computer's memory, are generated too quickly to be managed by a single computer, or take many different forms or formats. Figure 1.2 shows the interrelationship between AI, ML and Data Science.

The evolution and differences between big data analytics, predictive big data analytics and ML are sometimes used interchangeably in discussions of AI. ML is a subset of AI techniques that have allowed researchers to tackle

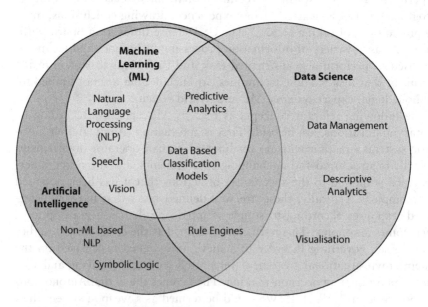

FIGURE 1.2   Overlapping Technologies.

Source: Adapted from Defence Artificial Intelligence Strategy 2022, June 2022. Available at: https://assets.publishing.service.gov.uk/government/uploads/system/uploads/attachment_data/file/1082416/Defence_Artificial_Intelligence_Strategy.pdf

many problems previously considered impossible, with many applications across national security and defence.[8] Quality data and clear judgement have made AI successful in the commercial world. These may not be available to the same degree for all military tasks. In the military, judgement includes command intentions, administrative management, rules of engagement and moral leadership. These functions cannot be automated with narrow AI technology. Increasing reliance on AI will make human beings' judgement even more important for military power. This incentivises adversaries to improve protection of their information systems and command institutions and intervention in the adversaries' information network and command set up. Instead of the idea that AI will lead to swift robot wars and significant changes in military power, we anticipate that conflicts involving AI will be characterised by environmental uncertainty, internal resistance and political disputes.[9]

AI is not just technology; it is a stack and an enabler more akin to inventions such as electricity or the internal combustion engine than the battle tank or fighter aircraft. It is a vast topic including Data and Data Science, how data is stored, transported and secured, which involves all three paradigms of computer science. It involves exquisite algorithms like GPT-3, applications, domain experts, integration and precision engineering. Talent for each one of these fields is different.

Dual-use technology

From the two World Wars to the Cold War era, most major defence-related technologies, including nuclear technology, the internet and the Global Positioning System (GPS), were first developed by government-funded research programmes before spreading to the commercial sector later. The private sector is now at the forefront of AI innovation. The progress made in the development and application of AI in both academic and commercial fields is faster than the pace at which the military can obtain and utilise the most recent advancements. Today, commercial companies are leading AI development, with armed forces adapting their tools for military applications later.

A lot of research on crucial AI applications and the level of human control over them have dual purposes. For example, the specifications for an autonomous drone used for package delivery are similar to those used for delivering explosives. The same image recognition software that is used to identify cats on YouTube could also be used by military drones to detect terrorist activity in Syria and Afghanistan.

AI and autonomous systems are attractive technologies to militaries for the following reasons:

- AI can outperform humans in automated military and civilian tasks and in more demanding tasks. AI systems have demonstrated superior performance to humans in dogfighting, board games and cancer detection tasks. AI has the potential to improve the precision of weapons, increase the accuracy of data analysis and provide earlier alerts of attacks.

- AI can process information much faster than humans. This speed is crucial as AI-enabled technologies become more integrated into military decision-making processes.
- AI algorithms may be embedded in and enhance automated systems, which carry out tasks where soldiers risk their lives neutralising mines or conducting intelligence, surveillance and reconnaissance in dangerous areas. Autonomous vehicles can perform these tasks without losing human lives.
- Autonomous systems can act as a force multiplier. They can perform tasks such as patrolling areas for extended periods without needing to rest or consume resources. This allows military personnel to be freed up to perform other duties.

Most of the Research and Development (R&D) on AI is conducted by the private sector. These efforts have significantly advanced this technology, identified new types of applications and decreased costs. Much of the foundational research is publicly available, and researchers have also created open-source tools to promote wide use. Though such research and tools are not focused on military applications, many of these capabilities are dual or multi-use across civilian and military contexts.

The significant recent breakthroughs in AI have been achieved by US software companies focusing on private sector customers. The best tools that could prove critical to the military, such as algorithms capable of identifying enemy hardware or specific individuals in videos, are built at companies like Google, Amazon and Apple or inside start-ups. These top companies compete fiercely to attract the best talent available, offering lucrative salaries, advanced technical toolsets, and a workplace culture with less formal processes and bureaucracy than a typical government agency or government contractor. Generally, the defence industrial base (DIB) consists of a relatively small number of companies specialising in serving the defence industry. If the armed forces want to maintain access to the companies that have the most experience deploying AI solutions and the technical experts at the cutting edge of AI research, they will have to collaborate with companies that do not think of themselves as defence contractors.

Several obstacles related to technology, process, personnel and culture hinder AI implementation in the military. The current procedures within the military, including those related to safety, performance standards, procurement, and intellectual property and data rights, create obstacles to integrating AI in military operations. The safety and performance standards for civilian and military use are not always similar or easily transferable. The tolerance level for failure in civilian AI applications may not be appropriate for military situations or vice versa. AI systems may encounter unexpected failures in complex environments like combat. These challenges may make it harder to apply commercially developed AI technology in the military.

For developing AI for military applications, there will be several non-technical and technical challenges:

- The leading AI companies may be reluctant to partner with the military due to the complexity of the defence acquisition process.
- Commercial technology companies are also often reluctant to partner with DOD due to intellectual property and data rights concerns. Intellectual property is the 'lifeblood' of commercial technology companies, yet the 'armed forces is putting increased pressure on companies to grant unlimited technical data and software rights or government purpose rights rather than limited or restricted rights.'[10]
- Non-technical challenges include investment, innovation and workforce challenges. Militaries will need to invest enough capital in research and development. The armed forces need to improve their ability to adapt and integrate technologies from the non-defence commercial sector and to find ways to attract the top AI experts, many of whom are offered higher pay in the private sector.

AI applications for defence

AI encompasses a wide range of technologies, many of which offer great promise in military applications. These capabilities are welcome developments in the eyes of many military operators. They provide prospects for dramatic increases in combat power and the ability to accomplish mission objectives faster and with less exposure to lethal threats. AI is being incorporated into many applications like ISR, logistics, cyberspace operations, information operations, command and control, semi-autonomous and autonomous vehicles and LAWS, AI-assisted decision-making, AI-assisted Common Operating Picture (COP), drones and swarms.[11]

Other reports on military AI have used slightly different categorisations and use terminology in different ways. For instance, the SIPRI report 'Mapping the Development of Autonomy in Weapon Systems' uses five categories to describe existing autonomous military systems: Air defence systems; active protection systems; robotic sentry weapons; guided munitions and loitering weapons.[12]

Many experts believe that AI will revolutionise warfare by allowing military commanders to supplement or replace their crewed served weapons with various unmanned systems. As warfare becomes increasingly fast-paced, battle commanders are likely to place greater reliance on AI-enabled machines to monitor enemy actions, evaluate the trove of information that is collected and initiate appropriate countermeasures. 'AI applications will help militaries prepare, sense and understand, decide and execute faster and more efficiently. In the future, warfare will pit algorithm against algorithm,' the National Security Commission on Artificial Intelligence (NSCAI) affirmed in its 2021 report.[13]

Some of the benefits of military application of AI are as follows:

- Increases speed of decision-making.
- Uses big data.
- Improves targeting and vision.
- Provides decision-making support.
- Mitigates manpower issues.
- Improves cyber defence.
- Improves accuracy and precision.
- Reduces labour and costs.
- Improves intelligence, surveillance and reconnaissance (ISR).
- Helps operate in anti-access/area denial (A2/AD) environments.
- Improves deception and information operations.
- Removes emotion from decision-making.

Although the military applications of AI are expected to yield a wide range of benefits, they also present significant risks. Some of these risks are:

- Decisions may be made too fast.
- Could result in increased escalation.
- Might be less accurate/precise than humans.
- Difficult to differentiate combatant from non-combatant.
- Difficult to differentiate anomaly from threat.
- Cannot adapt to programming.
- May result in automation bias.
- Data might be dirty, resulting in errors.
- Reduces attribution, encouraging bad behaviour.
- Could cause arms racing and proliferation.
- Systems might be fielded before full safety testing.
- Humans could lose essential skills.
- Might be vulnerable to spoofing/hacking.
- Behaviour might not be explainable.
- May exhibit emergent behaviour.
- It could lower the costs of war.

Intelligence, rurveillance and reconnaissance (ISR)

ISR is a complex big data process that includes sensors, people, platforms, networks and weapons. It also integrates all domains of surface, undersea, air, space and cyberspace. The ability to autonomously collect intelligence from drones, sensors in the terrestrial domain, space and cyberspace promises to increase the amount of data generated. The volume, velocity and variety of data will need to be analysed in part or whole by machines using AI. Some of that analysis will need to be done on ISR platforms deployed in the field, as bandwidth limitations would make it infeasible to transfer such large quantities of data. Much of the analysis will be done in intelligence processing centres. Wherever it is done, AI will enable dramatic improvements in the quality of intelligence derived from the masses of ISR data collected.

The proliferation of sensors, analytical tools, precision-guided munitions and non-kinetic payloads (i.e., cyber, directed energy) fundamentally alters the hider-finder contest. As sensors and analytical tools continue to develop and proliferate, hiding them in every domain, including space and undersea which have traditionally been the most opaque, will become increasingly difficult. AI-enabled intelligence, surveillance and reconnaissance platforms and indication and warning (I&W) systems will be critical for advanced warfighting capabilities. AI-enabled systems will optimise tasking and collection for platforms, sensors and assets in near-real time responding to dynamic intelligence requirements or environmental changes. At the tactical edge, 'smart' sensors will pre-process raw intelligence and prioritise the data to transmit and store, which will be helpful in degraded or low-bandwidth environments.

The following will happen:

- Autonomous vehicles will conduct ISR, collecting vast amounts of data and accessing areas too dangerous for human access.
- AI algorithms will consolidate and analyse data (for instance detect patterns of life, conduct human terrain mapping and social network analysis) from a number of sources and sensors and provide decision support for human operators.
- AI will be used for image recognition and target discrimination.
- AI will enhance early warning systems and serve as a virtual assistant (like Google Home or Alexa) to human operators.

Military drones for surveillance will:

- Channel remote communication, both video and audio, to ground troops and military bases.
- Track enemy movement and conduct reconnaissance in unknown areas of a war zone.
- Assist with mitigation procedures after a war by searching for lost or injured soldiers and giving recovery insights for a terrain.

Intelligence

AI is extremely useful in the intelligence acquisition process due to the large data sets available for analysis. AI will help intelligence professionals find needles in haystacks, connect the dots, discern trends and discover previously hidden or masked indications and warnings. AI-enabled capabilities will improve every stage of the intelligence cycle, from tasking through collection, processing, exploitation, analysis and dissemination. AI algorithms can sift through vast amounts of data to find patterns, detect threats, identify correlations and make predictions.

AI can automate the work of human analysts who now spend hours sifting through drone footage, satellite imagery, communications signals, economic indicators, social media data and other intelligence inputs from Open Source Intelligence (OSINT), Human Intelligence (HUMINT) etc., for actionable information, potentially freeing analysts to make more efficient and timely decisions based on the data.[14] For example, Project Maven would incorporate computer vision and AI algorithms into intelligence collection cells that would comb through footage from unmanned aerial vehicles (UAVs) and automatically identify hostile activity for targeting.

The intelligence community in the USA has sponsored AI research projects like developing algorithms for multilingual speech recognition and translation in noisy environments, geo-locating images without the associated metadata, fusing 2D images to create 3D models, and building tools to infer a building's function based on pattern-of-life analysis.[15]

An ideal example of the use of AI for intelligence collection is Project Maven of the Pentagon. The intelligence agencies are overburdened with masses of intelligence data in terms of surveillance and reconnaissance data, unmanned systems video, paper, computer hard drives, thumb drives and many more such resources collected by the intelligence agencies for operational use. Pentagon wanted AI to lead the hunt for Islamic State militants in Iraq and Syria by turning countless hours of aerial surveillance video into actionable intelligence. Project Maven aimed at using AI and machine learning algorithms to improve the analysis of drone footage. It used computer vision algorithms to automatically analyse drone video footage, enabling military analysts to quickly identify objects of interest and perform pattern recognition on the data. The Maven algorithms augmented or fully automated the object detection, classification, and alert tasks using computer vision, supporting the Defeat-ISIS campaign. Maven was a resounding success and surpassed expectations. By 2020, Maven was being applied across multiple conflicts,

TABLE 1.1 AI use in various tasks related to Intelligence

Tasking	• Sequencing and deconflicting the tasking of intelligence platforms efficiently across collection disciplines. • Detecting and prioritising targets of interest by analysing patterns that suggest opportunities to collect unique information or exploit vulnerabilities. • Assisting decision-makers in identifying information requirements and prioritising collection targets.
Collection	• Identifying potential opportunities for collection by finding gaps in technology-enabled counterintelligence and security systems as well as alternative collection pathways to an intelligence target. • Automating the validation process by cross-checking collection across all other reporting and collection disciplines. • Enabling smart sensors at the edge to improve collection fidelity and trigger collection when necessary.
Processing	• Transforming unstructured data into a structured, filterable, sortable and digestible data to aid analysis. • Employing natural language processing to transcribe, translate and summarise foreign language materials. • Summarising raw intelligence reporting with critical information highlighted and tailored for analysts.
Analysis	• Accelerating pattern matching and anomaly detection across intelligence disciplines and the intelligence record. • Generating visualisations to illustrate relationships, networks, geographies and time lapses. • Automating portfolio-specific indications and warning alerts for analysts.
Dissemination	• Tracking usage and impact of disseminated intelligence reporting and analysis. • Automating the creation and delivery of finished and raw intelligence to the appropriate users and analysts at any level of classification. • Streamlining classification downgrading to facilitate intelligence sharing with other government agencies, allies and the private industry.
Practices	• Systematising the auditing and approval process for routine business practices, such as accounting, as well as flagging anomalies for manual review. • Monitoring IT systems to provide predictive maintenance and upkeep requests. • Supporting the workflow and review requirements for contracting officer's technical representatives through a project's life cycle.

(Continued)

TABLE 1.1 (Continued)

Security	• Strengthening physical security measures by enhancing network video surveillance, trace detection and other intrusion detection systems. • Augmenting security clearance investigation and continuous evaluation. • Mapping supply chains of IC vendors and equipment.

Source: Mid-Decade Challenges to National Competitiveness, Special Competitive Studies Project, September 2022, https://www.scsp.ai/2022/09/special-competitive-studies-project-releases-first-report-sept-12-2022/.

Logistics

In the area of military logistics and maintenance, AI can create revolutionary cost-saving efficiency, which is why most militaries prioritise making progress on this front. Logistics may lead to the most radical changes in how armed forces 'do business.' AI systems can optimise the procurement process and automate supply chains. While minimising costs, they can forecast the need to repair equipment and order resupplies. AI can be used to post personnel by helping militaries figure out which soldier is best suited to what unit.

Many countries have started using AI to improve logistical issues like predictive maintenance, acquisition, personnel management and healthcare. Predictive maintenance is the most common logistical problem that governments are using AI to solve. Several Western countries have started using AI to monitor fleets of ships and aircraft to predict when each system will need maintenance. The US Army's Logistics Support Activity (LOGSA) has contracted IBM's Watson to develop tailored maintenance schedules for the Stryker Fleet based on information pulled from the 17 sensors installed on each vehicle. LOGSA started a second project to use Watson for analysing shipping flows for repair parts distribution, attempting to determine the most time and cost-efficient means to deliver supplies.[16]

National Mission Initiatives (NMIs) use AI to create efficiencies and reduce costs for maintenance by predicting in advance when a component might fail. This technique is known as predictive maintenance. Rather than waiting for a system or part to fail before fixing it or relying on a set, AI could provide a unit-based, specially tailored recommendation.

Cyberspace operations

In the cyber world, attackers and defenders will rely on AI, which may lead to unintended consequences. With the introduction of AI, the relationship between attackers and defenders could shift. AI would make the job of cyber defenders more difficult, with faster computers enabling increasingly complex attacks and more rapid network intrusions. Employing ML increases the chances of discovering vulnerabilities in code significantly. The AI software could then exploit autonomously, leading to greater efficiency and speed of offensive cyber operations. However, the same methods can be used to develop more robust defensive systems capable of automatically fending off malware or other adversarial intrusions into networks.

AI could help offensive cyber operations in the following ways:[17]

- Cyberattacks would be more precise and tailored.
- Spear-phishing abilities would be improved due to the rise of sophisticated natural language processing.
- To avoid detection and countermeasures, malware could mutate into thousands of forms once it is inside a network.
- The need to guarantee stable communication links between the different systems and components in the digital battlefield would increase the attack surface, rendering the entire ecosystem much more susceptible to adversarial cyberattacks.
- ML systems that learn from training data can be made vulnerable to data-poisoning attacks, in which the training data is manipulated or spoofed to influence the intended functioning of the system.
- Cyberattacks can be intentionally designed to trick or fool algorithms into making mistakes.
- AI-enabled bots can automate network attacks, making it difficult to extinguish the attacker's command and control channels.
- Patterns of pixels can be injected into images to change what an AI system sees; humans cannot detect the changes to the original images.

AI is crucial for improving military cyber operations by providing more advanced defence against hacking attempts. Traditional cybersecurity methods rely on finding previous instances of known malicious code, but hackers can easily bypass this by making small changes to the code. With AI, security systems can learn to identify abnormal patterns in network activity, creating a more robust and adaptable defence. Major reforms are required including extensive integration of AI-enabled cyber defences to match and neutralise offensive AI-cyber techniques.

For the defenders, AI could help in the following ways:

- AI could accelerate the detection of attackers inside a network.
- ML could help automate vulnerability discovery, deception and attack disruption.

Adversaries may launch AI-enabled cyber weapons with exceptional speed, accelerating the human capacity to exploit digital vulnerabilities. The defending state may have no time to evaluate the signs of an incoming attack; they have to respond immediately or risk disablement. The state may respond near simultaneously before the event can occur. This could lead to new forms of automated preemption or anticipatory self-defence and strain the legal and policy frameworks that guide government decision-making.

AI will enhance the magnitude, accuracy and longevity of adversarial information operations. It will provide significant power to nations that utilise it effectively. AI is making cyberattacks and misinformation campaigns more dangerous. Today it is possible for many autonomous agents to generate text snippets or short conversations to persuade a target audience to believe a particular narrative of geopolitical or military significance. AI can analyse the large amounts of data that people reveal about themselves online and gain an improved understanding of how to tailor specific messages to influence them.[18]

AI aggravates the problem of malign information in the following ways. AI can produce original text-based content and manipulate images, audio and video, including 'deepfakes', that will be very difficult to distinguish from authentic messages. AI can construct profiles of individuals' preferences, behaviours and beliefs to target specific audiences with specific messages. Further, AI can be embedded within platforms through ranking algorithms to proliferate harmful information. AI-enabled malicious information campaigns can send one powerful message to one million people and deliver a million individualised messages tailored to the targeted individuals' specific digital profiles, emotional states and social connections. While social media companies have significant resources dedicated to monitoring and managing the information on their platforms, collaboration between government agencies and these companies remains inconsistent and improvised. Cheap and commercially available AI applications ranging from deepfakes to lethal drones are becoming available to rogue states, terrorists and criminals. Defending against AI-enabled opponents acting at machine speeds without utilising AI is a recipe for disaster. Humans alone will not be able to match or defend against AI-enabled cyber or disinformation attacks without the help of AI-enabled machines.

AI is creating more convincing photo, audio and video manipulations, known as deepfakes, that can be used as part of information operations to spread false information, shape public opinion, undermine public confidence and potentially blackmail diplomats. Progress in machine learning and computer graphics has increased the ability of both government and non-government actors to create and disseminate highly realistic audio-visual content known as synthetic media and deepfakes. AI-generated deepfakes can now produce material indistinguishable from actual people, places and occurrences, posing a threat to national security. The technology is becoming advanced enough that it may soon be able to evade detection by forensic analysis tools.

AI may be utilised to construct complete 'digital patterns-of-life,' where an individual's digital behaviour is combined and matched with records of purchases, credit history, employment history and subscriptions to create a detailed behavioural profile of service members, suspected intelligence agents, government officials or private citizens. Like deepfakes, this information can then be used for specific influence operations or blackmail.

The spread of synthetic media has had a worrying outcome: Malevolent actors have labelled real situations as 'fake,' exploiting the new forms of deniability that have emerged with the erosion of trust in the era of deepfakes. As AI becomes more advanced and the cost of computer processing decreases, the difficulty that deepfakes present to online information environments during armed conflicts will only increase.

Deepfakes can be used for various aims, such as creating fake commands from military leaders, causing uncertainty among the public and military and providing legitimacy to conflicts and rebellions. Even though these strategies frequently fail, as they will continue to do so more often than not, the ability to influence an adversary's communication and messaging means that security and intelligence officials will inevitably employ them in various operations.

The ability to hack and manipulate data links, ISR platforms or force trackers will permit adversaries to deceive a state into going astray, seriously disrupting or derailing ongoing operations, and injecting uncertainty about the validity of reports, intelligence analysis or performance of vital systems. Even when armed forces have deepfake detectors, leaders will have to doubt the reports they receive and their understanding of the world around them. Leaders that cannot determine the truth cannot fight effectively.

The employment of AI would increase the potential impact of disinformation campaigns and other types of information warfare, enabling such efforts to become more efficient, scalable, and widespread.[19]

Command and control

AI plays an increasingly significant role in command and control systems, with uses ranging from data acquisition and analysis to presenting information to the operator. AI is proving highly useful in decision-making assistance. During military operations, it is crucial to quickly acquire information from the battlefield and use AI techniques to combine and analyse it for the commander. The commander needs this information to make crucial decisions quickly in high-stress situations. However, the volume of information can often be so large that there is a risk of overwhelming the commander. The issue arises when the information is not presented clearly, concisely and meaningfully, so that the commander can easily understand. AI-based tools are becoming increasingly important for reducing the workload of decision-makers. Situation assessment involves understanding and interpreting the operational scenario and potential threats to achieve tactical and strategic goals. While the experience and knowledge of the enemy and own forces are still crucial for properly assessing the situation, AI can assist by facilitating the initial assessment process through data fusion, pre-processing the analysis and presenting the information in a clear and structured manner.

The Common Operational Picture (COP) is the critical element of the presentation of information in command and control systems. The COP is often represented by a map that displays objects of interest. In traditional systems, these objects are manually keyed in by a human operator and then made available to authorised users of the COP. Today information to decision-makers comes in diverse formats from multiple platforms, often with redundancies or unresolved discrepancies. ML techniques automatically enable information to be added to the map through the analysis of satellite images or reconnaissance drone footage, for example. This allows objects of interest, such as enemy forces or critical infrastructure, to be detected, analysed and presented to decision-makers automatically.

C2 systems can potentially deliver critical system support when time is limited, when communications links are cut by an adversary or when the number of options is too large for people to be able to analyse alternative courses of action. Thus, the strategic importance of using AI at the tactical and operational levels can hardly be exaggerated.[20] Despite this, human judgement will significantly influence command and control decision-making. However, the increasing capabilities and rapid development of AI-based tools will allow humans to focus on tasks that still perform better than machines, rather than spend time on secondary tasks.[21]

Speed and decision-making

The high speed of processing huge data sets can reduce the time for decision-making in the planning and execution of operations significantly, as well as increase the efficiency of troop and weapons management. AI is a tactical and strategic tool for outpacing adversary decision-making processes, commonly known as the Observe, Orient, Decide, and Act (OODA) loop. It is believed that human-machine teams augmented by AI-enhanced technologies will act more quickly than opponents in a conflict, gaining decisive advantages that could enable victory. AI can help in military decision-making through image recognition across ISR, predictive analytics for maintenance and route planning, collecting valuable open-source information by web trawlers and increased situational awareness by AI-enabled sensing. However, today's AI capabilities carry common risks that may make emphasising speed less desirable. The importance of the OODA loop's speed ignores the limitations of the complexity of human-machine teaming, the importance of decision quality and attention to timing. Focusing on speed de-emphasises potential risks like un-explainability behind AI decisions, inadvertent escalation, legal or ethical concerns and training and data issues. Quicker decisions are not necessarily better. Getting disassociated from the adversary's OODA Loop may be less helpful. Technology cannot always plug precisely into human processes and may require humans to adapt to avoid automation bias.

The following issues are critical.[22]

- Fast-moving warfare and compressing decision-making will have strategic implications that cannot be easily anticipated.
- The line between the handoff from humans to machines will become increasingly blurred.
- Synthesis and diffusion of AI into the military decision-making process will increase the importance of human beings across the entire chain of command.

AI could lead to unintended escalation. A wargame conducted by RAND Corporation in 2020 found that widespread AI and autonomous systems with machine decision-making speeds leading to quicker escalation and weakening deterrence could lead to inadvertent escalation and crisis instability.[23] AI-enabled militaries could significantly speed up inadvertent escalation once some course of action is triggered. If forces are designed to respond at machine speeds, they may also escalate a conflict at machine speeds. Some of the critical issues to consider are:

- Is there adequate time and are there sufficient de-escalation procedures to avoid unintended escalation?
- What is the risk of not responding to an attack where decision-making is happening at machine speed?
- If both sides are risk-averse, are there incentives to seek a first strike advantage?
- AI relies heavily on data obtained in peacetime. How will the decision-making change once conflict begins?
- Will forces with slower, centralised decision-making during peacetime have faster, decentralised reactions during the conflict?
- What systems will keep humans in the loop?
- Will some processes become more autonomous, with significantly faster OODA loops?

Even as AI expands its wings, warfare will remain human-centric. Military decision-making processes in non-linear, complex and uncertain environments require much more than copious, cheap datasets and inductive machine logic. Commanders' intentions, the rules of law and engagement, and ethical and moral leadership are critical in military decision-making. As humans will remain the essential elements in the decision cycle for the conceivable future, effectively integrating human judgement and machine function with AI will become either a source of competitive military advantage or a liability.[24]

In high-intensity, fast-moving, data-centric, multi-domain human-teaming environments, pressures to make decisions might weaken the critical role of 'mission command,' which connects tactical leaders with the political-strategic leadership and is a decentralised, lateral and two-way communication between senior commanders and junior officers.[25] AI used to complement and support humans in decision-making might preclude the role of human 'genius' in mission command when it is most demanded. Commanders' intuition, latitude and flexibility will be demanded to mitigate and manage the unintended consequences, organisational friction and strategic surprise. The complex concepts of speed and time require a delicate balance. Often, the timelines are dominated by the time it takes to move equipment or people or even just the time that munitions are moving to targets. In these cases, it is important not to overstate the value of accelerating the decision process.

Command and ethics

Semi-autonomous and autonomous vehicles and weapon systems will enable vehicles and weapon systems to be capable of performance, speed and discrimination levels that exceed human capabilities. They could reduce the risk of accidental engagements, decrease civilian casualties, minimise collateral infrastructure damage and allow detailed auditing of the decisions and actions of operators and their command chains. All US military services are working to include AI into semi-autonomous and autonomous vehicles and weapon systems, including fighter aircraft, drones, ground vehicles and naval vessels.[26] Autonomous vehicles are perhaps the most prevalent autonomous military systems. They include unmanned aerial vehicles (UAVs), unmanned ground vehicles (UGVs) and unmanned underwater vehicles (UUVs).

Autonomous weapon systems combine AI software and combat platforms to attack enemy assets on their own. The US Defense Department defines such a device as 'a weapons system that, once activated, can select and engage targets without further intervention by a human operator.' Typically, this sort of weapon is enabled to conduct activities within certain parameters programmed into the software, such as the geographical space within which they can operate or the types of targets they can engage.[27]

Military advantages of semi-autonomous and autonomous vehicles and weapon systems are as follows.[28]

- Significant strategic and tactical advantages in the battlefield.
- Autonomous weapon systems act as a force multiplier. Fewer soldiers are required for a given mission, and the effectiveness of each soldier is greater.

- Autonomous weapon systems expand the battlefield, permitting combat to reach areas that were previously inaccessible.
- They can reduce casualties by removing soldiers from dangerous missions, like explosive ordnance disposal.
- Robots are better suited than humans for 'dull, dirty, or dangerous missions.' A dirty mission exposes humans to potentially harmful radiological material.
- Unmanned systems can be used to deliver supplies and evacuate casualties.
- They can strike lethally even when communication links have been severed.
- The judgements of autonomous weapon systems will not be affected by emotions such as fear or hysteria. The systems will be capable of processing much more incoming sensory information than human beings without distorting or discarding it to fit preconceived notions.
- By reducing the amount of equipment that soldiers have to carry, they improve awareness of the situation at the squad and platoon levels and acts as partners rather than just tools.
- They can free up personnel for other tasks, such as intelligence functions, training, education, planning and leadership. This can enable the soldiers to be deployed in areas that require more leadership and creativity.
- Introduce new operational concepts. It could saturate an operational area with small autonomous systems that force an adversary to move, be detected and be targeted by friendly forces. They can employ autonomous air, ground and naval systems to attack A2/AD systems before the introduction of ground forces.

There is currently a lack of scientific evidence that robots possess the necessary capabilities for precise target identification, understanding the situation or determining the appropriate level of force. Therefore, their use may result in a high level of accidental harm. It is not appropriate to leave decisions about the use of violent force to machines. The use of autonomous weapon systems goes against the 'Principle of Distinction,' which is a fundamental rule of armed conflict. These systems may have difficulty distinguishing between civilians and combatants, a task that is challenging even for humans. This means that using AI for targeting decisions is likely to lead to unintended harm to civilians and excessive collateral damage. There is a problem of accountability when autonomous weapon systems are used. International Humanitarian Law (*jus in bello*) holds that someone must be held responsible for civilian deaths. Any weapon or other means of warfare that makes it impossible to identify who is responsible for the casualties caused is not in compliance with *jus in bello* and should not be used in war.

Human judgement must be involved in decisions to take human life in armed conflict. The kind of involvement necessary for humans to remain accountable for the use of autonomous weapon systems will vary depending on the time-criticality of the situation as well as the operational context, circumstance and type of weapon systems involved. It is incumbent upon states to establish processes which ensure that appropriate levels of human judgement are relied upon in the use of AI-enabled and autonomous weapon systems and that human operators of such systems remain accountable for the results of their employment.

LAWS can identify and attack targets independently without human control. They use computer algorithms and sensors to determine if an object is a threat, decide when to attack and guide the weapon to the target. This allows them to function in situations where traditional weapons may not be able to operate, such as when communications are limited or unavailable.[29] Currently, there are no laws that prohibit the development of LAWS. However, an international group of government experts has begun to discuss the issue. About 30 countries have called for a ban on these systems because of ethical concerns, while others have called for regulations or guidelines for their development and use. The US Department of Defense Directive 3000.09 sets guidelines for the development and use of LAWS to ensure they follow the laws of war, treaties, weapon safety rules and rules of engagement. The directive states that all systems, regardless of their classification, must be created in a way that allows commanders and operators to control the use of force and comply with the department's weapons review process.[30]

There are potential operational risks to using LAWS. These risks could arise from hacking, enemy behavioural manipulation, unexpected interactions with the environment, or simple malfunctions or software errors. While such risks could be present in automated systems, they could be enhanced in autonomous systems, in which the human operator would not be able to physically intervene to terminate engagements. This could result in fratricide, civilian casualties or other unintended consequences.

Opportunities and challenges for AI in defence

AI and digital capabilities are a double-edged sword. They present a multitude of opportunities to improve defence and security functions, as well as a variety of new ways for criminals and terrorists to advance their methods of operation.

Opportunities

- AI technology is good at solving well-defined, narrow problems where the necessary data and feedback are fully available to the system.

- It can facilitate autonomous operations, lead to more informed military decision-making and will likely increase the speed and scale of military action.
- It can be used in military scenario planning at all levels of command to determine the requirements for a mission. AI ensures the most advantageous use of resources during a mission.
- Improved situational awareness and enhanced targeting and reconnaissance. AI can analyse large amounts of data from various sources, such as satellite imagery, unmanned aerial vehicles and sensors, in real-time and provide military leaders with a more comprehensive and up-to-date picture of the battlefield and identify potential targets and gather intelligence on enemy forces.
- Although humans will be present, their role will be less significant and the technology will make combat less uncertain and more controllable, as machines are not subject to the frailties that cloud human judgement, like being tired, frightened, bored or angry.
- Far superior at solving multiple control problems very quickly due to their ability to process massive amounts of information to detect patterns without suffering fatigue, recognition error, bias or emotional interference.
- Non-lethal activities such as logistics, maintenance, base operations, veterans' healthcare, lifesaving battlefield medical assistance, casualty evacuation, personnel management, navigation, communication, cyber-defence and intelligence analysis can benefit from AI.
- AI may play an important role in new systems for protecting people and high-value fixed assets and deterring attacks through non-lethal means.
- LAWS have the exceptional potential to operate at a tempo faster than humans can achieve and to lethally strike even when communication links have been severed.
- AI-powered intelligence systems provide the ability to integrate and sort through large troves of data from different sources and geographic locations to identify patterns and highlight useful information, significantly improving intelligence analysis.
- The judgements of autonomous weapon systems will not be affected by emotions like fear or hysteria. The systems will be capable of processing more incoming sensory information than soldiers without discarding or distorting it to fit preconceived notions.
- Between human and robot soldiers, the robots could be more trusted to report ethical infractions they observed than a team of soldiers would be, as they might close ranks. There is an ethical advantage in removing humans from high-stress combat zones in favour of robots.

Challenges

The ability of armed forces worldwide to adopt and integrate AI is hindered by various factors, such as the bureaucratic nature of the armed forces, outdated acquisition and contracting processes and a culture that is reluctant to take risks. These factors make it difficult for the armed forces to incorporate external innovation and move quickly towards widespread AI adoption. Armed forces are hardware-oriented towards ships, planes and tanks. Spending remains concentrated on legacy systems designed for the industrial age and Cold War. It is now trying to leap into a software-intensive enterprise.

Implementing AI in the military poses unique challenges, such as managing the large amounts of data required, ensuring data quality, and addressing data storage, access, classification and integration issues. Additionally, technical limitations in existing systems and technologies impede compatibility with how data is captured and processed, making it difficult to implement AI effectively. The effectiveness of AI algorithms is limited by the quality and nature of the data they are trained on. Obtaining large, high-quality datasets containing military information can be challenging due to its sensitive and classified nature. As a result, many military ML algorithms may rely on simulated data that may not accurately reflect real-world conditions. Additionally, the training process of ML algorithms can be vulnerable to manipulation through the introduction of false data, a tactic known as data poisoning.

Some of the disadvantages of AI systems in the military are:

- Armed forces lag far behind the commercial sector in integrating new and disruptive technologies such as AI into their operations. They largely remain hindered by antiquated technology, cumbersome processes and incentive structures that are designed for outdated or competing aims.
- Reliance on AI systems in military operations increases the dependence on technology and can reduce the decision-making abilities of military leaders.
- The data required for ML is currently isolated, disorganised or frequently discarded, and the platforms are not connected. The methods for acquiring, developing, and implementing AI are rigid, sequential processes that hinder early and ongoing experimentation and testing, which is essential for AI.
- AI systems can struggle to function reliably or correctly if the data inputs change or encounter uncertain or novel situations. AI systems can be designed and trained on biased data, leading to biased and discriminatory decision-making in military operations.
- Adversarial attacks on ML systems can target the algorithms or data that the system uses. Even minor modifications can lead to malfunction,

incorrect conclusions or other unforeseen issues. AI/ML systems cannot typically explain the reasoning behind their decisions, recommendations and actions. The lack of explainability significantly hinders building trust between human and AI teams.

- The complexity of AI algorithms and the black-box nature of many AI systems make it difficult to understand and predict their behaviour, leading to concerns about accountability and transparency in military operations.
- AI technology is unpredictable, vulnerable to unique forms of manipulation and challenges human-machine interaction.
- Enduring necessity for human presence on the battlefield alongside AI systems in some capacity is the principal restraining factor that will keep the technology from upending warfare.
- LAWS will inevitably produce errors. These errors will be difficult to correct or prevent from reoccurring. It is difficult to take corrective action without understanding how the weapon system behaves and why. Automation bias could create denial that an error has even occurred.
- AI systems used in military operations are vulnerable to cyberattacks and hacking, which could compromise the security of military assets and put troops at risk.
- The deployment of LAWSs might result in accidental and collateral damage risk causing mass fratricide, with large numbers of weapons turning on friendly forces and civilian casualties. This could be because of hacking, enemy behavioural manipulation, simple malfunctions, unexpected interactions with the environment or software errors.
- LAWS' errors would likely repeat with a consistent level of force until some external agent intervenes. Human error tends to be idiosyncratic and one-off, given a human operator's common sense, moral agency and capacity for near real-time consequence management.
- Waging war via robots is unethical. Because it spurs so much moral outrage among the population from whom the country needs support, robot warfare has serious strategic disadvantages.
- Service members at every level lack the technical education and experience to employ AI.
- Current capabilities in the military sphere are far from perfect, and mistakes may have serious consequences.
- Inability to multitask. A fairly simple set of tasks is currently impossible for an AI system to accomplish.

The use of AI in warfare should be approached with caution, and the potential disadvantages should be carefully considered and addressed in the development and deployment of military technology.

Human-machine collaboration (HMC), human-machine teams (HMT) and brain-computer interface (BCI)

One of the most important applications of AI to warfare will be human-machine collaboration (HMC), human-machine teams (HMT) and brain-computer interface (BCI) in which military personnel and intelligent machines work together to achieve warfighting tasks. HMT is the combination of three elements: The human, the machine and the interactions and interdependencies between them.

The idea behind HMT is that humans and machines possess different strengths and can complement each other's abilities. Humans are better than machines at tasks related to sensory perception, some forms of communication, and high-context tasks that rely on intuition and creative pursuits. Machines are better suited for tasks that require processing large amounts of data, precise calculations, memory retention and repetitive tasks. Combining the strengths of both humans and machines can lead to the creation of effective human-machine teams that outperform humans or machines individually in various tasks.

The effectiveness of AI depends on the capabilities of its creators and the expertise of those who utilise it to arrive at decisions. As operations procedures become more complex and the volume of data increases, leaders at all levels and across all domains will need to incorporate data-driven decision-making alongside their intuitive decision-making abilities. The aim is not to replace human intelligence with AI but to augment and amplify human decisions toward better, faster solutions. The partnership between humans and machines is critical.

In the military, delegating tasks to machines responsibly and effectively is just as critical as traditional skills like navigation and marksmanship. War fighters must be able to determine when it is appropriate to rely on their machines, considering factors like the machine's weaknesses and susceptibility to cyberattacks. This judgement is integral to their success, similar to understanding the abilities and traits of their human colleagues. Military personnel who possess programming skills, in addition to operating skills, will be highly sought after. HMC and HMT are not mutually exclusive and strictly demarcated concepts. There may be instances where both HMC and HMT are used together in various applications. Although autonomy is not equivalent to HMC and HMT, these concepts depend on autonomous technology for their effectiveness in some situations, particularly during intense conflicts. Under human supervision, autonomous systems will be indispensable in completing tasks and missions in situations where time is critical.

Robert Work, former US Deputy Secretary of Defense explained succinctly:

> The coin of the realm during the Cold War was armored brigades, mechanized infantry brigades, multiple launch rocket system battalions,

self-propelled artillery battalions, tactical fighter squadrons, among others. Now, the coin of the realm is going to be learning machines and human-machine collaborations, which allows machines to allow humans to make better decisions; assisted human operations, which means bringing the power of the network to the individual; human-machine combat teaming; and the autonomous network.[31]

The idea of distributed forces that are highly networked relies heavily on a widespread network of inexpensive sensors, satellites and reconnaissance tools, along with a multitude of unmanned systems that operate on the ground, sea and in the air, and can be sacrificed without significant consequences. This approach aims to diversify attack vectors and broaden areas of attack. However, to use this approach effectively, it is necessary to become skilled in HMC and HMT. These two areas have the potential to alter the nature of warfare significantly.

The brain-computer interface (BCI) technology is already being developed for commercial purposes, which will help determine its strengths and weaknesses. Even though BCI applications are still in the preliminary research stage, other military technologies like robotics and big data analysis will have to take into account the possibility of BCI becoming available in the future. The emergence of BCI, alongside other technological advancements that enable humans to become more closely linked with machines on the battlefield, has the potential to cause significant strategic and operational changes within the armed forces. It will raise ethical and organisational issues across the defence community. Precautions will have to be taken to alleviate vulnerabilities to military operations and institutions and to reduce potential ethical and legal risks of adopting BCI technologies.[32]

Employing low-cost and quicker-to-produce AI-enabled machines makes it possible to create new operational concepts that use autonomy to enable machines and operators to overcome complicated challenges. A group of low-cost, autonomous systems can offer more diverse uses than a single, expensive platform and can pose a lower degree of risk. If machines work together on tasks assigned by their human teammates, they could potentially overcome traditional defences, often with a smaller risk of human casualties compared to traditional offensive operations. Machines could also function as the 'eyes and ears' of human soldiers by assisting them in gathering more information about their surroundings and taking risks on their behalf.

HMT has the potential to bring about significant changes in warfare. The evolving dynamic between humans and machines can modify how wars are fought and ultimately won, from tactical to strategic levels. The private sector's research of HMT also offers various possibilities for developing dual-use technologies, which can expedite these changes.

Despite the potentially rapid changes that HMT can bring to warfare, numerous challenges humans encounter will persist. Armed forces personnel will continue to depend on their teammates and will experience anxiety over failure and death. It will remain crucial to adapt faster than adversaries. Achieving political goals will still involve limiting a human actor's options or compelling them to alter their risk analysis. Ultimately, military and political leaders will still need to comprehend human nature, decision-making and consequences.[33]

Military AI integration challenges

AI technology is a game changer. Applications of AI in the armed forces may become a transformational change. Armed forces must look closely at the effect of AI on doctrine, organisation, training, materiel, leadership and education, personnel and facilities (DOTMLPF). It would integrate innovative leaders, skilled soldiers and trained teams for the best application of AI technologies. Even if the armed forces have the right AI-ready technology foundations in place, they will still be at a battlefield disadvantage if they fail to adopt the right concepts and operations to integrate AI technologies. Integrating AI into the armed forces is a complex and ongoing process. It involves the development and implementation of new technologies, as well as the training and education of personnel to use them effectively. It is important for this process to be guided by clear goals and strategies and for ongoing communication and collaboration between different levels of leadership and different branches of the military. With the right effort and investment, AI has the potential to significantly enhance the effectiveness and efficiency of military operations across all domains.[34]

Armed forces have a rigid formal concept of operations for warfighting. With the introduction of AI technologies in the defence sector, the existing doctrines must be revisited. Vast amounts of data are available today. Different branches of the services have to ensure that the data is compatible with different users to share. The armed forces are already drowning in data, but it is in various formats that make it impossible for different users to share. The services must establish common technical standards to ensure their data is compatible.

The traditional methods of gathering and processing information in a command post are becoming increasingly inadequate in today's fast-paced and technologically advanced battlefield. With the rapid development of hypersonic missiles, anti-satellite weapons and cyber attacks, the battlefield can change in minutes, leaving staff officers struggling to keep up. The current process of gathering information through phones, emails and other means, then sorting and analysing it through chat rooms, sticky notes and verbal communication is slow and prone to errors. The cycle of OODA loop can take hours or even days, by which point the information may already be out of date. The current system cannot keep pace with the speed of modern warfare. The current manual staff processes rely heavily on having a large number of staff officers physically present in one location to communicate with each other. This requires significant resources such as diesel generators to power all the electronics and multiple antennae to receive reports, transmit orders and access video footage from drones. These visible, infrared and radio-frequency signatures that facilitate communication are easily identifiable by the enemy through the means of drones and satellites, making the command post a vulnerable target. This approach of manual staff process is becoming increasingly obsolete in twenty-first century warfare. AI can replace human beings doing the cognitive work, watch surveillance videos for enemy forces, track own forces' locations,assignments and supply levels, etc. Then, a single commander could see the whole battlefield at a glance. You could build a virtual hilltop for the commander to look down from. One can do it without the large footprint of a huge headquarters.

Any successful use of AI in military affairs begins in the mind of the decision-maker. There is a concern about the willingness of senior leaders to accept AI-generated analysis. The burden of starting the process is on the decision-maker, not the data scientist. Functional leaders and commanders must ask tough questions: What decisions do I make, what data do I rely on, and how can I have data presented to me in a time and place that is relevant?[35] To effectively integrate AI into military operations, involving individuals with practical expertise and real-world experience in specific domains is important, rather than just data scientists. The technology should be adapted to the needs of these domain experts, rather than expecting them to adapt to the technology. It is crucial to involve multiple communities of experts to plan for the integration of AI in data and warfighting. This approach will lead to a more efficient and effective use of AI in military operations.

Armed forces need more digital experts now or they will remain unprepared to buy, build, and use AI and associated technologies. They need new talent pipelines. AI expertise alone is insufficient for successfully integrating AI into military operations. To fully utilise AI capabilities, armed forces need experts in specific fields such as intelligence analysis, fire support and logistics in addition to AI expertise. For example, experts in intelligence analysis are needed to understand how human beings currently identify targets in drone video and satellite imagery. Experts in fire support are also necessary to determine the best aircraft or artillery unit to strike those targets. Experts in logistics are needed to schedule mid-air refuelling and ammunitions resupply runs. This approach will lead to a more comprehensive and effective use of AI in military operations.[36]

The military's ability to quickly deploy, use and update software, including AI models, will play a crucial role in determining its relative strength compared to its adversaries. In future conflicts, the side that can adapt and update software and AI models faster will have a significant tactical and operational advantage. Software, including AI models, is important in military operations, from detecting targets to decision-making, targeting and assessing battle damage. The quality of the software will determine a military's primacy in collecting and analysing information, developing an operating picture, thwarting enemy attacks, identifying opportunities in time and space to attack, and helping with target selection and servicing most effectively.[37] To ensure software advantage, the armed forces should make a new information architecture that will allow armed forces to be far more flexible, scale on demand, and adapt dynamically to changing conditions. This would include access to cloud computing and storage.[38]

Militaries that collect useful data, quickly draw lessons and integrate updates into their software at a more rapid pace than their competitors will have a clear advantage. Militaries will also benefit from denying their adversaries the ability to collect helpful data or use their software.

AI systems, particularly when combined with virtual and augmented reality technologies, can enhance individualised and customised training by adapting to human behaviour and creating customised training environments or scenarios. To fully realise the potential of AI in training, it is recommended to create new military training and occupational specialities to develop AI talent and to establish government fellowships and fast-track promotion options to attract and retain top technology professionals.

Adapting from a force that is focused on hardware and industrial-age methods to one that is focused on software and digital-age methods can be difficult. However, the biggest challenge to overcome is not technology but rather the cultural mindset and approach.[39]

AI and the nature and character of warfare

The nature of war describes what war is and the character of war describes how it is fought. War's nature is violent, interactive between opposing wills and fundamentally political. War's character is influenced by technology, law, ethics, culture, methods of social, political, and military organisation and other factors that change across time and place.

The increasing use of AI in military operations will alter how wars are fought, leading to new tactics, strategies and approaches to conflict. The character of warfare changes in concert with the available tools and how they influence how militaries organise themselves to fight wars. AI systems can potentially increase the speed with which countries can fight. Even if humans are still making final decisions about the use of lethal force, fighting at machine speed can dramatically increase the pace of operations.

Some potential impacts of AI on the character of war include:

- Increased speed and efficiency of decision-making and military operations.
- Development of new weapons and systems that rely on AI for targeting and autonomous operations.
- The ability to analyse vast amounts of data in real-time leads to improved situational awareness and intelligence.
- The potential for autonomous weapon systems to operate without human intervention leads to ethical and legal questions about using force.

There is near unanimity that the character of warfare is changing. But is it changing the nature of war? The impact of AI on the nature of war will depend on several factors, including the ethical and legal frameworks governing its use, the level of technological development and the actions of states and other actors.

Some potential impacts of AI on the nature of war include:

- More precise and efficient targeting and weapon systems.
- The ability to analyse vast amounts of data in real-time leads to improved situational awareness and intelligence.
- The potential for increased automation and autonomy in military operations, reducing the need for human soldiers on the front lines.
- The ability to simulate and analyse war scenarios, leading to improved war planning and decision-making.

Autonomy will change the nature of war in several ways. These are:

- It could weaken the role of political direction by forcing response delegation to lower echelons for faster forms of attack. Autonomy can lessen the ability of governments to gain the support and legitimacy of their populations while making it easier for foreign governments to manipulate their adversary's population.
- Deep learning forms of AI will augment the intuition and judgement of experienced commanders.
- Automated technologies could reduce popular support for professional military institutions, which paradoxically could free governments to employ force more readily, since the political consequences are reduced.

- As with the earlier ages, friction and uncertainty will endure. The age of autonomy can introduce new forms of friction while reducing human factors in tactical contexts.[40]

The nature of war has changed since the Cold War. Globalisation, emergence of violent non-state actors, spread of identity wars, non-trinitarian (irregular) wars and Information Technology (IT) have fundamentally altered the nature of war. Former US Deputy Secretary of Defense, Bob Work, identified AI and human performance enhancements as potential breakthroughs in defence technology: 'We believe we are at an inflection point at artificial intelligence and autonomy. I am starting to believe very, very deeply that it is also going to change the nature of war.'[41]

Some argue that AI does not change the nature of war. It emphasises the essential element behind the deep nature of war: Its human component. Psychology, ethics, politics, passions and the proximity of pain and death are what war is all about. It reminds us that war is a very intimate expression of our humanity and something we cannot delegate. AI is severely limited and cannot achieve any strategic task nor be used efficiently in any war. Just like any other kind of weapon, AI-augmented weapon systems have limitations and could be manipulated by the enemy , and it could well backfire against its users. Human biases are always present in the decision-making process of military organisations. AI is neither completely bias-free nor totally separated from the human touch.

The chaos of war, its fog, friction, and chance will likely never be deciphered, regardless of what technology is thrown at it. While AI-enabled technologies will be able to gather, assess and deliver heretofore unimaginable amounts of data, these technologies will remain vulnerable to age-old practices of denial, deception and camouflage.[42] Potential military applications of emergent technology should not be viewed as a panacea for the chaos, friction and chance that will continue to define war in the future. Regardless of the possibilities offered by technological advances, success in the next conflict is likely to be just as reliant on human genius.[43]

Over the years, Ex-US Secretary of Defense Jim Mattis, an avid reader of military history and theory, has cultivated a reputation for deep thinking about the nature of warfare. He came to a few conclusions about the fundamental nature of combat. He said, 'It's equipment, technology, courage, competence, integration of capabilities, fear, cowardice – all these things mixed together into a very fundamentally unpredictable fundamental nature of war. The fundamental nature of war is almost like H_2O, ok? You know what it is.' However, his opinion about the nature of war has changed. He said,

I'm certainly questioning my original premise that the fundamental nature of war will not change. You've got to question that now. I just don't have the answers yet... The character of war changes all the time. An old dead German [Carl von Clausewitz] called it a "chameleon." It changes to adapt to its time, to the technology, to the terrain, all these things... If we ever get to the point where it's completely on automatic pilot and we're all spectators, then it's no longer serving a political purpose. And conflict is a social problem that needs social solutions.[44]

AI and recent wars

First artificial intelligence war

Israeli military action, the Operation 'Guardian of the Walls' in May 2021 in the Palestinian territory of Gaza, has been described as the first artificial intelligence war. Israel relied heavily on ML and data gathering. A senior Israeli officer involved in digital conflict said, that this is the first time AI was used broadly across an operation. Israel used AI to coordinate attacks on Gaza. The AI has been fed sufficient raw data related to warfare and intelligence related to the conflict. The Israel Defense Forces (IDF) used resources such as signal intelligence (SIGINT), visual intelligence (VISINT), human intelligence (HUMINT), geographical intelligence (GEOINT); data collected through satellites, aerial reconnaissance vehicles, field agents, and ground intel. The Israeli military gathered the military intelligence reports by adapting readily available AI technologies. Vast amounts of data were analysed for precision attacks. Drawing on satellite imagery, sensors and other sources, Israel was able to obtain 3D geographical information about Gaza and identify the locations of rocket launchers that Hamas installed. Algorithms and 3D modelling helped them to be precise and accurate.[45]

IDF set up a multi-disciplinary centre to help produce relevant targets, aiding the military in effective planning and targeting.[46] An advanced AI technological platform was created that centralises all data on the militant groups in the Gaza Strip on one system to help with the analysis and extract intelligence crucial for the operations.

According to a report by Jerusalem Post, it was IDF's elite intelligence officers in Unit 8200, an Intelligence Corps unit of the IDF that specialises in code decryption and signal intelligence, who created advanced programs named Alchemist, Gospel and Depth of Wisdom, which were developed and used during the conflict. The Gospel helped the IDF's military intelligence unit to enhance the intelligence recommendations for the officers and helped them identify 'quality targets,' which were then passed to the air force to launch the strikes. Unit 8200 created multiple algorithms using signal, geographical and human intelligence to pinpoint strike targets, which were then passed to command to order a strike.[47]

The unit used AI algorithms to predict the enemy's rocket launches, and their place and time, helping soldiers on the ground to defend against them effectively. For example, the IDF's J6/C4i Directorate's Lotem Unit specifically creates software for AI and uses ML for real-life applications. A part of this technology has been deployed in Israel's Iron Dome. The Lotem Unit created an app that helped the soldiers predict rocket launch locations, time of launch, and the target for such missiles by studying the data from field sensors, etc. Another app that the Lotem unit created was used to identify suspicious objects from videos and write down the description, which helps save manpower.

Hamas's underground tunnel network was heavily damaged in several nights of airstrikes. Israelis could map the network, consisting of hundreds of kilometres under residential areas, where they knew almost everything about them. The mapping of the underground network was done by a huge intelligence-gathering process that was helped by technological developments and the use of Big Data to fuse all the intelligence. Once mapped, the IDF could have a full picture of the tunnel with details, such as the tunnels' depth, thickness, and the nature of the routes. With that, the military could construct an attack plan that was used during the operation. Over the years, IDF Unit 9900's satellites have gathered GEOINT. They were able to detect changes in terrain in real-time automatically. During the operation, the military could detect launching positions and hit them after firing. Using satellite imagery, Unit 9900 troops were able to detect 14 rocket launchers located next to a school. Using the data gathered and analysed through AI, the IAF could use the suitable munitions to hit a target, be it an apartment, a tunnel or a building. The IDF used supercomputing and a swarm of AI-guided drones to comb through data and identify new targets within the Gaza Strip. It is believed that this is the first time a swarm of AI drones has been used in combat.

Ukraine war

AI technology has been utilised in various ways during the Ukraine conflict, including facial recognition software to identify enemy soldiers and ML to enhance the efficiency of military and aid supply chains. Additionally, AI has been used for propaganda and information warfare, with deepfake videos blurring the boundary between real and artificially generated content. Unmanned systems proved crucial to stopping Russian tank columns, firing missiles, providing pinpoint accuracy for Ukrainian artillery and missile strikes, sinking Russia's Black Sea flagship and striking Russia's naval bases in Crimea. Autonomous systems are deployed in Ukraine using the concept of loitering munitions, operating singly and increasingly in swarms.[48] Electronic countermeasures against remotely operated weapons are pushing both sides to increase the level of autonomy of those weapons. Russia is using smart cruise missiles like Iranian Shahed missiles to attack predefined targets such as administrative buildings and energy installations. They can fly low along rivers to avoid detection and circle an area while they await instructions. Shahed missiles are good at hitting stationary targets but not so much against mobile targets that can quickly manoeuvre away from the original position.

Significantly, swarms of missiles are used to overwhelm air defence systems and minimal radio links to avoid detection. It has been reported that Shahed missiles are now being equipped with infrared detectors that allow them to home in on nearby heat sources without requiring target updates communicated from controllers by radio. This could be an important step towards full autonomy. As jamming systems turn out to be the norm, radio control becomes more difficult and autonomous weapons become more attractive. Military analysts say the longer the war in Ukraine lasts, the more likely drones will be used to identify, select and attack targets without human help.

The AI and ML algorithms are used to process massive amounts of data, which is crucial for making decisions. One of the examples of using AI for battlefield information processing is by an American company called Primer. Reportedly, it is used to capture, process, analyse, transcribe and translate Russian military communications that are often unencrypted. AI and ML algorithms have been used for information operations for the production of deepfakes and in authentic and fake social media accounts that Russia has employed on social media platforms like X (formerly known as Twitter), Facebook and Instagram. Overall, AI and ML are being used in this war between Ukraine and Russia. However, their use is still limited and still cautious.

Future prospects

The future of AI in warfare will be formed by the decisions made by governments, military leaders and technology experts. In military contexts, the responsible development and deployment of AI will be crucial to ensure that it contributes to stability and security rather than fuelling new conflicts or aggravating existing ones. AI can be copied easily and run on relatively small machines once trained. With present technology, detecting their presence or verifying their absence is difficult or impossible. AI will be ubiquitously acquired, mastered and employed. The imposition of restraints on weaponising AI or achieving a collective definition of restraint will be extremely difficult. This is an important area for further technical research and policy development.[49]

By 2025, digital technologies such as robotics, autonomy, connectivity, secure cloud data and AI will have a transformative effect on confrontation and conflict. These technologies will lead to armed forces that are composed of a combination of human operators and autonomous systems, rather than just people operating equipment. This process will begin by enhancing how armed forces currently organise, operate and train. As technology advances and more experience is gained, it will become as disruptive as the impact of companies like Airbnb and Uber on their respective industries. The digital age will bring about the most significant change in how nations engage in confrontation and conflict, and it will be a long-term competition where success will depend on the ability to rapidly adapt and innovate, while those who are slow to change will fall behind.

One example of rising AI technology system is ChatGPT (Generative Pre-training Transformer), which is a state-of-the-art language model that is trained to generate human-like text. ChatGPT is a recently released chatbot from the company OpenAI. ChatGPT is a powerful tool that can revolutionise the way we work, communicate, process information and live. ChatGPT may replace jobs that are repetitive, routine and predictable. It may impact jobs that require human intelligence and a high level of education – journalism, writing, translation, law, education, computer coding/engineering and research. ChatGPT can write code in various computer languages such as Python, C++ and JavaScript. It can review and identify errors within written codes. This raises a question regarding the future of programmers and developers.

The Large Language Models (LLMs) like ChatGPT have national security implications. At a strategic level, nations will try to utilise AI models for economic, military and national security advantages.[50] ChatGPT could be the interface through which people access a varied range of AI-powered tools, including robotics and computer vision. Such intelligent tools could transform the I&War scene through AI-generated predictive insights in a national security context. It would enable semi-autonomous systems to train and interact with helpful robots and support tasks ranging from logistics and supply chain management to the vetting of export control applications. ChatGPT can empower a more robust and faster Intelligence Community across all intelligence disciplines and phases of the intelligence cycle across all levels of war. A ChatGPT with external search capabilities and expert-informed training data would improve open-source intelligence (OSINT) and analysis capabilities. With the right combination of AI, data, modelling, simulation capabilities and predictive insights, models can game out various policy options and provide that analysis to human decision-makers. There are limitations to how much of a role ChatGPT should play in semi-autonomous systems that make operational decisions. However, these could provide a real-time, data-informed perspective on policy or war options.

ChatGPT could generate unique texts at speed and scale that evade existing filtering systems. People may question whether what they read on social media was developed by AI. We may be moving towards a world where most of our content is AI-generated, allowing state and non-state actors to shape the information landscape. However, the recent proliferation of synthetic content detection capabilities like OpenAI Text Classifier, Copyleaks AI Content Detector, GPTRader, etc., will provide tools for national security professionals and informed citizens to identify synthetic media. ChatGPT is only as good as the data upon which it is trained. The use of such technologies will have associated risks and must be wisely considered by humans. Nevertheless, ChatGPT and other LLM tools will proliferate quickly and people will integrate tools like ChatGPT into their daily lives.

As AI capabilities are integrated into national security apparatus, ChatGPT and the like will be put to the test against those of our adversaries. Advanced countries will have to leverage the competitive advantages of their private sector to ensure that their own equivalents to ChatGPT are better than others. As we increasingly rely on AI systems, we need a way to explain and understand them. We cannot query an AI application to discuss its thought process. Breakthroughs in this space would establish trust in AI systems by allowing human users to understand why an AI system made a particular decision. ChatGPT and other LLM/AI-powered tools will affect national security at a fundamental computer science level via coding. Techniques are now available to detect and prevent Zero Day attacks and can help to harden code. These technologies can also lower the barrier to entry for malicious cyber actors. Governments will find it challenging to keep pace with this rapidly developing technology for developing AI regulations or integrating AI capabilities into the workforce.

In 20 years' time, we may have swarms of unmanned systems of airborne drones, ground systems surface vessels, etc. Different units operating and carrying out attacks together, which would require a high level of AI-enabled autonomy. This would drive an arms race and some actors may be forced to adopt a certain level of autonomy because human beings would not be able to deal with autonomous attacks as fast as would be necessary. Speed is a crucial issue here.

Because of the speed and autonomous systems encountering other autonomous systems, we could find ourselves in situations where the systems react with each other in a way that is not wanted. Situations like these are called 'Flash Wars,' where you have an attack or even think there is an attack and autonomous systems react to that. In turn, another autonomous system by the opponent reacts to the attack. The space for deliberate diplomatic negotiations will decrease, increasing the risk of miscalculation and misunderstanding and leading to the possibility of rapid, unintentional and accidental escalation. We may have an accidental military conflict that we did not want. We are moving into a world where systems would be more autonomous. We must ensure that we minimise the risk of unwanted escalation of lethality decided by machines without human control.

Conclusion

AI will play a significant role in future military applications. Militaries worldwide are incorporating AI and autonomy into their organisational processes, command and control systems, logistics systems and weapon systems to leverage current developments from the commercial world. It has many potential areas where it can improve efficiency, reduce the workload for human operators and perform tasks more quickly. Continued research will further develop its capabilities, transparency and robustness. The military cannot ignore the potential of this technology. However, the full potential of AI in integrated warfare has yet to be realised. Regarding using AI in military operations such as targeting or other tactical aspects of warfare, the armed forces still have a long way to go in terms of deploying the technology and addressing issues related to trust and ethics.[51]

While the potential benefits of AI in military applications are significant, we must exercise caution in its deployment. Placing vulnerable AI systems in contested environments and entrusting them with critical decision-making poses a considerable risk of negative consequences. At this time, humans must retain ultimate responsibility for key decisions. Given the likelihood of AI systems being targeted by attacks and the current limitations in AI technology resilience, the most viable areas for investment in military AI are those that operate in uncontested environments. AI tools closely supervised by human experts or with secure inputs and outputs can provide value to the military without raising concerns about vulnerabilities. To fully harness the potential of AI in the military, traditional acquisition practices will need to be reformed or discarded. Collaboration between government, academia and industry is crucial for successfully integrating AI into the military.

AI can potentially revolutionise the way wars are fought, but it also poses significant ethical and security challenges. Even after extensive training, advanced algorithms have been known to make considerable mistakes in identifying objects or people and scientists still do not understand how these systems arrive at their decisions. AI-enabled machines are vulnerable to hacking and sabotage, like other cyber-dependent systems.[52] The development of Military AI poses various risks that need to be addressed. There is no agreement on an AI development timeline, but professionals agree that there will be a steady increase in the use of AI in military systems. International competition could encourage countries to hasten the development of military artificial intelligence without adequate attention to safety, reliability and humanitarian consequences. The former chief executive officer of Google, Eric Schmidt suggested 'even those powers creating or wielding an AI-designed or AI-operated weapon may not know exactly how powerful it is, or what it will do in a given situation.'

Notes

1 '"Whoever Leads in AI Will Rule the World": Putin to Russian Children on Knowledge Day,' *RT*, 1 September 2017, available at: https://www.rt.com/news /401731-ai-rule-world-putin/, accessed on 1 February 2023.

2 Major-General P.K. Mallick, VSM (Retd.), 'Artificial Intelligence in Armed Forces: An Analysis,' *Centre for Land Warfare Studies (CLAWS) Journal*, (Winter 2018), pp. 63–79.

3 'Final Report,' National Security Commission on Artificial Intelligence, 19 March 2021, available at: https://www.nscai.gov/wp-content/uploads/2021/03 /Full-Report-Digital-1.pdf, accessed on 20 January 2023.

4 'National Security Strategy,' October 2022, available at: https://www.whitehouse .gov/wp-content/uploads/2022/10/Biden-Harris-Administrations-National -Security-Strategy-10.2022.pdf, accessed on 29 January 2023.

5 'Summary of the 2018 Department of Defense Artificial Intelligence Strategy,' available at: https://media.defense.gov/2019/Feb/12/2002088963/-1/-1/1/ SUMMARY-OF-DOD-AI-STRATEGY.PDF, accessed on 24 January 2023.

6 'AI Principles: Recommendations on the Ethical Use of Artificial Intelligence by the Department of Defense,' *Defense Innovation Board*, available at: https:// media.defense.gov/2019/Oct/31/2002204458/-1/-1/0/DIB_AI_PRINCIPLES _PRIMARY_DOCUMENT.PDF, accessed on 30 January 2023.

7 'Artificial Intelligence and National Security, R45178, Congressional Research Service,' 10 November 2020, available at: https://crsreports.congress.gov/prod-uct/pdf/R/R45178/10, accessed on 16 and 31 January 2023.

8 Ben Buchanan, 'The AI Triad and What It Means for National Security Strategy,' Center for Security and Emerging Technology, p. iii, Georgetown University, August 2020, available at: https://cset.georgetown.edu/publication/the-ai-triad -and-what-it-means-for-national-security-strategy, accessed on 6 January 2023.

9 Paul Scharre, *Army of None: Autonomous Weapons and the Future of War* (New York: W.W. Norton, 2018); Michael C. Horowitz, 'When Speed Kills: Lethal Autonomous Weapon Systems, Deterrence and Stability,' *Journal of Strategic Studies*, vol. 42, no. 6 (2019), pp. 764–88, DOI: 10.1080/01402390.2019.1621174, accessed on 5 January 2023.

10 Kelley M. Sayler, 'Emerging Military Technologies: Background and Issues for Congress,' *Congressional Research Service*, 1 November 2022, available at: https://crsreports.congress.gov/product/pdf/R/R46458/12, accessed on 5 January 2023.

11 'Artificial Intelligence and National Security, R45178, Congressional Research Service,' 10 November 2020, available at: https://crsreports.congress.gov/prod-uct/details?prodcode=R45178, accessed on 10 January and 2 February 2023.

12 Vincent Boulanin and M. Verbruggen, *SIPRI Mapping the Development of Autonomy in Weapon Systems* (Solna: SIPRI, 2017).

13 'National Security Commission on Artificial Security (NSCAI),' Final Report, p. 22, 2021, available at: https://www.nscai.gov/wp-content/uploads/2021/03/Full-Report -Digital-1.pdf, accessed on 10 February 2023.

14 'Applications in Maintaining the Intelligence Edge: Reimagining and Reinventing Intelligence Through Innovation,' CSIS Technology and Intelligence Task Force, pp. 8–22 (13 January 2021), available at: https://csis-website-prod.s3.ama-zonaws.com/s3fs-public/publication/210113_Intelligence_Edge.pdf, accessed on 22 January 2023.

15 Patrick Tucker, 'What the CIA's Tech Director Wants from AI,' *Defense One*, 6 September 2017, available at: http://www.defenseone.com/technology/2017/09

/cia-technology-director-artificial-intelligence/140801, accessed on 5 February 2023.

16 Marcus Weisgerber, 'Defense Firms to Air Force: Want Your Planes' Data? Pay Up,' *Defense One*, 19 September 2017, available at: http://www.defenseone.com/technology/2017/09/military-planes-predictive-maintenance-technology/141133/, accessed on 15 February 2023.

17 Micah Musser and Ashton Garriott, 'Machine Learning and Cybersecurity: Hype and Reality,' Center for Security and Emerging Technology, June 2021, available at: https://cset.georgetown.edu/publication/machine-learning-and-cybersecurity, accessed on 24 January and 23 February 2023.

18 Emily Falk and Michael Platt, 'What Your Facebook Network Reveals About How You Use Your Brain,' *Scientific American*, 9 July 2018, available at: https://blogs.scientificamerican.com/observations/what-your-facebook-network-reveals-about-how-you-use-your-brain/, accessed on 10 February 2023.

19 M. C. Horowitz et al., *Artificial Intelligence and International Security* (Center for a New American Security, 2018), pp. 5–6, available at: https://s3.amazonaws.com/files.cnas.org/documents/CNAS-AI-and-International-Security-July-2018Final.pdf, accessed on 21 February 2023.

20 'Generating Actionable Understanding of Real-World Phenomena with AI,' *DARPA*, 4 January 2019, available at: https://www.darpa.mil/news-events/2019-01-04, accessed on 12 February 2023.

21 'Artificial Intelligence in Command and Control Systems,' *GMV*, available at: https://www.gmv.com/en/node/213/printable/print#:~:text=AI%20plays%20an%20increasingly%20significant,presenting%20information%20to%20the%20operator, accessed on 19 January 2023.

22 Owen Daniels, 'Speeding Up the OODA Loop with AI: A Helpful or Limiting Framework? Institute for Defense Analyses May 2021 in Joint Air & Space Power Conference 2021,' available at: https://www.japcc.org/essays/speeding-up-the-ooda-loop-with-ai/, accessed on 5 February 2023.

23 Forrest E. Morgan, Benjamin Boudreaux, Andrew J. Lohn, Mark Ashby, Christian Curriden, Kelly Klima, Derek Grossman, 'Military Applications of Artificial Intelligence, Ethical Concerns in an Uncertain World,' *RAND Corporation*, 2020, available at: https://www.rand.org/pubs/research_reports/RR3139-1.html, accessed on 13 February 2023.

24 Y. H. Wong, J. Yurchak, R. Button, A. Frank, B. Laird, O. Osoba, R. Steeb, B. Harris, and S. Joon Bae, *Deterrence in the Age of Thinking Machines* (Santa Monica: RAND Corporation, 2020), available at: https://www.rand.org/pubs/research_reports/RR2797.html, accessed on 11 February 2023.

25 James Johnson, 'The AI Commander Problem: Ethical, Political, and Psychological Dilemmas of Human-Machine Interactions in AI-enabled Warfare,' *Journal of Military Ethics*, vol. 21, nos. 3–4 (2022), pp. 246–71, DOI: 10.1080/15027570.2023.2175887.

26 Bill Canis, 'Issues in Autonomous Vehicle Deployment, CRS Report R44940,' pp. 2–3, available at: https://crsreports.congress.gov/product/pdf/R/R44940/9.

27 For extensive background on such systems, see Scharre, *Army of None*.

28 Major-General P.K. Mallick (Retd), a talk delivered on AI and the Transformation of Warfare: Perspectives from South Asia and Beyond at PRIO Organized International Conference, Kolkata on 3–4 March 2022, available at: https://indianstrategicknowledgeonline.com/web/Latest%20AI%20APPLICATIONS%20IN%20DEFENCE.pdf.

29 Nathan J. Lucas, 'Lethal Autonomous Weapon Systems: Issues for Congress, CRS Report R44466,' 14 April 2016, available at: https://crsreports.congress.gov/product/pdf/R/R44466/4, accessed on 17 January 2023.

30 Paul Scharre, 'Autonomous Weapons and Operational Risk, Center for a New American Security,' February 2016, available at: https://s3.amazonaws.com/ files.cnas.org/documents/CNAS_Autonomous-weapons-operational-risk.pdf, accessed on 19 January 2023.

31 William Eliason, 'An Interview with Robert Work,' *Joint Force Quarterly*, vol. 84 (26 January 2017). Available at: https://www.rand.org/content/dam/rand/pubs/research_reports/RR2900/ RR2996/RAND_RR2996.pdf.

32 A.F. Salazar-Gomez, J. Del Preto, S. Gil, F.H. Guenther, and D. Rus, 'Correcting Robot Mistakes in Real Time Using EEG Signals,' *2017 IEEE International Conference on Robotics and Automation (ICRA)*, Singapore, July 2017, https:// gil.seas.harvard.edu/sites/projects.iq.harvard.edu/files/gil-engineering/files/pub _correcting_robot_mistakes_in_real_time_using_eeg_signals.pdf.

33 Bajraktari, 'Human-Machine Teaming in Warfare,' 30 June 2022, https:// scsp222.substack.com/p/human-machine-teaming-in-warfare, accessed on 28 January 2023.

34 Sydney J. Freedberg Jr., 'Building JADC2: Data, AI & Warfighter Insight,' *Breaking Defense*, 13 January 2021, available at: https://breakingdefense.com/2021/01/ building-jadc2-data-ai-warfighter-insight/, accessed on 24 January 2023.

35 Michael Groen, 'Uncle Sam Needs AI, ASAP: DoD Artificial Intelligence Chief,' *Breaking Defense*, 11 January 2021, available at: https://breakingdefense.com /2021/01/uncle-sam-needs-ai-asap-dod-artificial-intelligence-chief/, accessed on 14 January 2023.

36 Freedberg, 'Building JADC2,' accessed on 2 February 2023.

37 Nand Mulchandani and N.T. John, Jack Shanahan, 'Software-Defined Warfare: Architecting the DOD's Transition to the Digital Age,' *Center for Strategic and International Studies*, available at: https://www.csis.org/analysis/soft- ware-defined-warfare-architecting-dods-transition-digital-age, accessed on 20 January 2023.

38 'Final Report,' National Security Commission on Artificial Intelligence, 19 March 2021, available at: https://www.nscai.gov/wp-content/uploads/2021/03 /Full-Report-Digital-1.pdf, accessed on 20 January 2023.

39 Groen, 'Uncle Sam Needs AI, ASAP,' accessed on 27 January 2023.

40 Kareem Ayoub and Kenneth Payne, 'Strategy in the Age of Artificial Intelligence,' *Journal of Strategic Studies*, vol. 39, nos. 5–6 (2016), pp. 793–819, available at: https://doi.org/10.1080/01402390.2015.1088838.

41 Sydney J. Freedberg Jr., 'War Without Fear: DepSecDef Work on How AI Changes Conflict,' *Breaking Defense*, 31 May 2017, https://breakingdefense .com/2017/05/killer-robots-arent-the-problem-its-unpredictable-ai/.

42 Major-General P.K. Mallick, 'Is Artificial Intelligence (AI) Changing the Nature of War?' Vivekananda International Foundation, 18 January 2019, available at: https://www.vifindia.org/article/2019/january/18/is-artificial-intelligence -changing-the-nature-of-war, accessed on 19 January 2023.

43 Mallick, 'Artificial Intelligence in Armed Forces: An Analysis.'

44 Aaron Mehta, 'AI Makes Mattis Question "Fundamental" Beliefs About War,' 17 February 2018, available at: https://www.c4isrnet.com/intel-geoint/2018 /02/17/ai-makes-mattis-questionfundamental-beliefs-about-war/, accessed on 1 February 2023.

45 Takeshi Kumon, 'The First AI Conflict? Israel's Gaza Operation Gives Glimpse of Future,' *Nikkei Asia*, 28 June 2021, available at: https://asia.nikkei.com/ Politics/International-relations/The-first-AI-conflict-Israel-s-Gaza-operation -gives-glimpse-of-future, accessed on 12 February 2023.

46 Anna Ahronheim, 'Israel's Operation Against Hamas was the World's First AI War,' *The Jerusalem Post*, 27 May 2021, available at: https://www.jpost.com /arab-israeli-conflict/gaza-news/guardian-of-the-walls-the-first-ai-war-669371, accessed on 16 February 2023.

47 Ahronheim, 'Israel's Operation Against Hamas was the World's First AI War.'

48 Peter W. Singer, 'One Year In: What Are the Lessons from Ukraine for the Future of War?' *Defense One*, 22 February 2023, available at: https://www.defenseone .com/ideas/2023/02/what-ukraine-has-changed-about-war/383216/, accessed on 25 February 2023.

49 Eric Schmidt, 'AI, Great Power Competition & National Security,' *Dædalus*, vol. 151, no. 2 (2022), pp. 288–98, https://doi.org/10.1162/DAED_a_01916, accessed on 19 February 2023.

50 Ylli Bajraktari, 'The "ChatGPT Moment" and National Security, Special Competitive Studies Project,' 8 February 2023, available at: https://scsp222 .substack.com/p/the-chatgpt-moment-and-national-security?utm_source=post -email-title&publication_id=652508&post_id=101645894&isFreemail=true &utm_medium=email, accessed on 8 February 2023.

51 'Artificial Intelligence is the Future of Warfare (Just Not in the Way You Think),' *Modern War Institute*, available at: https://mwi.usma.edu/artificial-intelligence -future-warfare-just-not-way-think/, accessed on 6 February 2023.

52 Morgan et al., 'Military Applications of Artificial Intelligence, Ethical Concerns in an Uncertain World.'

2
DEFENCE APPLICATIONS OF ARTIFICIAL INTELLIGENCE

Ajey Lele

Introduction

AI commonly gets identified as the simulation of human intelligence processes by machines. The main focus here is on information and communications technologies (ICT). The overall AI system involves specific hardware, software, and algorithms. Data is the heart of any AI system. Generally, AI systems work by feeding large amounts of data and undertaking pattern analysis based on that data. Here, both quantity and quality of data are equally important. Today, AI is found impacting practically every sector of society. It is changing business processes radically. There are various AI-enabled devices, which are found bringing in a silent revolution in various fields from science research, including climate research, industry, agriculture, education, and medicine, and critical infrastructure management. It is also becoming an integral part of various security establishments. Various military applications of this technology are becoming evident. The defence industry is found working towards designing and developing various AI-enabled military systems. This chapter takes a broad overview of the strategic applications of AI know-hows. It discusses various AI-related military developments, identifies both the offensive and defensive areas where AI has already made inroads in the defence sector, and traces the ongoing and proposed research and developments happening in this field and relevant investments made by major states. The chapter ends with a discussion on facets of legality, arms control, disarmament, and ethics.

The idea of Artificial Intelligence (AI) is not new. However, this technology still remains an evolving technology and there is a requirement for much innovation in order to obtain the perceived benefits of this technology. The

DOI: 10.4324/9781003421849-3

scientific community is continuously working towards developing various applications of this technology. Currently, much work is being done towards developing various defence applications of AI. Actually, some armed forces are already inducting this technology in their overall defence architecture. From a military perspective, AI has both offensive and defensive dimensions. For some years now, there has been an ongoing debate on the efficacy of Lethal Autonomous Weapon Systems (LAWS) in modern-day conflict. This debate, which mainly revolves around the aspects of ethics and morality, is still evolving. This chapter discusses various AI-related military developments.

Context

Technology shapes warfare. It defines, governs or circumscribes warfare and is known as the instrumentality of warfare. It is viewed that technology shapes warfare; and, conversely, war (not warfare) shapes technology. Also, military technology is not deterministic but opens the doors for further experimentation, and modern-day wars are the best example of this.[1] The first two decades of the twenty-first century have seen an explosion in technological advances affecting cultures, economies and governments around the world. Militaries and the realm of warfare have been no different. Technological innovation is known to modify the way armies engage with each other, and in doing so, it has been observed that it re-maps the boundaries of how conflicts are fought. Seismic changes in warfighting have always happened owing to the introduction of new technologies. From the horse-drawn armies of the 1910s, through the mechanised forces of the 1930s and jet-propelled aircraft of the 1940s, to the nuclear age, it has been a technology that has dictated the processes of warfare. General Richard Barrons – a former head of Britain's Joint Forces Command – argues that: 'We are seeing changes on the battlefield as profound as anything in the past 150 years as data, artificial intelligence and connectivity become the new, key components of warfare. But we are still in the foothills of what is coming.'[2]

John McCarthy, a professor emeritus of computer science at Stanford, is credited with having coined the term 'artificial intelligence.' McCarthy was a giant in the field of computer science and a seminal figure in the field of AI. While at Dartmouth in 1955, McCarthy authored a proposal for a two-month, 10-person summer research conference on 'artificial intelligence' – the first use of the term in publication.[3] He defined AI as the science and engineering of making intelligent machines like intelligent computer programs. Important advances in AI in recent times have led to many debates about the potential social, economic and security impact of AI. However, little sustained attention has been paid to the impact of AI on international relations or how the technology impacts the work of the country's ministries, government leaders and policymakers. The initiation of deep learning and neural networks in the late 1990s brought a new wave of attention towards AI. In AI, foreign relationships and global security are contemporary hotbeds. The development of AI technology is central to the economic aspect, with great market potential for the future.

The AI revolution is measured to be more powerful than the Industrial Revolution since it has shaped the personal lives of the people apart from affecting the industries. In this era of Industry 4.0, it is observed that the Industrial Revolution had a great impact on international relations and trade, making industrial machines more autonomous, including the defence industry. The states during the First Industrial Revolution strove to control resources like coal and oil. While during the 1970s and 1980s, with the IT Revolution, electronics and microprocessors fetched a new trend of innovation that gave birth to the internet, GPS, etc. (remember these were all military projects). The search for AI has voyaged through stages of hope and desolation since the 1950s.[4] But, post-2000, AI has leapfrogged. It has been a much-debated topic today and is projected to emerge as one of the major technologies of the twenty-first century that could influence geopolitics, geo-economics and geostrategy for the present and the future.

AI progressively facilitates the way humans interact with others. In a relative sense, AI remains an emerging technology and there is still scope to appreciate the nuances of this technology from various dimensions, from physics to philosophy and from military to morality. International relations in regards to developing the context for AI are crucial. There exists a correlation between AI and international power distribution. But, theoretical clarity is missing towards identifying the power dimensions of AI. There is a need to identify AI's impact on the international system at the strategic level.

It is important to appreciate the possible world structure emerging from interactions of AI with social, economic and strategic global structures. Earlier research has delineated that interactions between research, knowledge, innovation and technology create a new perspective on productivity and competitiveness. It has been argued that Science and Technology (S&T) has been a neglected issue in the international relations research agenda. Attempts are ongoing to study international relations (IR) and AI and to formulate theoretical positions in this regard. Some use a theoretical and scenario-planning approach, but there is also a view that much progress cannot be made owing to the absence of empirical data. Some studies have examined specific states and their challenges, military matters, strategic studies, war and cyberspace.[5] AI has the potential to make sweeping changes in the process the human society operates presently. Hence, there is an ongoing quest to evaluate its likely influences on international stability.

AI being a disruptive technology, it is supposed that it affects nearly all aspects of international security, from diplomacy, intelligence and defence, to conflict management. Obviously, a number of countries are increasingly getting interested in growing their capabilities in AI, in order to achieve dominance and collateral hegemony in international relations. There is a possibility of the AI emerging as a 'new nuclear power' revamping the existing global order. This could lead to all states seeking to have it with its dual usage, civilian and military.[6] Investment of militaries in AI is expected to impact global security.

AI systems are being tested to solve a range of complications in transportation, economy and finance, education, health care, intelligence analysis and cybersecurity. Many prominent scientists are demanding a ban on the use of this technology for military purposes.[7] AI technology has the potential to bring in comprehensive changes for the military-industrial complex (MIC) of various states. There could be some national security applications of AI, which could be beneficial for militaries to ensure peace, like in the field of cyber security. Some security-related applications of AI could improve international stability. At the same time, it is important to realise that the current AI systems have substantial limitations and vulnerabilities too. Hence, there is a need to be careful towards making AI-related investments in the military field. Ethics and military investments do not always go together. Military applications of AI pose risks to international stability. Some security-linked applications of AI could be destabilising, and competitive dynamics among the states could lead to damaging consequences such as a 'race to the bottom' on AI safety.[8] But similar logic could be offered for various other weapon systems too. It is unlikely that the growth in AI-assisted weapon systems would halt. Military technologists have identified some important areas for development of AI-based military systems and mainly technologically developed countries are found working on the military aspects of AI.

At present, the militaries are trying to understand what AI means to them and what they can possibly achieve with the induction of this technology. Basically, militaries are trying to find solutions to the modern-day challenges they are being faced as a result of the continuously evolving nature of warfare. Generally, many of the present-day militaries are required to remain ready to fight both conventional and asymmetric warfare. Some of the militaries have nuclear deterrence architecture in place and they are looking at AI to help them to strengthen their nuclear triad mechanisms. Additionally, there are concerns about cyber warfare and the possibility of space warfare, hence some defence establishments are looking at AI to offer them some solutions to improve their capabilities in such emerging domains of warfare.

For military technologists it is also important to relate with existing technological developments in the field of AI. Artificial General Intelligence (AGI) is a theoretical application of generalised AI in any domain, for solving problems. Normally, AI is divided into two types: Narrow AI and General AI. Narrow AI (ANI) is created to solve one given problem, for example, a chatbot. ANI occasionally gets referred to as 'weak AI,' which is any AI system that can beat a human in a narrowly defined and structured task. It is designed to perform a single function like an internet search, face recognition, or speech detection under various limitations. Owing to such constraints, it gets identified as 'narrow' or 'weak.' The commonly known Narrow AI techniques include Machine Learning (ML), Natural Language Processing and Computer Vision. Amazon, Spotify and Netflix all use Narrow AI algorithms to endorse their products and services. Their algorithms work using data to profile human behaviours and find matching attributes from other users or products.[9]

Narrow AI systems can perform what they are taught. The decision-making capability is restricted and directly related to their design parameters. AGI also involves a system with comprehensive knowledge and cognitive capabilities. Here the purpose is to achieve performance which is indistinguishable from that of a human, although its speed and ability to process data are far greater. Such a system known as Strong AI, is theoretical at present and expert opinions differ as to whether it is possible to create one. Some experts believe that an AI system would need to possess human qualities, such as consciousness, emotions and critical-thinking to reach the level of AGI.[10] At present, various military applications are being worked on in the vicinity of Narrow AI.

Military applications

The twenty-first century is witnessing major investments in the field of military equipment development and production. In fact, it was the 1991 Gulf War which showcased the importance of modern technologies like Information and Communication Technologies (ICT) and space technologies. This war also came to be designated as the first space war. Obviously, many states in the world have learnt the lessons from this war and are depending more on technologies to improve their defence architecture. Apart from military platforms like fourth/fifth generation fighter aircraft, new battle tanks, ships and submarines, states are increasingly investing in drone technologies, radars, sonars, electronics, signals and communication warfare equipment. AI has relevance for all these military paraphernalia including various weapon delivery platforms. Globally, various defence establishments are very enthusiastic about the potential of AI and are establishing various research nodes for developing and fusing these technologies within their armed forces.

The progress in AI is expected to bring new potential to defence technology and hence is expected to make changes towards the defence industrial setups. AI can help in military operations and assist in improving the performance of individual military units and overall defence forces. This could be achieved by using deep learning machines, which require big data and artificial neural networks. In addition, there is a need to develop teamwork between humans and machines, where AI-powered machines could assist humans in taking precise and appropriate command decisions. But all these depend on the skill and expertise of scientists and engineers to develop and create technologies and applications involving AI. AI systems require quality data in large quantities. Such data needs to be routinely collected from the everyday use of various sensors, satellites, ships and aircraft. Data for AI can also be created through physical defence workouts or training, war games and digital simulation.[11] It is important to note that AI is hugely data-driven, and qualitative and quantitative aspects of the database are important. There are some challenges in regard to the availability of the data in the military domain. At times, despite having a good algorithm in place, the data limitations could possibly hamper the development of the solutions.

From a defence perspective, AI has a long history. It could be a coincidence, but within two years after the term AI was coined in 1956, the US Department of Defense established the Advanced Research Projects Agency (which was later renamed DARPA) to facilitate research and development of military and industrial strategies. Arthur Samuel, an IBM computer scientist and a pioneer in computer gaming and AI, coined the term 'machine learning' and much focus on such ideas started taking root during the period 1959-1962. During the 1960s, the US Department of Defense began training computers to mimic basic human reasoning. During 1964–66, ELIZA, an early natural language processing computer program, was created at MIT's AI Laboratory. While Shakey the Robot, developed by SRI International,[12] was the first symbol demonstrating the strength of AI. By 1991, the US military started using the DARPA-funded Dynamic Analysis and Replanning Tool (DART), an AI program, to schedule the transportation of supplies or personnel and to solve other logistical problems. It was using intelligent agents to aid decision support systems located at the US Transportation and European Commands. This system is known to have saved the military millions of dollars right after its launch. For all these years, DARPA has been involved in supporting various AI-based projects, including the development of autonomous cars. It has been continuing its focus on the research, innovation and development of autonomous weapon systems. The US Department of Defense's Project Maven uses computer vision: A combination of machine- and deep-learning (neural networks), to autonomously find objects of interest within massive amounts of video or still imagery from drones.[13] Apart from the USA, many other states, including the European states, have been researching AI's military ability for a long time. Particularly, the efforts and investments made by states like the United Kingdom, Russia and China are noteworthy.

Today, there are numerous military applications of AI on which various military agencies are researching. However, it is important to note that AI would not replace any existing military system directly, but would help to improve upon the existing military systems or assist in the development of new military systems. It could be said that AI would be helping the systems to perform tasks that normally require human intelligence, such as visual perception, speech recognition, decision-making and translation between languages.[14] At present, the development in such fields is mostly at a near-nascent stage, and further development would increase the efficiency and accuracy of these systems.

About the possible applications of AI in matters military, it might be useful to get answers from the AI itself. In November 2022, an AI-based application called ChatGPT was launched which can offer (generate) answers for various questions, although accuracy is an issue. As per the available information, ChatGPT is a prototype AI chatbot developed by OpenAI which specialises in dialogue. The chatbot is a large language model fine-tuned with both supervised and reinforcement learning techniques. It is a fine-tuned version of a model in OpenAI's GPT-3.5 family of language models. To check the strength of this application, a question about the defence uses of AI was asked to ChatGPT (the application available online). The purpose of undertaking this exercise was to assess how detailed and focused the answer generated by ChatGPT was on the aspects of defence applicability of AI. Broadly, it could be argued that these are still early days and much needs to be learnt in regards to the exact applicability of AI for defence. It is well understood that there could have (possibly) been more detailed answers depending on the choice of words while framing the question. Following is the text of the answer received for a specific question.[15]

Artificial Intelligence (AI) has the potential to revolutionise many aspects of defence and national security. Some potential applications of AI in defence include:

Predictive Maintenance: AI can be used to predict when equipment is likely to fail and schedule maintenance accordingly, reducing downtime and improving readiness.

Target identification and tracking: AI can be used to analyse surveillance data and identify potential targets, as well as track their movements.

Military Simulation and Training: AI can be used to create realistic simulations for training purposes, allowing soldiers to practice their skills in a safe and controlled environment.

Autonomous weapon systems: AI can be used to develop semi-autonomous or fully autonomous weapon systems, such as drones or ground vehicles, which can operate without human intervention.

Cybersecurity: AI can be used to monitor network traffic and identify potential threats, as well as to protect against cyberattacks.

Intelligence Analysis: AI can be used to analyse large volumes of data and identify patterns and trends that may not be immediately apparent to humans.

Overall, the use of AI in defence has the potential to improve the efficiency and effectiveness of military operations, as well as to reduce the risk to human lives. However, it also raises ethical and legal questions, such as the extent to which AI should be used to make decisions about the use of force, and the potential for AI to be used for malicious purposes.

At present, advanced analytics and other AI technologies are showing much growth prospects. These technologies are expected to improve military effectiveness. Information Technology (IT) has been a saviour to the world in many walks of life during the Covid-19 crisis. Today, the world is finding creative ways to use and innovate the use of AI in various domains of IT. At the same time, the Ukraine-Russia War has changed the geostrategic calculus. States which were shying away from making major investments towards upgrading their military structures are redefining their defence policies. States are keen to expand on their defence R&D and innovation setups. AI is getting viewed as a technology which could allow the defence industry to leapfrog. There are various military applications (and various possible applications) of AI getting discussed.[16] Following are some of the important fields, where AI applicability is becoming noticeable.

Robots

Combat systems such as weapons, sensors, navigation, aviation support and surveillance can employ AI in order to make operations more efficient and less reliant on human input.[17] Some important developments are happening to move towards using AI applications for the development of combat systems and weapon delivery platforms. There is a realisation that the developments in AI and ML have the potential to revolutionise how humans interact with technology. These advances are driven primarily by the massive quantity of information that can be found, making a major impact towards developing military-capable systems. For some time now, there has been an increasing focus towards using AI in the military domain for the purposes of training, simulation and cybersecurity. Also, AI-enabled land combat systems are also under research. However, a major development appears to be happening in the arena of drone technologies.

Over the years, much progress has taken place towards developing combat drones. The unmanned systems, also called Unmanned Aerial Vehicles (UAV) or, when they are designed as weapon systems, as Unmanned Combat Aerial Vehicles (UCAV), are getting used in the battlefield as remotely operated or autonomous systems. These systems have now become a part of the defence inventory for various states and are fast becoming a significant part of the combat forces, apart from being used for Intelligence, Surveillance and Reconnaissance (ISR). During September–November 2020, the 44-day war between Armenia and Azerbaijan over the disputed Nagorno-Karabakh region demonstrated how effectively AI-driven drone technology could be used on the battlefield. It was a war which witnessed the use of conventional and AI-embedded weapon systems. The war witnessed the heavy use of missiles, drones and rocket artillery. Drones of Russian, Turkish, Israeli and homegrown designs undertook both reconnaissance missions to provide fire support to artillery and also participated in strike missions. It was Azerbaijani drones that emerged as the winner, and post-war assessment indicated that these systems were game-changers. They provided important advantages in ISR, besides long-range strike capabilities. They helped in finding, fixing, tracking and killing targets with precision.[18]

It could be said that by now drone systems have made their place as a flighting platform. The next step is using AI in a dual role: 1) to assist the fighter aircraft/bomber in multiple ways; and 2) AI is being taught to fly warplanes, indicating that fighter aircraft would be totally controlled by AI with the possibility of removing the pilot altogether. The US Department of Defense has given a project to DARPA called the Air Combat Evolution (ACE) programme. Here, the idea is to develop a fighter plane equipped with AI. AI would allow the aircraft to execute tighter turns, take greater risks, and make better shots than human pilots. All this would transform the human pilot's role. The focus of the programme is to ensure that the AI will fly the plane in partnership with the human pilot. This would make the pilots a battle manager. Eventually, all this would help the military establishment to have smaller, less expensive units. In the future, this will lead to the development of an armed force waging combat with fewer humans and a larger number of expendable machines. DARPA calls this mosaic warfare.[19]

Overall, naval warfare involves three different types of warfare: Underwater warfare, surface warfare and aerial warfare. Today, AI is being used in all these three arenas. Unmanned surface vehicles are one of several new tools the navies are pursuing that readily use AI. Accordingly, AI and ML are critical to the navies' new-fangled maintenance initiative called Condition-Based Maintenance Plus. Once fully operational, this competence will leverage AI to sort through a ship's data and use predictive analytics to identify issues with systems and subsystems that need to be addressed to resolve problems at sea. For this, a robust digital infrastructure has been put in place for harnessing the available data. In 2020, the US Navy launched a system named Jupiter. This system deals with the vast amount of information collected. It has been designed on the Pentagon's data analytics tool known as Advana, and it helps to translate everyday data into insights, decisions and outcomes which the US Navy can then act on (applications ranging from administrative tasks to warfighting capabilities). This is done by making data more 'discoverable, accessible, understandable and usable.'[20]

The twenty-first century is witnessing the global naval powers preparing a strong road map for future underwater warfare. Historically, silent submarines are known as the authentic source of surveillance and data gathering. Submarines are important instruments towards forming a strong defence strategy with the deterrent power to stop any adverse naval situations. During April 2016, the US Navy had launched the Sea Hunter, a 40-meter unmanned and completely autonomous warship designed (a DARPA design) for conducting anti-submarine warfare. The entire navigation of this vehicle is controlled by AI with zero crew onboard. During April 2021, a design by MSubs for Britain's Royal Navy was launched. This was the UUVs (Unmanned Underwater Vehicles), an exclusive research prototype for XLUUV (Extra Large Unmanned Underwater Vehicle). Today, global powers are working towards introducing XLUUV for its capability to handle various underwater challenges.[21]

UAVs and UUVs have unique design problems. However, the most difficult system to design is an unmanned ground vehicle (UGV). This is mainly because the nature of obstructions a system encounters when undertaking operations on the ground is far more than in the air or underwater. The unmanned systems in full autonomous roles as such require a lot of processing power. The design becomes more complicated in the case of UGVs. Analysing large amounts of data to complete complex deep learning algorithms requires significant processing capabilities. A large number of devices like cameras and sensors are in use for collecting and processing data. With more devices collecting data, there are issues associated with the bandwidth. As such, in the case of unmanned systems of any kind, the size of the unit is also important. High-performance embedded computing (HPEC) helps in this regard. HPEC computing, coupled with AI capabilities, brings forth a powerful computing solution in size, weight, and power (SWaP) in an optimised form. All this fuels AI and deep learning and offers unmanned systems access to on-demand intelligence during the operational phase.

High-performance embedded edge computing is critical to the success of AI missions. For operating in a contested environment, which basically means a scenario involving restricted bandwidth and degraded communications, the UGVs would be in trouble. Here, the tactical use of cloud-based computing and AI becomes a liability. It has been realised that the computational processing capability must reside on-premise to ensure the low latency and near real-time speed demanded of AI-based applications. Hence, various recent developments in commercial off-the-shelf (COTS) technologies are known to assist with practical embedded edge computing use in unmanned vehicles. The high-powered embedded computers provide the processing power needed to run various new AI algorithms for keeping UAVs, UUVs and UGVs on the move.[22]

The military industry has realised that juxtaposing AI on existing robots could allow these systems to make their own decisions. There are both advantages and limitations in regards to putting such systems in the battlefield. Robotics involves developing a system with the assistance of various branches of engineering including electrical, mechanical, electronics, information and communication technologies. Present-day AI involves intelligent decision support systems, which allow the systems to think and decide on their own. Such systems take inputs on a range of issues from weather conditions to terrain conditions to capabilities of the attacker, nature of possible incoming attack location, type and nature of counterattack to be launched, situational awareness for possible target engagement, etc.

Various robotic systems assisted by AI for the purposes of military are getting designed and developed. Such systems could be both defensive systems and offensive systems. Here, defensive systems essentially indicate non-lethal systems. Such systems could be reconnaissance systems, which are robotic and autonomous in nature. Important applications could be perimeter fencing systems, where cameras could respond to the happenings in their area of operation.

For transporting logistical load, the US agencies have developed a mechanical mule,[23] which carries the load to the inhospitable region in order to reduce the burden and cater for the risk to human life. Such a robotic system (BigDog) was developed to be capable of navigating rough terrain. This mule could follow humans automatically for 24 hours and cover a distance of around 32km with a load of 180 kg in one go. However, the US military has most likely shelved development of this robotic pack mule owing to performance issues, it being apparently too noisy to make a good soldier (a loud robot that's going to give away their position).[24] In this case, there are possible issues related to mechanical engineering aspects of the system development that require improvement. Also, the AI training for the system would need to be improved on.

Lethal autonomous weapon systems (LAWS)

One of the controversial AI-based weapon systems, which is called by few as Slaughterbots or killer robots, is the lethal autonomous weapon system (LAWS). Here, AI is used for the identification and selection of the target, and the system decides and acts on its own to destroy the target. Such systems operate without any human intervention. At present, such systems are mostly so-called 'fire-and-forget' systems, which, once activated, select and engage targets on their own without any guidance from humans. Like any other military system, LAWS also has its own pluses and minuses. Allowing machines (robots) to use their own intelligence (AI) to kill has raised various moral concerns[25] and prominent scientific and industrial community leaders are arguing against the development of such systems. LAWS are both offensive as well as defensive systems.[26] At present, the only deployed fully autonomous systems belong in the defensive category. Humans are yet to achieve technological expertise to make machines which can decide on their own when to start a war or a conflict. Hence, LAWS, as they are deployed today, could be considered as defensive weapon systems, which are programmed to respond to incoming threats. Almost all the prevailing autonomous weapon systems (mainly used in missile and air defence roles) are designed as point defence or area defence weapon systems. They respond to incoming missile threats, but are not capable of launching an attack independently. The US has operated armed ground robots like the Special Weapons Observation Reconnaissance Direct-action System (SWORDS[27]), which was deployed in Afghanistan for detecting and disabling improvised explosive devices. SWORDS was the first weaponised UGV. Such a robotic system has limited inbuilt inherent intelligence and is remotely operated by a soldier. Such systems indicate that similar systems, with a capability for firing without human intervention or oversight, could be designed and developed. Such systems would be categorised as offensive LAWS.

The 1991 Gulf War saw successful usage of the US Navy's Tomahawk missile. Tomahawk has been recognised as one of the important autonomous systems in recent times. Also, some other systems used by the US, mostly in the post-Cold War era, could be said to belong to the autonomous category. They are artillery-launched weapons, the Brilliant Anti-armour sub-munition (BAT) and Terminally Guided Warhead. In addition, systems like sentry robots, systems mostly involving ground robots like PackBot, TALON, and MARCBot, supersonic and stealthy drones also belong to the autonomous category with AI being at the centre of such developments.

At present, the only fully autonomous weapon systems that are completely operational are counter-rocket, artillery and mortar systems, such as the so-called 'Iron Dome,' anti-missile systems, such as Terminal High Altitude Aerial Defence (THAAD) and anti-aircraft systems such as S-400/S-500.

In addition, there are systems based on robotic technologies, like drones and unmanned ground and underwater vehicles, which are able to navigate, but cannot select and engage targets autonomously; they only respond to the incoming target. A further example could be the sentry robot SGR-1, which has been developed by Samsung Techwin. Presently, this system has been used by the Republic of Korea along its border with the Democratic People's Republic of Korea.[28] This robot can detect targets from a distance of around 3.5 km, however, at present, the final order to fire is given by a human operator.

A Close-in Weapon System (CIWS) is a point-defence system used for defence against short-range anti-ship missiles. These systems are also useful for engaging enemy aircraft which have successfully infiltrated outer defences to approach with high speed targets (normally, a battle ship or tanker ship). Land-based CIWSs can also address threats like shell bombardment and rocket fire. All major maritime forces in the world are equipped with CIWS. These systems could also be used on land to protect military bases. Such systems have both gun-based and missile variants. The gun-based system comprises of multiple-barrel, rotary rapid-fire cannons placed on a rotating gun mount. Both variants require various types of passive and active radar units for providing terminal guidance.

The Iron Dome[29] has proved its effectiveness for short-range applications. It is a system conceptualised by Israel and jointly funded by the USA. The Iron Dome is a counter-rocket, artillery and mortar system capable of intercepting multiple targets from any direction. The Iron Dome uses an autonomous guidance and control system capable of intercepting specific targets which represent a high-priority threat according to the system configuration. The Tamir Adir system is a sea-based variant of an Iron Dome missile battery, developed as an autonomous maritime missile interception system. Israel conducted a successful test of this system in May 2016. The Tami Adir system is capable of engaging and destroying airborne targets from a moving platform.[30] The Israeli Defence Forces (IDF) has an operational Naval Iron Dome system which was tested aboard Sa'ar 5 corvette INS Lahav. The integration of the Iron Dome air defence missile system aboard surface vessels follows the need of Israel to protect strategic assets at sea, such as oil/natural gas rigs. Israel's hydrocarbon fields and drilling installations in the East Mediterranean Sea require continuous security cover.[31]

THAAD[32] is intended to defend against short and medium range ballistic missiles. This system claims to have a 100% intercept test success rate. The entire system architecture depends on support from radars and satellites. The system operates in a fully autonomous mode whereby an infrared satellite detects an incoming missile's heat signature and sends all collected real-time tracking data to the ground-based system through a communications satellite (announcing early warning). When the threat is confirmed, based on an assessment carried out from inputs received from various early warning systems, a suitable command is conveyed to the sensors and weapon systems (such commands put the weapon system into active mode). Subsequently, the long-range radar detects and tracks the missile for some time to further improve the accuracy. The tracking data helps to compute the trajectory of the incoming threat missile. Among the group of batteries available to address the threat, the most effective interceptor battery is engaged and carries out the interception. The entire process of identifying, engaging and destroying the missile is fully autonomous in nature and known to have very high efficiency.

Some LAWS, which are still either at the drawing board level or in the realm of theoretical possibilities, also warrant attention. These include space-based autonomous systems, which could be used to target space-based systems as well as targets on Earth.[33] There exists a possibility that, with the overall growth happening in the technology sector globally, some capable states could seek to develop such systems in the near future.

The nature of warfare is ever evolving. An increasingly automated battlefield is expected to add another dimension to warfare, which will have a mixed impact on militaries. States are bound to develop countermeasures (and counter-countermeasures) to LAWS. In general, LAWS are likely to continue to have relevance both as tactical and strategic weapon systems. It is important to note that autonomy cannot be thought of in absolute terms; there may be either low or high levels of autonomy. Arguably, militaries will be required to keep these weapons under their effective control and decide about the contexts in which they can be deployed, and the nature and degree of autonomy allowed for any given deployment. Militaries will also be required to effectively navigate the various legal challenges, arms control considerations and moral issues related to LAWS, in order to continue to keep these weapons in their arsenals. Today, LAWS provide both opportunities and vulnerabilities for militaries. Hence, it is necessary for militaries to incorporate such weapons into their warfighting doctrines with due diligence.

Swarming

Drone swarms are one technology that is quickly taking root in the military architecture of the defence forces of various states. Such systems are known for their military effectiveness. Here, all the drones in the swarm are in communication with each other and perform preprogrammed functions. In case of any drone from the swarm becoming unserviceable, the other units take over its functions. Broadly, swarming is known as a system of autonomous networked Small Unmanned Aircraft Systems (SUAS) functioning collaboratively to achieve shared intentions with an operator on or in the loop. Here, each platform (an individual drone in the swarm) is controlled jointly with others. All drones operating in a swarm are interlinked and in continuous communication with each other. The purpose is to share information from their sensors and take AI-driven collective decisions toward the achievement of a single goal. This datalink and the AI software are thus crucial for its functioning. However, it may be noted that every system gets designed and programmed as per the expected military aim of the mission. Each single drone forming a swarm is just a minor, but important component playing a specific role in a larger system. Some such drones are designed to use their sensors to locate and track targets and share information with other drones in the system. The swarm gets programmed to react dynamically to changes in the battlespace and AI technologies play a major role in that effect.

In general, swarming holds immense potential and capability to revolutionise warfare. It can be applied to different types of military tasks. Swarms can patrol large areas with greater efficiency and shorter reaction times than human personnel. Swarms allow savings in various fields from time factor to saving human lives. They can be put to use for both defensive and offensive missions. Intelligence gathering, search and destroy missions against enemy air defences, submarines or mobile missile launchers, over-the-horizon targeting, air combat, and anti-access/area denial (A2/AD) and counter-insurgency missions could be some of the missions where swarms could be used.[34]

The US defence establishment has major interests in armed drone swarms. The Perdix system consists of autonomous drones operating as cooperative swarms of 20 or more flying units. The drones are launched to achieve a specific goal, are expected to engage in collective decision-making and are known to possess swarm self-healing abilities whereby, in the case one or more drone units are forced to drop out, the entire system reconfigures itself automatically for mission completion. Work on this system began in 2013. During this time, several testing missions have been launched and the system's software is currently being upgraded. The US Army is finalising Project Convergence, which is a campaign of learning to evaluate how the army will fight with modernised equipment and advanced capabilities. Here the idea is to have a low-cost swarm capability with the ability to identify and engage threats with the use of a single controller.[35]

C4ISR

Command, Control, Communications, Computers, Intelligence, Surveillance and Reconnaissance (C4ISR) solutions are known to simplify the complex – turning data into knowledge and knowledge into action.[36] Presently, states are found employing either totally or partly the Revolution in Military Affairs (RMA) as a part of their modernisation process. Owing to the presence of information technology-based tools, availability of secure communication satellites and significant assistance from satellites, various armed forces across the world are opting for developing structures, doctrines and strategies for engaging in Net Centric Warfare (NCW). All this has led to the development of C4ISR architecture. Such architectures are normally based on the ICT tools and various inputs (real-time or otherwise) from space-based systems. Also, various applications involve navigational feedbacks based on Global Positioning Satellites (GPS), Inertial Navigation Systems (INS) systems and Geographic Information Systems (GIS). Big data and inputs from various radar/telescopic systems make the system more proficient. All this reduces the dependency of commanders on conventional techniques used for operational planning like paper maps and sand models.

In a network-centric environment, AI assists soldiers to access and share information throughout the entire network almost in real-time. For this, ISR are the key elements of battlefield management. The change in the battlefield scenario over the years has also led to changes in techniques of identification, surveillance and reconnaissance. Improvements in technologies in various related spheres are presenting sensors with high sensitivity, small sizes and better visibility. Particularly, AI is contributing significantly to the fields of ground, aerial, space and underwater ISR capabilities. Noticeably, UAVs are found using AI-based techniques for the purposes of intelligence, surveillance and reconnaissance. This is bringing transparency to the battlefield and allowing dynamic visualisation of an almost actual (three-dimensional model) battlefield. AI is increasingly integrated in areas like internet search, military applications, terrain and atmospheric conisations content optimisation and robotics. In strategic terms, AI has the potential to change the way armed forces execute their command, training, logistics, and force deployment. Various surveillance platforms, UAVs, Synthetic Aperture Radars, Airborne Early Warning Systems and Satellites etc. continuously generate vast volumes of data and imagery, around the clock. Using such data, AI-based systems develop approaches to aid the battlefield commanders in identifying hidden patterns in the swathes of data with the highest precision and accuracy. It could be a standalone input given by an AI-based system or such information could be used for augmenting the human ability to analyse and act.[37] In the near future, such and other AI-based technologies are likely to become embedded in various other technologies and help provide autonomous decision-making on behalf of military commanders. AI is expected to have a major role in ICT, including both traditional telecommunications as well as various communications-enabled applications.[38] AI also allows devising Intelligent Decision Support Systems, which actually leads to the transformation of human decision-making to machine-based decision-making or provides computational assistance to commanders to make decisions. C4ISR structures are also expected to be aided by blockchain technologies and the Internet of Military Things (IoMT).

Cybersecurity

Present-day ransomware attacks are very sophisticated. It has been observed that in recent times, though the overall number of attacks has dropped, their efficiency and success rate have risen. At present, the focus of hackers is more on businesses than individuals. The key ransomware statistics indicate that during the Covid-19 period, schools, universities and hospitals were also targeted globally. Possibly, around every 11 seconds there were ransomware attacks during 2022, and businesses suffered the attacks every 40 seconds. The overall global cost per year is estimated to be $20 billion yearly. Every year, ransomware possibly generates an estimated $1 billion in revenue for cybercriminals.[39]

AI is expected to transform this activity. Owing to the presence of AI, such cyberattacks and responses are expected to become faster, more precise and more disruptive. AI could take the corrective measures against any cyberattack in the shortest possible time. AI is already being employed to verify code and identify bugs and vulnerabilities. The worth of AI in cybersecurity is predicted to reach $18 billion by 2023.[40] Modern-day armed forces are already facing major cyber threats. It is expected that with increasing dependence on information technology tools, the challenges from cyber threats are going to be more severe in coming years. Hence, it is important for the armed forces to factor AI as a tool in their cyber defence architecture.

AI is a trending topic for many industries now. AI is getting used by a variety of organisations for operational functions like automated tasks, natural language processing, deep learning and problem-solving. All such tasks have military relevance too. Today, with the increasing number of cyber threats and attacks, AI security serves as a crucial element and demands critical attention. Security in AI falls into two different areas: Using AI for cybersecurity and using cybersecurity for AI. Distinct capabilities of AI serve huge benefits for cybersecurity since it holds the ability to analyse and mitigate large sets of potentially malicious data without any human interaction and also while providing suggestions for future threats. At the same time, AI programs themselves are vulnerable to direct cyberattacks as well.[41] Military agencies while investing in AI need to be aware of all such aspects related to AI applicability and limitations. They are also required to find/work on solutions for various AI-associated challenges before inducting AI technologies.

Conclusion

For some decades now, much of research, development and innovation has been happening in military-related AI technologies. Various weapon systems are getting designed, with some achieving operational status, while some are still in the initial level of development. Major powers like the USA, Russia and China are making major investments in this field. This technology is showing vast potential for future use. The current focus is on using this technology mainly for the purposes of logistics and training. There are various ethical issues associated with the use of this technology, which the militaries have to grapple with. Mainly, the technology is still not ripe for total autonomy during deployment. As such, humans would not like the machines to make the decision to kill a human. Hence, even with the autonomous weapon systems, there is a broad understanding that the human should always remain in the loop. Broadly, AI allows the military leadership to get suggestions based on systematic assessment of vast data. It is found to make the process of decision making, from early stages of planning to actual operations, more effective and logical. AI-based weapon systems have great advantages and are currently are proving to be force multiplayers for the militaries. All in all, AI is expected to revolutionise military operations, and as technology matures, there would be endless possibilities for its usage in the military.

Notes

1 Alex Roland, 'War and Technology,' 27 February 2009, available at: https://www.fpri.org/article/2009/02/war-and technology/#:~:text=Technology%20de fines%2C%20governs%2C%20or%20circumscribes,is%20that%20it%20 changes%20warfare, accessed on 23 September 2022.
2 'Technology and the Future of Modern Warfare,' 16 December 2021, available at: https://www.information-age.com/technology-and-future-of-modern-war fare-19223/, accessed on 24 August 2022.
3 https://news.stanford.edu/news/2011/october/john-mccarthy-obit 102511.htm l#:~:text=McCarthy%20created%20the%20term%20%22artificial,and%20i nvented%20computer%20time%2Dsharing, accessed on 2 November 2022.
4 Preethi Amaresh, 'Artificial Intelligence: A New Driving Horse in International Relations and Diplomacy,' 13 May 2020, available at: https://diplomatist.com /2020/05/13/artificial-intelligence-a-new-driving-horse-in-international-rela tions-and-diplomacy/, accessed on 7 December 2022.
5 Oscar M. Granados and Nicolas De la Peña, 'Artificial Intelligence and International System Structure,' *Revista Brasileira de Política Internacional*, vol. 64, no. 1 (2021), e003, available at: https://www.redalyc.org/journal/358 /35866229002/html/, accessed on 7 December 2022.
6 'Artificial Intelligence and Global Security,' available at: https://www.ia-forum .org/Content/ViewInternal_Document.cfm?contenttype_id=0&ContentID =9255, 11 December 2022, accessed on 11 December 2022.
7 During 2015, Elon Musk and Stephen Hawking, along with hundreds of AI researchers and experts, have called for a worldwide ban.
8 https://hai.stanford.edu/events/ai-and-international-security, accessed on 11 Dec. 2022.
9 Zoe Larkin, 'General AI vs Narrow AI,' 16 November 2022, available at: https://levity.ai/blog/general-ai-vs-narrow-ai#:~:text=What%27s%20the%20differ ence%20between%20Narrow,any%20problem%20that%20requires%20AI, accessed on 15 December 2022.
10 https://www.techtarget.com/searchenterpriseai/definition/narrow-AI-weak -AI#:~:text=Narrow%20AI%2C%20also%20known%20as,intelligence%20for %20a%20dedicated%20purpose, accessed on 15 December 2022.

11 'Artificial Intelligence in Defence Technology,' available at: https://www.innefu
.com/blog/artificial-intelligence-in-defence-technology, accessed on 6 December
2022.

12 Is a 75-year pioneering research institute (formerly Stanford Research Institute)
involved in various research area including robotics, bio medical sciences and AI.
See 'Shakey the Robot' April 1984, http://ai.stanford.edu/~nilsson/OnlinePubs
-Nils/shakey-the-robot.pdf

13 https://militaryembedded.com/ai/machine-learning/artificial-intelligence-time-
line, accessed on 14 November 2022.

14 'The Most Useful Military Applications of AI,' available at: https://sdi.ai/blog/
the-most-useful-military-applications-of-ai/, accessed on 6 December 2022.

15 https://chat.openai.com/chat, accessed on 18 December 2022.

16 For this section, the author has used some information he has previously
published. Please refer to A. Lele, 'Artificial Intelligence (AI),' in *Disruptive
Technologies for the Militaries and Security, Smart Innovation, Systems and
Technologies*, vol. 132 (Singapore: Springer, 2019). https://doi.org/10.1007/978
-981-13-3384-2_8.

17 'The Most Useful Military Applications of AI.'

18 Shaan Shaikh and Wes Rumbaugh, 'The Air and Missile War in Nagorno-
Karabakh: Lessons for the Future of Strike and Défense,' 8 December 2020,
available at: https://www.csis.org/analysis/air-and-missile-war-nagorno-kara-
bakh-lessons-future-strike-and-defense, accessed on 14 June 2022.

19 Sue Halpern, 'The Rise of AI Fighter Pilots,' 17 January 2022, available at:
https://www.newyorker.com/magazine/2022/01/24/the-rise-of-ai-fighter-pilots,
accessed on 25 November 2022.

20 Mikayla Easley, 'Surface Navy Building Digital Infrastructure to Harness AI,'
18 March 2022, available at: https://www.nationaldefensemagazine.org/articles
/2022/3/18/surface-navy-building-digital-infrastructure-to-harness-ai, accessed
on 19 December 2022.

21 Amit Das, 'Submarine Warfare & Artificial Intelligence,' 26 July 2022, available
at: https://www.financialexpress.com/defence/submarine-warfare-and-artificial
-intelligence/2606457/, accessed on 20 August 2022.

22 Jamie Whitney, 'Artificial Intelligence and Machine Learning for Unmanned
Vehicles,' 26 April 2021, available at: https://www.militaryaerospace.com/
unmanned/article/14202040/artificial-intelligence-and-machine-learning-for
-unmanned-vehicles, accessed on 20 August 2022.

23 Jack Linshi, 'U.S. Military Takes Robotic Mule Out for a Stroll,' 15 July 2014,
available at: https://time.com/2987608/us-military-robotic-mule/, accessed on
30 November 2022.

24 Erico Guizzo, 'Boston Dynamics Wins Darpa Contract to Develop LS3 Robot
Mule (It's a Bigger BigDog),' 1 February 2010, available at: https://spectrum
.ieee.org/automaton/robotics/military-robots/boston-dynamics-ls3-robot-mule,
accessed on 14 October 2017; James Vincent, 'US Military Says Robotic Pack
Mules Are Too Noisy to Use,' 29 December 2015, available at: https://www
.theverge.com/2015/12/29/10682746/boston-dynamics-big-dog-ls3-marines
-development-shelved, accessed on 15 October 2017.

25 Ronald Arkin, 'Lethal Autonomous Systems and the Plight of the Non-
Combatant,' *Georgia Institute of Technology, AISB Quarterly*, no. 137 (July
2013), pp. 1–6.

26 It is important to note that the defensive systems which are discussed in the section on robotics are different than in this case. In the context of robotics, the
defensive systems are non-lethal, whereas in case of LAWS, it is a lethal autonomous system which is considered to be used in defensive mode. This section

is based on the information readily available on the internet and some of the author's earlier writings.

27 Kim Jones, 'Special Weapons Observation Remote recon Direct Action System (SWORDS),' Special Weapons Observation Remote recon Direct Action, available at: https://www.sto.nato.int › MP-AVT-146-36, accessed on 12 December 2022.

28 This was indicated to the author during one of his visits to Seoul by the officials there during informal talks.

29 https://www.rafael.co.il/worlds/air-missile-defense/short-range-air-missile -defense/, accessed on 11 December 2022.

30 https://www.timesofisrael.com/israel-successfully-tests-shipborne-iron-dome -missile-interceptor/, accessed on 7 December 2022.

31 https://navyrecognition.com/index.php/news/defence-news/2017/november -2017-navy-naval-forces-defense-industry-technology-maritime-security-global -news/5749-idf-declares-naval-iron-dome-operational-with-sa-ar-5-corvette .html, accessed on 7 December 2022.

32 https://www.lockheedmartin.com/en-us/products/thaad.html, accessed on 8 December 2022.

33 I.A. Nesnas, L.M. and R.A. Volpe, 'Autonomy for Space Robots: Past, Present, and Future,' *Current Robot Report*, vol. 2 (2021), pp. 251–63. https://doi.org/10 .1007/s43154-021-00057-2.

34 Alessandro Gagaridis, 'Warfare Evolved: Drone Swarms,' 10 June 2022, available at: https://www.geopoliticalmonitor.com/warfare-evolved-drone-swarms/, accessed on 3 December 2022.

35 https://www.defensenews.com/unmanned/2022/03/01/us-army-to-demo-offen- sive-drone-swarms-in-next-project-convergence/, accessed on 12 December 2022.

36 https://www.lockheedmartin.com/en-us/capabilities/c4isr.html, accessed on 28 November 2022.

37 https://idsa.in/issuebrief/the-global-race-for-artificial-intelligence_msharma _230218.

38 'Artificial Intelligence in Military Application Information Technology Essay,' 23 March 2015, available at: https://www.ukessays.com/essays/information -technology/artificial-intelligence-in-military-application-information-tech- nology-essay.php, accessed on 1 October 2017 and http://www.businesswire .com/news/home/20160812005152/en/Artificial-Intelligence-Communications -Applications-Commerce-Internet-Things, accessed on 11 October 2017.

39 Ivana Vojinovic, 'Ransomware Statistics in 2022: From Random Barrages to Targeted Hits,' 5 December 2022, https://dataprot.net/statistics/ransomware -statistics/, accessed on 28 Nov. 2022.

40 Mariarosaria Taddeo, 'Regulate Artificial Intelligence to Avert Cyber Arms Race,' 16 April 2018, https://www.nature.com/articles/d41586-018-04602-6, accessed on 27 September 2018.

41 'Artificial Intelligence, a New Chapter for Cybersecurity?' 10 November 2022, https://www.tripwire.com/state-of-security/artificial-intelligence-new-chapter -cybersecurity, accessed on 20 December 2022.

3

ARTIFICIAL INTELLIGENCE AND MILITARY AVIATION

A.K. Sachdev

Artificial intelligence (AI), the distributary of science that deals with computer-related technologies pursuing the development of intelligent machines, is yet to receive a universally accepted definition but is being hailed as the third revolution in warfare. While the first two revolutions (gunpowder and nuclear weapons) were arms, AI does not – strictly speaking – fit that definition. Nevertheless, it is common to read or hear references to the 'AI arms race.' AI is making significant inroads into weaponry in all domains of warfare – land, sea, air, space and cyber – and all military powers are engaged in endeavours to harness AI to their advantage over perceived rivals.

At this stage, it would be good to differentiate between 'automated' and 'autonomous' in the context of computer-driven technologies. Automation implies that the computer carries out a task without human assistance within a rule-based structure deterministically, i.e., producing consistently identical results for a particular set of inputs, while an autonomous system operates stochastically in as much as it makes predictions about outcomes based on inputs so as to arrive at a course of action probabilistically. Thus, an autonomous system, even when given the same inputs repeatedly, would produce a range of decisions/actions. That is where AI comes in. According to USDOD,

> AI refers to the ability of machines to perform tasks that normally require human intelligence — for example, recognizing patterns, learning from experience, drawing conclusions, making predictions, or taking action — whether digitally or as the smart software behind autonomous physical systems.[1]

DOI: 10.4324/9781003421849-4

In other words, AI describes a computer's competence for natural language recognition, pattern identification, experiential learning, predictions/conclusions based on inputs, visual perception, decision-making and decisive action; essentially, these represent imitation of human intelligence. However, there is a fundamental problem in making computers mimic the human brain. We still do not understand comprehensively how the brain functions, and thus cannot design computers to impersonate it.

The term Common AI (CAI) refers to those tasks that computers can do as well as, and indeed faster than, human beings, while General AI (GAI) pertains to the higher level, cognitive functions of the brain, generally termed as 'thinking.' The Holy Grail of AI is to achieve singularity – a state of technology wherein computer programs become so advanced that AI transcends human intelligence, potentially erasing the boundary between humanity and computers (hence the term 'singularity'). Erudite estimates expect that to happen in another two decades.[2] Meanwhile, current research endeavours to elevate CAI to GAI by understanding how the brain functions and then having AI machines mimic the processes. While we move towards that milestone, AI is still a significant handmaiden for military apparatuses and the militarisation of AI is well under way with leading military powers investing large resources in emerging AI-related technologies.

AI has linkages with many current, emerging and future fields of science. Some of them are quantum computing, nano technology (size, weight and power consumption [SWaP] is critical to embedded AI in avionics, and technology is obsessed with reducing this set of parameters to make AI more usable and deployable), brain machine interface, natural language processing, human augmentation, robotics, data science (data science combines multiple fields, including statistics, scientific methods, and data analysis, to extract value from data), big data analytics, cloud computing, edge computing, small (and big) satellites, 5G/6G communications, 3D printing, cyber technologies, Internet of Things, digital twins, augmented/virtual/ mixed reality and metaverse (Meta is developing its own AI supercomputer, which it predicts would be the fastest in the world). This article looks at the major developments in AI, in conjunction with these domains, in the technology-intensive field of military aviation.

Modern combat aerial platforms

To locate the inroads of AI into combat aircraft, it would help to know that as technologies provided improvements during their evolution, the significant milestones gradually came to be known as generational changes. There is no unanimous agreement on airtight definitions of these generations, but generally, the jet fighters that appeared at the end of the Second World War (mid 1940s) are referred to as the first generation, while significant technological improvements in design and speeds led to the second generation in the 1950s. The second-generation aerial vehicles were the fighters operating during the Korean War, many with swept-wing designs. Supersonic speeds and advanced engines characterised the third generation through the 1950s and 1960s. The fourth generation came in during the 1970s and was differentiated by significant improvement in avionics and automation, for example Fly By Wire (FBW) and Full Authority Digital Engine Controls (FADEC). These had sophisticated levels of automation (essentially 'if... then...' capabilities) but no autonomy nor AI yet. A 4.5th (also sometimes 4+ or 4++) generation is also listed by some sources to refer to fourth-generation fighters with some fifth-generation features. Numerically, the majority of current-day fighters are essentially fourth-generation, although they are being constantly retrofitted with features (improvements upon the fourth generation) to enhance their original capabilities. In the 1990s, the fifth generation emerged with a significant amount of stealth, some of the other features that characterised this generation being fitted with Active Electronically Scanned Array (AESA) radars, super cruise, plug-and-play electronics, and automation permitting a fair amount of autonomy supported by fledgling AI in some areas of operations.

While the fifth generation was typified by high stealth (its defining and emphatic feature), the distinctive feature of the sixth generation is AI. Some fourth-generation fighters carried a Weapon System Operator (WSO) to help the pilot, but fifth-generation fighters are all single seaters with AI sharing the pilot workload. In the sixth generation, AI is expected to move from sharing the cockpit with a human pilot to manning it independently; an 'optionally manned cockpit' design is thus slated to be a key attribute of the sixth-generation fighter. Its other attributes are expected to be modular design, virtual/augmented/mixed reality and the capability to control drone swarms.

The USA, Russia, China, the UK, Japan and France are engaged in sixth-generation fighter programmes, with the USA, expectedly, in the lead but China not far behind. USA has two sixth-generation programmes: F-X, also known as Next Generation Air Dominance (NGAD) or Penetrating Counter Air (PCA) for the US Air Force (USAF) and F/A-XX, a replacement for the US Navy's F/A-18E/F (also referred to as NGAD by the US Navy). The projected date of induction for both NGAD is 2030. The NGAD is often projected as a 'family of systems' with a manned platform working with multi-mission UAVs called Collaborative Combat Aircraft (CCA) which will carry out roles including probing, suppressing and destroying enemy air defences, surveillance and weapon delivery in high-risk areas. In September 2022, American defence firms Kratos and General Atomics announced arriving at a significant milestone – that of providing autonomous capability using an AI pilot on the UAV Avenger MQ-20A.[3] The 'family of systems' may later include long-range aircraft for catering to distant theatres. While the USA has been going all alone with the two NGAD programmes till recently, indications from reported statements by USAF Secretary Frank Kendall are that the US may be willing to discuss collaboration with Australia, Japan and the UK, especially about the Collaborative Combat Aircraft (CCA) unmanned complement of the NGAD.[4] However, such partnerships may not be easy to forge due to the UK and Japan already being involved in other sixth-generation programmes.

Europe has two sixth-generation programmes afoot. The first is the Future Combat Air System (FCAS) or Tempest being developed by the UK, Sweden and Italy, none of which have a fifth-generation programme. Japan has its own F-X sixth-generation programme but has been partnering on some parts of the Tempest programme since 2020. In December 2021, it was announced that the UK and Japan are jointly going to produce an engine for Tempest and F-X.[5] The collaboration could extend to other areas of development (of the Tempest) too. In September 2022, Defense News reported that the Italian Air Force Chief would travel to Japan to hold talks with his Japanese counterpart about collaboration on sixth-generation technology.[6] Besides the stealth and data fusion features of fifth-generation, the Tempest is expected to have significant AI to assist the pilot and for management of teamed unmanned platforms. It is expected to enter service in the mid-2030s and, according to the UK Ministry of Defence (UKMOD), the demonstrator aircraft would fly by 2027.[7]

The second European sixth-generation fighter is being developed by a consortium comprising France, Germany and Spain. Its official name is also FCAS, but to differentiate it from the UK programme, it is often referred to by its French abbreviation SCAF (standing for *Système de combat aérien du futur*). It aims at producing a Next Generation Fighter (NGF) which will connect through the cloud to a variety of UAVs for offensive and surveillance roles. This programme was initially slated to do its first test flight in 2027 and go operational by 2045. However, some problems about work share are rumoured, and there may be delays.[8] While the workshare problems appear solvable in the near future, the business issues involving the various governments and OEMs are also impediments to surmount, and so the final aircraft may not be inducted before 2050, according to Dassault's CEO Eric Trappier.[9]

Russian news agency Telegrafnoye Agentstvo Sovetskogo Soyuza (TASS) reported in 2016 that Russia was working on a sixth-generation aircraft.[10] Later, some unconfirmed reports called the programme *Perspektivnyj Aviatsionnyj Kompleks Dal'nego Perekhvata* or PAKDP (which translates to Prospective Air Complex of Long-Range Interception) and gave it the name MIG-41, projecting for it a first flight in the mid-2020s and entry into service after 2030. More recently, reports indicate that the systems meant for the sixth-generation fighter are being tested on the fifth generation Su-57 on which the new fighter is based.[11] Very few details have been officially revealed about the MIG-41, but it is expected to have substantial AI and be hypersonic and capable of operating in near space in the satellite hunter role. An unmanned variant may also emerge from the programme. However, the Ukraine War has put a huge question mark on its future.

In December 2021, a model of this Chinese fighter was exhibited in Shanghai at the First Science and Technology Conference of the airborne Cockpit System Division; a cockpit simulator was also on display. Earlier, in June 2018, a Chinese model called 'Dark Sword' was unveiled as the future sixth-generation fighter. More recent reports have attributed to China a programme that is keeping pace with USNGAD[12] and indeed, according to General Mark Kelly, head of USAF Air Combat Command, the USA's lead is just about a month.[13]

As can be expected, the cost of a sixth-generation fighter, whenever it fructifies, would be huge and already there is a debate even within the USA, the most affluent of sixth-generation fighter design nations, about the affordability of such an aircraft. US fifth-generation fighters F-22 and F-35, despite their matchless performances, have seen minimal sales due to their unaffordable price tags. The humongous cost of fifth-generation fighters has scared away many other nations. Some of the nations with current sixth-generation programmes have leapfrogged the fifth. There is even a thought process of shunning fifth and sixth-generation fighters, rejuvenating old fourth-generation programmes and giving them some additional features of later generations – at least those that can be retrofitted without major design changes. This thought process points towards the emergence of a 4.5th generation, which achieves a compromise between cost (far less than a fifth/sixth-generation aircraft) and capability (a lot better than fourth-generation). The single major differentiating design change would be AI.

Already, elements of AI are being retrofitted on some older battle-proven aircraft. According to one report: 'The Air Force has over 600 projects incorporating a facet of artificial intelligence to address various mission sets.'[14] The B-2 bomber, more than four decades old already, has been given updates of onboard memory and its networking speeds are being enhanced considerably so as to enable essential AI functions (data base access, information organising, scanning, viewing, organising targeting, radar warning, images and video). AI-aided autonomous fusing sensor information has been introduced and is helping to lower pilot workload related to flying and navigation, thus leaving pilots' cognitive faculties free for mission-related tasks. The F-15 is being given AI and ML algorithms to enhance the capabilities of its EW systems[15] under the 'cognitive EW' concept. The EA-18G Growler is also getting AI injected into its systems to imbibe enhanced Electronic Warfare (EW) capabilities: An illustration is a system called Reactive Electronic Attack Measures (REAM).[16] Northrop Grumman was awarded a contract for the development of this capability of transitioning Machine Learning Algorithms (MLAs) to the EA-18G's airborne electronic attack suite to achieve capabilities against agile, adaptive and unknown hostile radars or radar modes. REAM technology is expected to join active USN fleet squadrons around 2025.[17]

Many of the AI applications mentioned above are already available in varying degrees of sophistication and could be grafted onto existing or evolving aircraft types without much difficulty. As in-cockpit AI permits aircraft to perform more and more vital combat functions at speeds far exceeding those that human pilots are capable of, combat aircraft could move from AI sharing cockpits with human pilots (as the U-2 illustration mentioned above) to manning them independently. Thus, an 'optionally manned cockpit' design is a key attribute of the sixth-generation fighter. As is also evident from the discussion on sixth-generation fighters above, a quintessential element of ongoing sixth-generation fighter programmes is the AI-driven capability to team with UAVs for the performance of sundry offensive and defensive roles. This teaming is discussed later in this chapter.

AI as cockpit crew

On 20 August 2020, in the finale of the AlphaDogfight trials under US Defense Advanced Research Projects Agency's (DARPA's) auspices, an algorithm developed by Heron Systems and installed in a simulator as a simulated F-16, comprehensively defeated an experienced US Air Force fighter pilot in five rounds of mock combat. Almost exactly a year ago, DARPA had emphatically stated that, 'No AI currently exists, however, that can outduel a human strapped into a fighter jet in a high-speed, high-G dogfight.'[18] Deeper analysis would show that the simulation trial conditions were biased in favour of AI. The human pilot was not in an actual aircraft which he was trained to fly by the seat of his pants, and had only a Virtual Reality (VR) headset to give him situational awareness. However, the demonstration was significant inasmuch as it ratified the all-important construct of AI replacing a pilot in the cockpit for something as dynamic as a dogfight.

Civil aviation of course has already had models of AI in the cockpit. The Autonomi system from Garmin enables, in case of a pilot incapacitation, for on-board systems to take over either automatically or on activation by the pilot or a passenger, stabilise the aircraft and then autoland it, if required. Interestingly, this system, now certified by US Federal Aviation Administration (FAA), is not projected as an AI system but as automation. The patent application for Emergency Autoland System[19] did not use the term 'artificial' or 'intelligence' anywhere.

However, military aviation requirements are much more demanding and stringent. On 15 December 2020, the USAF flew a single-seat U-2 on a simulated missile strike mission[20] with an AI algorithm called ARTUμ as a working crew member. A human pilot flew the aircraft and coordinated with ARTUμ, which was responsible for sensor employment and tactical navigation. The system is designed to be transferable to another type of aircraft with ease and has the potential to be a cockpit occupant in modern fighters as well as future ones. The implication is that the pilot workload of flying and fighting can be reduced by AI in the cockpit taking over complex tasks of managing data from multiple sources and sensors. An ARTUμ (or similar system) could also manage autonomous systems teamed with the fighter it is located in (manned-unmanned teams are discussed later in detail). There have been reports of other programmes for unmanned combat aircraft. For example, Japan is reportedly developing an unmanned fighter to be operational by 2035.[21]

Current technological prowess encourages optimism about military aviation accommodating incremental AI participation in cockpit workloads. The AlphaDogfight series of demonstrations is part of DARPA's Air Combat Evolution (ACE) programme, which aims to progressively automate aerial combat and foster confidence in unmanned cockpits. Evolving fighter designs, prevalently clubbed into sixth-generation fighters, stand to benefit immensely from the inexorable tiptoe of AI into modern cockpits. A promising example is DARPA's Aircrew Labour In-Cockpit Automation System (ALIAS), which is aimed at developing a flexible, drop-in, removable kit that would permit appending high levels of AI-driven automation into an existing aircraft; the ultimate aim is to execute an entire mission from take-off to landing[22] with just AI working as a cockpit crew. This forms the basis for the Manned-Unmanned Teaming (MUM-T) concept, now in an advanced stage of maturity.

Manned-unmanned teaming

Rotary wing advancements in MUM-T preceded those in fixed wing combat aircraft by at least a decade and proved the concept for adoption by fixed wing platforms. In 1997, the US Army initiated four concept evaluation programmes which fruitioned into two operational systems involving the Apache AH-64D and the Apache AH-64E respectively linking with UAVs.[23] In October 2020, these programmes demonstrated a spectacular accomplishment wherein a live missile was fired through cooperative engagement between an Apache AH-64E helicopter, a Shadow RQ-7BV2 Block 3 Tactical Unmanned Aircraft System (TUAS) and an MQ-1C Gray Eagle Extended Range (GE-ER) UAS.[24]

Thus, the basic framework already existed for fixed-wing MUM-T to adopt, although the comparatively high speeds of fixed-wing fighters and the UAVs teamed with them required higher and more immersive levels of

AI. The 'Loyal Wingman' project of USAF Research Laboratory (AFRL) involves a manned command aircraft with an unmanned aircraft serving as an adjunct flying platform, scouting for threats and taking them on, if required (thus inviting the term loyal wingman). Initial live demonstrations were with F-16s – both manned and unmanned – and were impressive[25] enough to envisage the concept being grafted onto the sixth-generation programmes. According to a paper from Joint Air Power Competence Centre by[26] NATO's conglomerate of analysts, while the F-16 tests were conducted in a semi-autonomous mode based on predetermined parameters, future plans include 'flocking' and 'swarming.' In flocking, more than one unmanned wingman would operate under the manned command aircraft, which would give more abstract commands than exercise direct control over them, while 'swarming' would obviate the possibility of the command aircraft having situational awareness over each of a huge swarm of UAVs but permit it to exercise command over the swarm as an aggregate. A demonstration related to this concept was made in October 2016 with three F/A-18 Super Hornets releasing a swarm of 103 autonomous micro drones, which then proceeded to exhibit 'collective decision-making, adaptive formation flying, and self-healing.'[27] The swarming UAVs are especially alluring as they represent myriad possibilities. Some UAVs could have offensive roles, some could be sheer decoys, some could carry out Intelligence, Surveillance and Reconnaissance (ISR) tasks, some could carry EW payloads, some others could aid in target recognition and tracking, and yet others could be loitering munitions with possibly air-to-air capability, so that they can sacrifice themselves in defence of their command aircraft. Thus, the possibilities are myriad.

As can be seen, UAVs present a persuasive alternative to exposing increasingly expensive (like the almost unaffordable F35!), scarce, manned aircraft to high threat environments. They can be used again and again like manned aircraft and are comparatively economical, so greater risks can be taken with them, with no danger of losing a valuable human pilot. They are referred to as Low Cost Attritable Aircraft (LCAA) or Attritable/Reusable (A/R) UAVs,[28] designed for short life spans (in comparison to manned fighters), recoverable (if not attrited during a mission), low cost, and usable with multiple manned aircraft. The crucial AI technologies that are lodged in these UAVs enable them to be part of MUM-T programmes like the US Skyborg, one of three 'Vanguard programmes' launched by AFRL.

Skyborg was launched in 2018 by AFRL's Strategic Development Planning and Experimentation (SDPE) and AFRL calls it 'an autonomy-focused capability that will enable the Air Force to operate and sustain low-cost, teamed aircraft that can thwart adversaries with quick, decisive actions in contested environments.'[29] Accompanying UAV programmes include the Kratos-built XQ-58 Valkyrie which completed a successful flight in March 2019 under AFRL's Low Cost Attritable Aircraft Technology (LCAAT) project. Skyborg related contracts have also been awarded to General Atomics, Northrop Grumman and Boeing. Skyborg's eventual loyal wingman could well be an unmanned fighter (as in the 2018 trials of manned/unmanned F-16s flying together).

In May 2019, DARPA launched the Air Combat Evolution (ACE) programme to adopt AI in individual and team aerial combat tactics and to develop Air Combat Manoeuvring (ACM) algorithms for visual 1-versus-1, 2-versus-1 and 2-versus-2 engagements with a broad spectrum of performance. The AlphaDogfight trials mentioned earlier were a part of ACE which embraces manned and unmanned aerial combat. While ACE is aimed at developing AI software capable of close combat by unmanned platforms autonomously, Skyborg has modern capabilities (Beyond Visual Range [BVR] missiles and long-range sensors) in mind for loyal wingmen. However, as both programmes look at unmanned wingmen, they may be merged at some future point in time as the programmes mature. That will remove the dividing line between close aerial combat and BVR engagements by AI-enabled UAVs toiling as wingmen. Already, the X-62A, a modified F-16, is being used by both Skyborg and ACE.[30]

Boeing, which has a stake in the US Skyborg programme, is also designing and developing, in collaboration with the Royal Australian Air Force (RAAF), the Airpower Teaming System (ATS) – its first unmanned system in Australia. On 1 March 2021, Boeing announced[31] the first test flight of its loyal wingman, a UAV slightly smaller than an F-16 (its length is 11.6 metres) which can fly like a fighter (as loyal wingman) with an F-18 or F-35 class of aircraft (up to six could fly with one manned aircraft), or fly roles like EW and ISR along with P-8 Poseidon or E-7 Wedgetail aircraft. This UAV was later named MQ-28A Ghost Bat. Australia had bought six last year and plans to buy seven more[32] over the next two to three years.

Moving to Europe, the decision to develop FCAS was taken in 2017 by Germany and France, but they signed an investment agreement only in February 2020, with Spain joining up in December that year. Within the programme, the Next Generation Weapon System (NGWS) aims at teaming sixth-generation manned fighters (possibly NGFs) with UAVs (called Remote Carriers or RCs under the programme) which will be appendable in a scalable and flexible manner. RCs from 200 kg to several tons All Up Weight (AUW) are envisaged under NGWS, with EW, ISR, Target Acquisition, Air to Ground Strike and Suppression of Enemy Air Defences (SEAD) as the possible roles. Airbus Defence and Space was working with US-based ANSYS, an engineering simulation company, but reportedly, there are some problems being faced by the programme[33] and its future is uncertain.

Russia has also been working on developing new UAVs to function as autonomous loyal wingmen; reportedly 'Grom' (meaning thunder), developed by Kronstadt Group was presented at the Army 2020 Arms Expo in Moscow. Grom is designed to operate in conjunction with Su- 35/Su-57 fighters for reconnaissance and missile firing (500 kg payload) under their command.[34] It resembles the XQ-58 Valkyrie but is longer by half a length (13.8 m to Valkyrie's 8.8 m). According to TASS, Russia's state news agency, Kronstadt has stated that the Grom will be able to control a swarm of 10 smaller drones called 'Molniya' (or lightning).[35] Another UAV, the S-70 Okhotnik-B (Hunter) designed by Sukhoi Design Bureau and Russian Aircraft Corporation MiG, first flew in 2019 and is also designed to be a loyal wingman capable of carrying 2 tons of internal payload including missiles and bombs, or to be installed with electro-optic targeting, communication, and reconnaissance equipment. The actual AI capability and content of these two UAVs is not ascertainable as all information about them is from Russian state media (and hence susceptible to scepticism).

Sixth-generation programmes appear intent on MUM-T capabilities, but not all of them will survive the onslaught of shrinking defence budgets. Already, there is a mild clamour about revitalisation of fourth generation aircraft with additional AI retrofitted to give them capabilities transcending fourth generation. Without a doubt, there would be built into the MUM-T UAVs the modular capabilities to dovetail into partnerships with aircraft hierarchically lower than the sixth generation designs they are being developed for originally. An interesting consequence of this happening could be the realisation that a UAV is as effective in MUM-T roles with sixth generation aircraft as it is with less-than-sixth ones.

AI on UAVs

From a humble beginning, and driven by commercial motivations, drones or UAVs have grown tremendously. Their military uses have proliferated impressively too. In 1969, Israel's Military Intelligence Directorate used toy drones mounted with 35 mm still cameras programmed to click pictures every ten seconds to harvest details of trenches, signals cables and deployments over Egyptian territory along the Suez. Since then, the military use of UAVs has grown remarkably in terms of autonomy for myriad roles and tasks. UAVs have been used in combat during military conflicts for a long time; a US Predator first fired a missile in 2001.[36] More recently, Azerbaijan and Armenia saw wide (and decisive!) use of UAVs in the latter half of 2020.

Autonomous UAVs can now execute accurate strikes against ground targets, carrying out target selection and firing premised on AI (which is either on board or on a platform they are in data communication with). The MQ-9 Reaper has been used to test Agile Condor, an AI pod designed to detect, categorise and track potential objects of interest. This AI pod has the potential to identify targets and determine priorities for engagement. The loyal wingman UAV developed by Boeing for RAAF has similar target detection, prioritisation and engagement capabilities, as claimed by Boeing. This is applicable to loitering munitions as well, AI enabling them to carry out their suicidal missions without an operator in the loop but investing them with enough autonomy to recover safely if no worthwhile target is unearthed. An interesting European project is nEUROn, a demonstrator for the development, integration and validation of European technologies for next-generation combat aircraft and UAVs. Russian UAVs Okhotnik and Grom (both with AI capability for working with fighters) have been mentioned earlier; Altius, Sirius and Orion are three other Russian UAVs with autonomous/semi-autonomous capabilities.

AI has been quietly and surreptitiously making its way into UAV operations and each episode of success for UAVs strengthens the motivation to embed AI in diverse forms to enhance their capabilities. The USN's X-47B has demonstrated not only deck launches, landings and go-arounds, but even an aerial refuelling, all carried out autonomously.[37] The QF-16, an unmanned platform derived by converting old F-16s, has the capability to fly autonomously on predetermined routes for decoy roles or used as aerial targets for testing air-to-air missiles and guns. Looking ahead, it is easy to imagine UAVs that will be primed for fully autonomous missions with their designs hybrid to accommodate MUM-T errands when tasked to. Automatic target detection using optical, thermal and/or electromagnetic sensors would empower them for offensive roles while long-range sensors would render them invaluable for autonomous operations, including in hostile environments. A demonstration in 2020 showed a USAF MQ-9 Reaper engage a cruise missile using an AIM-9X air-to-air missile, thus revealing another aspect of AI-enabled UAVs' offensive potential.

In 2020, the US Army's Combat Capabilities Development Command (CCDC) had released a Request For Information (RFI) for Air Launched Effects (ALE)[38] which would be a new family of UAVs air launched from scout and assault helicopters, which would have multifarious capabilities including scout role, EW, attack, decoy, suicide munitions and swarm capabilities. The RFI stipulates that these UAVs need to be capable of semi-autonomous and fully autonomous operations and specifies that the camera systems on the UAVs would also be paired with AI-driven ML algorithms to automatically identify potential targets of interest. The steady rise of AI technologies is matched by the wants and needs of military aviation for a wide spectrum of AI capabilities.

Several spectacular demonstrations of UAVs operating in swarms have caught military attention in the recent past. Their small size (and Radar Cross Section or RCS), low cost (permitting large numbers to be used together), and AI to connect them together to seemingly think like one, makes UAV swarms objects of desire for the military. Dramatic multi-UAV attacks like the one on Saudi Aramco oil facilities at Abqaiq and Khurais in September 2019 have helped keep that interest alive (those attacks were termed 'swarm drones' by media but did not have the interconnecting AI that defines a swarm). DARPA's Offensive Swarm-Enabled Tactics (OFFSET)[39] programme has contracted with nine companies to develop AI technology that will enable 250 small air and/or ground units to collaborate; aimed primarily at facilitating operations in dense urban environments, the programme showcases the promise of swarming UAVs. The interest in swarming UAVs is not restricted to the US military only. China is making significant strides in that direction as is the European Union, where several joint military programmes are underway to produce AI-driven swarm technologies; the DRONEDGE-E project, for example, is aimed at designing an edge computing platform for the autonomous control of UAV swarms in real time with automatic generation of algorithms through AI. Details of some other swarm related projects can be found in a recent document called 'Artificial Intelligence in European Defence: Autonomous Armament?'[40] authored by GUE/NGL, the left-wing group of the European Parliament. Roborder is another European project that is aimed at developing AI-piloted UAVs which would be tasked to patrol European borders autonomously. Working in swarms, these UAVs will corroborate and coordinate information among air, ground, and sea-borne platforms.

In the context of high speed, highly manoeuvrable swarm UAVs operating in close proximity to each other (and enemy UAVs) and collaborating tactically, cloud computing is too slow for their cooperative communication. Edge computing[41] eliminates the time taken for devices/UAVs to communicate through cloud computing (which connects through centralised data centres or servers) and processes data on the spot at the 'edge' of the network, i.e., at or near the source of the significant data. This munificence of the information technology is quintessential to the progression of swarm UAVs; so is their employment of AI. In a demonstration by Lockheed Martin, four rotary-wing UAVs flew coordinated missions while connected to two Verizon 5G network nodes. These UAVs detected and geolocated a target which was transmitting a low-power RF signal through real-time detection data being transmitted over the 5G network and then being processed by signal processing algorithms employing edge computing resources.[42] Their copious use in future military kinetic action can be predicted.

Conclusion

From the foregoing account, some strands are discernible in the context of AI's inexorable inroads into military aviation. Firstly, AI is a force multiplier that has the potential to enable faster, smarter and safer military operations using minimal resources. Most AI applications are dual use, and a clear distinction between civil and military applications is rather difficult. Thus, international law may not have much inhibiting leverage over the use of AI for military aviation purposes. AI is changing the character and conduct of war into 'hyperwar' – combat waged under the influence of AI, where human decision-making is almost entirely absent from the Observe, Orient, Decide, and Act (OODA) loop. In the future, some military aviation missions will be consummated without a human in the loop. We have already seen optionally manned cockpits, unmanned wingmen, unmanned, autonomous UAVs and swarming UAVs collaborating with each other in real time through edge computing to eliminate delays due to the time taken by data/communication to travel from one entity to a satellite and back to another entity. The ethical issues involved are being debated hotly, but it is unlikely that AI-crewed aerial platforms will be kept away from any aerial action in the future.

AI is harshly disruptive. It is changing how air power theorists look at the use of combat aircraft, aerial combat, long-range strikes, air defence and support to surface forces. The importance given to strategic bombers appears set to diminish with much smaller craft taking over strategic roles (including some not possible by bombers, targeted assassinations for example). Sixth-generation combat aircraft will 'battlespace' into the bounds of space and mission decisions may get delegated to AI in orbit, although the overall control, for philosophical reasons and ethical ones, may remain with humans on the ground or in the air or even space.

AI is taking militaries towards no-contact war, with long-range strikes and BVR aerial combat. This would be possible as confidence in AI transcends current apprehensions about their reliability and lack of cognitive faculties. Without putting a timeline to it, that confidence appears achievable and feasible as advancements of technology promise. Attritable, disposable, economical UAVs are proving to be more efficacious than fixed-wing aircraft, although the disappearance of fixed-wing fighters is still quite distant. The speed of operations is going up, and so is the speed of decision-making, which is being redefined through AI, which enables taking enormous amounts of data from many dissimilar sensors, organising that data, and performing analytics on it to solve problems, make determinations and recommend courses of action. Ultimately, autonomous operations could become the norm because even after all the handling and analytics of information is done by AI, the decision-making time at the hands of a human may be unacceptably long. The pace of AI's march towards singularity will decide how soon that happens.

Notes

1 *Summary of the 2018 Department of Defense Artificial Intelligence Strategy* (Washington, DC: 2019), p. 5, available at: https://media.defense.gov/2019/Feb /12/2002088963/-1/-1/1/SUMMARY-OF-DOD-AI-STRATEGY.PDF, accessed on 3 October 2022.

2 Dom Galeon, 'Separating Science Fact from Science Hype: How far off is the Singularity?' *Futurism*, 31 January 2018, available at: https://futurism.com/separating-science-fact-science-hype-how-far-off-singularity, accessed on 8 October 2022.

3 Joe Saballa, 'Kratos, General Atomics Make Progress on AI-Driven Combat Aircraft,' *The Defense Post*, p. 1, Para 2, available at: accessed on 5 October 2022 https://www.thedefensepost.com/2022/09/22/kratos-general-atomics-ai-aircraft/#:~:text=General%20Atomics%20said%20in%20a,in%20challenging%20real%20world%20scenarios, accessed on 5 October 2022.

4 Valerie Insinna, 'On Next Generation Air Dominance Program, US Eyes Cooperation with Allies,' *Breaking Defense*, 3 October 2022, p. 1, available at: https://breakingdefense.com/2022/10/on-next-generation-air-dominance-program-us-eyes-cooperation-with-allies/?utm_campaign=BD%20Daily&utm_medium=email&_hsmi=228213778&_hsenc=p2ANqtz-8O9BjlBuaGg3qbMkjkJKsWs_6Ik2hdpfF5UAFqCdjLZCdzQ89sWO78QICnwRS0SzbJhuXmIkr-2S777pvkENJsrb-JfkQ&utm_content=228213778&utm_source=hs_email, accessed on 5 October 2022.

5 Anon, 'UK and Japan to Develop Future Fighter Jet Engine Demonstrator,' *MoD Press Release*, 22 December 2021, p. 1, available at: https://www.gov.uk/government/news/uk-and-japan-to-develop-future-fighter-jet-engine-demonstrator, accessed on 5 October 2022.

6 Tom Kington, 'Italy Air Force Chief Heads to Japan to Talk Next-Gen Fighter Jets,' *Defense News*, 23 September 2022, p. 1, available at: https://www.defensenews.com/global/europe/2022/09/22/italy-air-force-chief-heads-to-japan-to-talk-next-gen-fighter-jets/, accessed on 3 October 2022.

7 Andrew White, 'Tempest's Fate: UK Says Next-Gen Fighter Demonstrator to Take off Within 5 Years,' *Breaking Defense*, 18 July 2022, available at: https://breakingdefense.com/2022/07/tempests-fate-uk-says-next-gen-fighter-demonstrator-to-take-off-within-5-years/, accessed on 8 October 2022.

8 Kington, 'Italy Air Force Chief Heads to Japan to Talk Next-Gen Fighter Jets,' p. 1.

9 Christinq Mackenzie, 'Mired in Politics, Franco-German Next-Gen Fighter "Likely Headed for the 2050s",' *Breaking Defense*, 12 September 2022, p. 1, available at https://breakingdefense.com/2022/09/mired-in-politics-franco-german-next-gen-fighter-likely-headed-for-the-2050s/, accessed on 5 October 2022.

10 Anon, 'Russia's Sukhoi Comes up With first Designs of 6th-Generation Fighter --- Deputy PM,' *TASS*, 2 March 2016, p. 1 available at https://web.archive.org/web/20160303013236/https:/tass.ru/en/defense/860142, accessed on 5 October 2022.

11 Anon, 'Russia Tests Sixth-Generation Fighter Elements on Fifth-Generation Jet,' *TASS*, 14 July 2016, available at: https://tass.ru/en/defense/887593, accessed on 8 October 2022.

12 Thomas Newdick, 'China is Working on Its Own Sixth-Generation Fighter Program: Official,' *The Drive*, 28 September 2022, p. 1, available at: https://www.thedrive.com/the-war-zone/china-is-working-on-its-own-sixth-generation-fighter-program-official, accessed on 3 October 2022.

13 Gabriel Honrada, 'China racing for 6th-Gen Fighter Edge Over US,' *Asia Times*, 1 October 2022, p. 1, available at: https://asiatimes.com/2022/10/china-racing-for-6th-gen-fighter-edge-over-us/, accessed on 9 October 2022.

14 Kris Osborn, 'Air Force has Plans to Turn the B-2, F-35 and F-15 Into Even More Deadlier Killers. Here's How,' *National Interest*, 31 July 2018, p. 1, available at: https://nationalinterest.org/blog/buzz/air-force-has-plans-turn-b-2-f-35-and

-f-15-even-more-deadlier-killers-here%E2%80%99s-how-27327, accessed on 5 March 2021.

15 John Keller, 'Air Force Asks Industry for Artificial Intelligence (AI) Cognitive Electronic Warfare (EW) for F-15 Jets,' *Military & Aerospace Electronics*, 15 March 2021, p. 1, available at: https://www.militaryaerospace.com/computers/article/14199230/electronic-warfare-ew-cognitive-artificial-intelligence-ai, .

16 US Department of Defense, 'Contracts for April 25, 2018,' Sixth Item Under NAVY, 25 April 2018, available at: https://www.defense.gov/Newsroom/Contracts/Contract/Article/1503297/, accessed on 23 September 2022.

17 John Keller, 'Vadum to Support Electronic-Warfare Project to Counter Waveform-Agile Enemy Radar with Machine Learning,' *Military & Aerospace Electronics*, 20 February 2019, p. 1, available at: https://www.militaryaerospace.com/sensors/article/16722184/vadum-to-support-electronicwarfare-project-to-counter-waveformagile-enemy-radar-with-machine-learning, accessed on 20 September 2022.

18 DARPA Outreach, 'Training AI to Win a Dogfight,' 5 August 2019, p. 1, available at: https://www.darpa.mil/news-events/2019-05-08, accessed on 1 October 2022.

19 United States Patent Application Publication No US 2017/0249852 AI dated 31 August 2017; available at: https://patentimages.storage.googleapis.com/04/81/09/55bcc0207e50fd/US20170249852A1.pdf, accessed on 2 October 2022.

20 Secretary of the Air Force Public Affairs, 'AI Copilot: Air Force achieves first military flight with Artificial Intelligence,' Official US Air Force website, 16 December 2020, p. 1, available at: https://www.af.mil/News/Article-Display/Article/2448376/ai-copilot-air-force-achieves-first-military-flight-with-artificial-intelligence/, accessed on 2 October 2022.

21 Smriti Chaudhary, 'Japan to Deploy Unmanned Fighter Jets by 2035 with Aim to Counter Rising Chinese Military Might,' *Eurasian Times*, 1 January 2021, p. 1, available at: https://eurasiantimes.com/japan-to-deploy-unmanned-fighter-jets-by-2035-with-aim-to-counter-rising-chinese-military-might/, accessed on 9 October 2022.

22 Stuart H. Young, 'Aircrew Labor In-Cockpit Automation System,' DARPA Official Site, Undated Paper, p. 1, available at: https://www.darpa.mil/program/aircrew-labor-in-cockpit-automation-system, accessed on 7 October 2022.

23 Lieutenant-Colonel Steven G. Van Riper, 'Apache Manned-Unmanned Teaming Capability,' Association of the United States Army Site, p. 2, available at https://www.ausa.org/sites/default/files/apache-manned-unmanned-teaming.pdf, accessed on 23 September 2022.

24 Becky Bryant, 'Aerial MUM_T reaches New Heights at DPG,' *US Army Site*, 27 October 2020, p. 1, available at: https://www.army.mil/article/240283/aerial_mum_t_reaches_new_heights_at_dpg, accessed on 9 September 2022.

25 Joseph Trevithick, 'This Is What the US Air Force Wants You to Think Air Combat Will Look Like in 2030,' *The Drive*, 26 March 2018, see section entitled 'Loyal Wingmen,' available at: https://www.thedrive.com/the-war-zone/19636/this-is-what-the-us-air-force-wants-you-to-think-air-combat-will-look-like-in-2030, accessed on 4 September 2022.

26 Andy J. Fawkes and Lieutenant Colonel Martin Menzel, 'The Future Role of Artificial Intelligence,' *Joint Air Power Competence Centre Journal*, Edition 27, Autumn/Winter 2018, p. 73, available at: https://www.japcc.org/wp-content/uploads/JAPCC_J27_screen.pdf, accessed on 2 September 2022.

27 US Department of Defense, 'Department of Defense Announces Successful Micro-Drone Demonstration,' US DoD Release 9 January 2017, p. 1, available at: https://www.defense.gov/Newsroom/Releases/Release/Article/1044811

/department-of-defense-announces-successful-micro-drone-demonstration/, accessed on 3 September 2022.

28 Colonel Mark Gunzinger and Lukas Autenreid, 'The Promise of Skyborg,' *Air Force Magazine,* 1 November 2020, p. 1, available at: https://www.airforcemag.com/article/the-promise-of-skyborg/, accessed on 4 September 2022.

29 Air Force Research Laboratory, 'What is Skyborg?', p. 1, available at: https://afresearchlab.com/technology/vanguards/successstories/skyborg, accessed on 23 September 2022.

30 Sakshi Tiwari, US Skyborg Program ready for take off with a heavily modified F-16 Fighter Jet Redesignated As X-62A Aircraft,' *Eurasian Times,* 25 August 2022, p. 1, Para 18, available at: https://eurasiantimes.com/us-skyborg-program-ready-to-take-off-with-a-heavily-modified-f-16-jet/, accessed on 7 October 2022.

31 Boeing Media Room, 'Boeing Loyal Wingman Uncrewed Aircraft Completes First Flight,' 1 March 2021, p. 1, available at: https://boeing.mediaroom.com/news-releases-statements?item=130834, accessed on 5 September 2022.

32 Inder Singh Bisht, 'Australia Announces $ 317 Million Loyal Wingman Drone Investment,' *The Defense Post,* 19 May 2022, p. 1, Para 2, available at: https://www.thedefensepost.com/2022/05/19/australia-loyal-wingman-drone-investment/, accessed on 7 October 2022.

33 Justin Bronk, 'FCAS: Is the Franco-German-Spanish Combat Air Programme Really in Trouble?' *RUSI Commentary,* 1 March 2021, p. 1, available at: https://rusi.org/commentary/fcas-franco-german-spanish-combat-air-programme-really-trouble, accessed on 9 October 2022.

34 Anon, 'Russia's Top Long-Range Attack Drones,' *Air Force Technology,* 27 November 2020, p. 1, available at: https://www.airforce-technology.com/features/russias-top-long-range-attack-drones/, accessed on September 3, 2022.

35 Anon, 'Russia's Latest Combat Drone to Control Swarm of Reconnaissance UAVs,' *TASS News Site,* 15 March 2021, p. 1, available at: https://tass.com/defense/1265961, accessed on 9 September 2022.

36 Lee Ferran, 'Early Predator Drone Pilot: I had Bin Laden in my Crosshairs,' *ABC News Network* site, 18 November 2014, p. 1, available at: https://abcnews.go.com/blogs/headlines/2014/11/first-pilot-to-fire-missile-from-predator-drone-breaks-silence, accessed on 24 September 2022.

37 Andy J. Fawkes and Lieutenant-Colonel Martin Menzel, 'The Future Role of Artificial Intelligence,' *Joint Air Power Competence Centre Journal,* Edition 27, Autumn/Winter 2018, p. 72, available at: https://www.japcc.org/wp-content/uploads/JAPCC_J27_screen.pdf, accessed on 7 September 2022.

38 Joseph Trevithick, 'The Army Has Unveiled its Plan for Swarms of Electronic Warfare Enabled Air-Launched Drones,' *The Drive,* 20 August 2020, p. 1, available at: https://www.thedrive.com/the-war-zone/35726/the-army-has-unveiled-its-plan-for-swarms-of-electronic-warfare-enabled-air-launched-drones, accessed on 10 September 2022.

39 DARPA Outreach, 'OFFSE Awards Contracts to Advance Swarm Tactics for Urban Missions, Enhance Physical Testbeds,' 13 April 2020, p. 1, available at: https://www.darpa.mil/news-events/2020-04-13#:~:text=DARPA%20has%20awarded%20contracts%20to,operate%20in%20dense%20urban%20environments, accessed on 12 September 2022.

40 Christoph Marischka, *Artificial Intelligence in European Defence: Autonomous Armament?,* (THE LEFT Group In The European Parliament: Brussels, 2020), pp. 28–30, available at: https://documentcloud.adobe.com/link/track?uri=urn%3Aaaid%3Ascds%3AUS%3A1884c966-f618-4110-a5f3-678899e4c8ee#pageNum=1, accessed on 8 October 2022.

41 Paul Miller, 'What Is Edge Computing?,' *The Verge*, 7 May 2018, p. 1, available at: https://www.theverge.com/circuitbreaker/2018/5/7/17327584/edge-computing-cloud-google-microsoft-apple-amazon, accessed on 7 October 2022.
42 Kumar Natasha, 'Lockheed and Verizon are Testing Swarms of 5G-Enabled Reconnaissance Drones for ISR Missions,' *The Times Hub*, 3 October 2022, p. 1, available at: https://thetimeshub.in/lockheed-and-verizon-are-testing-swarms-of-5g-enabled-reconnaissance-drones-for-isr-missions/51903/, accessed on 8 October 2022.

4

AI-CYBER NEXUS

Impact on deterrence and stability

Kritika Roy

Introduction

Artificial intelligence (AI) is a growing field that finds application in various sectors. It has often been termed as 'general/purpose technology' that enhances or adds functionality when integrated into systems.[1] The major idea behind the use of AI within machines is to allow some form of autonomy with limited or no human assistance or supervision.

Autonomy is not a new concept, especially in weapon systems where its employment has been witnessed since at least the era of the Second World War. For instance, the Norden Bombsight and V-1 buzz bomb were controlled by sensors linked to computer systems allowing the control and application of lethal force while the final decision-making control always remained with the operator.[2] Autonomy depends on a diverse array of technology but primarily software. The viability of autonomy relies on (a) 'the ability of software developers to formulate an intended task in terms of a mathematical problem and a solution'; and (b) 'the possibility of mapping or modelling the operating environment in advance.'[3] Autonomy can be created or enhanced by AI and ML.[4] The current advancement in autonomy is because of the increasing computing power and continuous development in ML which has made the development of commercially viable autonomous systems, such as AI voice assistants, autonomous vehicles or autonomous weapons a possibility.[5] AI can have different levels of autonomy based on:[6]

A) Human involvement in the system's task execution:
   • Human in the loop: Semi-autonomous systems which remain under the active control of a human operator.

DOI: 10.4324/9781003421849-5

- Human on the loop: Human-supervised systems that remain under the oversight of a human operator.
- Human out of the loop: Fully autonomous systems that carry out operations without any interference or oversight of a human operator.

B) Sophistication of the system, that is, how well the system can govern its own behaviour and handle uncertainties in its operational environment.

- Reactive Systems: Follow condition-action rules, commonly referred to as 'if-then' rules.
- Deliberative Systems: Employ a model of the world (information on how the world functions and how actions by the system are responded to by the world), a value function (which offers information about the intended goal), and a set of alternative rules to search for and plan ways to accomplish the goal.
- Learning systems: Become more effective over time by gaining experience.

C) The number and type of functions automated.

- Operational tasks include 'mobility, health management (fault detection), etc.'
- Mission tasks include 'target identification and selection, explosive detection, etc.'[7]

AI has also been segregated into three branches: Narrow (or weak) AI, General Intelligence (AGI) and AI Super Intelligence (ASI). ASI is said to be a futuristic capability where the machines would surpass the intelligence of humans who have the brightest and most gifted minds. AGI has been linked to an advanced level of intelligence that would at least match, if not surpass, that of a human. It is supposed to be so advanced that it would be able to comprehend the environment around it and also make decisions and predictions based on this understanding, something more akin to what was portrayed in the movie *The Terminator*. However, such a development is still a distant future. To date, we have a narrow phase of AI today. This phase has improved a lot over time and is continuously developing, as machines are able to perform tasks such as object identification or language translation, but as per the user-defined inputs and parameters. Often these parameterised inputs are broadly defined but allow a substantial range of autonomy by establishing some limits on the AI's behaviour. For instance, AlphaZero – a computer program, developed by DeepMind – mastered the game of chess. Playing chess allowed the technology to initiate moves as per its own understanding (providing freedom to navigate on the chess board) but within the rules of the game. Hence, AI in the current context is defined as 'purpose-built, problem-specific and context-driven.'[8]

Understandably, the introduction of AI within weapon systems could facilitate significant improvements in the military domains, for instance, optimising solutions for operations that are data-intensive like logistics and maintenance. Speedy and concise analysis using AI would provide policy-makers with better information and more time to make decisions. However, these AI systems are also brittle, that is, there is a high plausibility that these systems may fail in unpredictable situations or behave in an unpredictable manner that could further snowball into escalation and deter the current stability matrix among nations. Cyberspace, in this context, is the foundation on which AI is grounded and hence, cybersecurity is a necessary pre-condition for the secure deployment of AI systems for all social, economic and military purposes. The shortcomings and vulnerabilities of cyberspace would intrinsically be linked to the implementation of AI, which would further exacerbate the deterrence equation.

Against this backdrop, this chapter will first analyse the relevance of the topic, employing the ongoing Ukraine-Russia War as a case in point that has witnessed a number of cyber operations feeding into kinetic warfare. Further, this chapter will examine the degree of autonomy showcased by the weapons employed in the Russian-Ukraine War by analysing several deployments of new and untested semi-autonomous weapons (with a promise of full-autonomous capabilities). This examination is important as it will highlight how the current Russia-Ukraine War will set certain precedents for the future of warfare, where disruption of networks and cyberattacks on critical information infrastructure may become a common phenomenon, as well as mass usage of autonomous systems in various roles. Furthermore, the paper will also investigate the nexus of AI-cyber tools that could not only enhance the efficacy of the weapon systems but also make them susceptible to vulnerabilities and errors. These advances could subsequently impact the current stability matrix and deterrence equation among nation states.

Russia-Ukraine War: The role of cyberwar and autonomous systems

The ongoing war between Ukraine and Russia has brought to focus the significance of cyberspace as well as brought forth the reality of autonomous weapons. Russia and Ukraine have consistently targeted each other's cyberspace with Distributed Denial of Service (DDOS) attacks, varied malware deployment and leakage of personal information.[9] These cyberattacks may not be as lethal as their kinetic counterparts but have a force multiplier effect, that is, they create more chaos in an already frenzied environment. In one instance, the Ukrainian communication satellite KA-SAT network owned by Viasat was targeted, which had a significant impact on communications on the ground. The Viasat attacks also had spillover effects on other European nations, including Germany, where around 2,000 wind turbines were disconnected from the internet.[10]

These instances of ongoing conflict reveal the significance of cyber operations running alongside and feeding into kinetic operations. Although analysts have noted that the conflict's cyber dimension was limited, nevertheless, there are serious concerns about the instability of the international security environment.[11] These risks are further exacerbated by the potential spillover effects of cyberattacks targeting Ukraine into other countries, which may have a systematic impact on cyberspace and beyond.

The current conflict also saw the use of drones with at least some autonomous capabilities. Two major drones with autonomous capabilities employed by Russia in the war include Lancet Kamikaze and KUB-BLA. KUB-BLA produced by ZALA Aero, is a small kamikaze drone aircraft that smashes itself into enemy targets and detonates an onboard explosive. This drone proclaims to operate autonomously with AI capabilities such as target identification, selection and engagement. However, there is no certainty whether the drone did operate fully autonomously in the current Russia-Ukraine War. An executive from Rostec, a part of Russia's government-owned defence industrial complex, indicated that KUB-BLA is a domestic analogous of Israel's Orbiter 1K drone.[12] This Israeli counterpart has a ground control station where an operator monitors the incoming videos using the sensors of the drone and manually selects the target. This even stands true for KUB-BLA, wherein the target selection and engagement were undertaken by a human operator and not the machine alone.[13]

Ukraine used Turkish Bayraktar TB2 known for autonomous flights and features 'laser guided smart ammunition.'[14] The USA declared that it would send 700 Switchblade and 'Phoenix Ghosts' with GPS (Global Positioning System) tracking and object recognition software to Ukraine. Switchblade is a small US single-use drone that comes equipped with 'explosives, cameras, and a guided system.' This loitering munition comes with some autonomous capabilities but needs an operator to lock on the target.[15]

Germany announced in November 2022, that it would supply '14 tracked and remote-controlled infantry vehicles for support tasks' as part of 2022's 1.64 billion USD spent on military support for Kyiv.[16] These unmanned vehicles are said to rely on an upgraded version of similar technology that was used for landmine disposal in the Afghanistan and Iraq Wars.

Milrem Robotics, an Estonian contractor and developer of the Tracked Hybrid Modular Infantry Systems (TheMIS) – a UGV – provided Ukraine with these units designed for evacuating casualties and for finding and removing landmines. However, analysts believe that often these technologies are introduced for a specifically defined purpose but may also find additional applications in combat zones. For instance, the TheMIS also has a 'follow-me' capability that makes it an excellent 'battle mule' for transporting ammunition and other equipment.[17]

In the ongoing war, there were also noted instances of the use of AI such as the employment of Clearview AI's facial recognition software by Ukraine's Defence Ministry. Ukraine has received free access to powerful search engines under a 'special operation' campaign to identify killed Ukrainians and Russian soldiers on the battlefield, match fingerprints or allow authorities to potentially screen people of interest at checkpoints, among other applications. The software works even if there is facial damage.[18] Although there are some promising applications of the technology such as reuniting missing people with families, there is also the fear of misidentification or unfair arrests. The so-called 'well-intentioned' technology could backfire, thereby harming the same people it is meant to help. Another obvious fear that stems from facial recognition's battlefield deployment is that it will allow Clearview the chance to perform stress tests and provide valuable data, instantly transforming Clearview AI into a major tool for military use.[19] If the AI is able to identify both live and dead enemy soldiers, it could be integrated into systems that use automated decision-making to direct lethal force. This perilous precedent set by battlefield testing and refinement of facial-recognition technology could be integrated into autonomous killing machines in the near future.

The Ukraine-Russia War is serving as a testing ground for cutting-edge, yet unproven technologies. Although these technologies are currently being employed in 'support' roles rather than as fully autonomous combat machines, even if that's what they were initially designed to do. Developing wartime technology is complicated because of the chaotic environment. Despite the best assurances of the manufacturers, there is simply no way to know what could possibly go wrong until a given tech sees actual field use. Nonetheless, the deployment of these machines in war zones is worrisome. First, there is limited control over how it is being used. Second, it would continuously be deployed in all conflict situations. Last, with each iteration these autonomous systems would only get better and smarter.

These illustrations of using cyber and AI tools can be seen as a 'novelty and an added complication,' as these may tamper with or potentially generate new conflict dynamics.[20] In the near future, these technologies would become more advanced because of the data being gathered along the way, which is a primary requisite for improving the effectiveness of these technologies.

Cyber and AI: A lethal combination

Over the past few decades, researchers have made enormous strides in developing computer programs that draw insights from immense volumes of data owing to cheaper access to machines that can process gigantic data sets quickly. With the advancement of AI, it is now possible for machines to accurately translate languages, compose poetry and possibly even aid in the discovery of new drugs.[21]

The introduction of AI within weapon systems has led to significant improvements in many processes as well as eliminating redundant yet time-consuming tasks. Major applications of AI have been noted in detection capabilities and analysing voluminous data from different sensors, early warning systems, intelligence surveillance and reconnaissance, inventory management precision and prediction. These systems can also be used to conduct remote sensing operations in areas that are hardly accessible by manned systems. Many of these tasks are currently being undertaken with a human in the loop. For instance, AI can enable systems to sift through large amounts of data to find relevant information, fuse data from different sensors and observe whether situations have changed or not; finally, it is the analyst who decides the next steps to be taken based on the concise and comprehensible data.

For cyber operations as well, AI could act as a force multiplier for both defensive and offensive cyber weapons. In 2016, the Defense Advanced Research Projects Agency (DARPA) organised a cyber challenge wherein a lethal machine dubbed Mayhem was deployed. Its features include 'automatically finding vulnerabilities in commercial-off-the-shelf software without developer participation' and 'automatic patching of its own vulnerabilities.' Many researchers also anticipate that an advanced version of Mayhem could be capable of 'fighting dynamically against attackers.' It is said to be 'smart for both offense and defense purposes.'[22]

In the field of cybersecurity, AI is being increasingly used in systems that prioritise threats and carry out automatic remediation. Examples include anomaly detection algorithms that can identify malicious traffic or user activity in real time. The development of new technologies that can conduct complex cyber campaigns with innovative traits, such as malware capable of tactical adaptability and value calculations vis-à-vis strategic objectives, is currently the subject of intense research worldwide. OpenAI has recently released a ChatGPT tool that allows users to interact with an 'intelligent' chatbot. A user can also ask it to converse or finish an assignment. A researcher experimented with the chatbot to see if it would be able to solve a simple capture the flag challenge. Surprisingly, the model not only identified the vulnerability within the code but also wrote a small code that allowed the exploitation of the vulnerability. This is to note that as the machine becomes more sophisticated, it could also be employed to write malicious code or find vulnerabilities in the adversary's network. This discovery is not without shortcomings, as OpenAI researchers have observed that the model can produce vulnerable or misaligned code and that while 'future code generation models may be able to be trained to produce more secure code than the average developer,' getting there 'is far from certain.'[23] There is also no clarity over how the model is addressing cybersecurity concerns or the risks posed by AI.

Thus, there is a general indication that AI could have a transformative impact on both defensive as well as offensive cyber weapons. However, the likely impacts of introducing AI-cyber tools within weapon systems may range from being a positive reinforcement to a destructive scenario. From a defensive perspective, the combination of AI-cyber could help to enhance the performance of weapon systems and detect irregularities and deviations. The combination could help build a robust network against any form of cyber subversion. Several cybersecurity firms are cognisant of the growing advantages of implementing AI solutions to provide rapid insights and cut through the noise, thereby reducing the response time during eventuality. From a national security perspective, as per the United States' strategic report, 'AI can enhance our ability to predict, identify, and respond to cyber and physical threats from a range of sources.'[24] The USDOD's Project Maven is believed to 'employ AI to decipher video footage and is expected to, among other things, increase the precision of drone strikes.'[25] It is also being expanded to other areas such as speeding up the reading of hard drives obtained from non-state actors or terrorist entities.[26]

From an offensive perspective, the introduction of AI within weapon systems may make the systems vulnerable to different types of cyberattacks and data manipulation. For example, an adversary could deploy malware, install a backdoor to take control of the system or manipulate/fool the behaviour of autonomous systems. Furthermore, the data being fed to the autonomous systems could be poisoned, wherein intruders feed an algorithm with altered data to modify the system's behaviour according to their own aims. Offensive attacks such as data manipulation or spoofing would be relatively easy to execute, but very difficult to detect, attribute or effectively counter. After witnessing the capabilities of Mayhem, Pentagon's startup-centric office Defense Innovation Unit Experimental (DIUx) initiated a project dubbed Voltron which is said to offer Mayhem-like capabilities to several different defence agencies to find coding flaws in both operating systems and custom programs used by the US military.[27]

The introduction of AI-cyber tools within weapon systems highlights a prevalent strategic paradox wherein, on the one hand, these technologies are making operations easier and eliminating redundant tasks. While on the other hand, the same technology would make systems more prone to cyberattacks, vulnerability exploitation and an increased probability of technical glitches.

Is integration of AI within weapon systems viable?

There have been some efforts to integrate AI within weapon systems. However, these integrations are not without challenges. The first challenge emanates from models trained through ML being susceptible to adversarial attacks where 'an adversary can craft a malicious input that reliably causes a trained model to produce the wrong behaviour (for example misclassify an object in an image or take an inappropriate action in the environment).'[28] Despite years of intensive investigation and knowledge about this vulnerability, no reliable fix has yet been discovered. Nevertheless, the urge to implement potentially vulnerable systems may arise given the promise of new capabilities that ML and automation offer.[29]

The next biggest challenge is the absence of quality data. AI-based systems require voluminous data for training models for different tasks such as target selection and engagement, path navigation and data relationship identification. To identify a particular object in an image (such as a cat or dog), a system may need to be trained with millions of pictures of that particular object. Equally crucial is the quality of the data used to train the systems. If the training data set is not representative, the system may not work as intended, may perform badly, and may introduce new biases or reinforce existing ones. This may lead humans to make incorrect decisions and take incorrect actions. In the case of combat zones, acquiring adequate training data as well as quality data has been a considerable challenge for AI development projects. The absence of quality data could lead to the failure of autonomous systems in a complex and unpredictable manner.

Furthermore, making systems autonomous also means allowing room for error. As witnessed in the incident of 1983 where a machine misidentified sunlight reflecting off high altitude clouds as incoming missiles. However, Lieutenant-Colonel Stanislav Petrov of the Soviet Union decided to follow his own instincts rather than the machine's indication, calling it a 'false alarm' and thus averting a nuclear crisis.[30] In a more recent incident, a man lost his life because the sensors of his autonomous car failed to distinguish between a large white 18-wheel truck and a trailer crossing the highway against a bright spring sky.[31] Vulnerabilities within systems or technical flaws may still persist or occur despite rigorous testing and proofing of the AI systems. There have been noted cases wherein AI employed for routine operations is plagued by flaws that were discovered only at the deployment stage. Consequently, the result of such oversights may be irreversible and may cause irreparable harm to military operations by compromising personnel, equipment and/or information.

Another issue is that ML algorithms that depend on deep neural networks are opaque and operate like a black box. The input and the output are observable but the computational process leading from one to the other is difficult for humans to comprehend. In other words, it is easier to train a system using the technique of deep learning and neural networks but trying to decipher what the machine understood is complex.[32] These learnings are not stored in a digital memory for an individual to unlock and decipher, rather the information is diffused in a way that is exceedingly difficult to untangle the machine's learning.[33] Simply put, this indicates that it might not be possible to fully comprehend how a trained AI comes to its conclusions or predictions. Therefore, using it in sensitive contexts such as contested border regions or conflict zones could pose a risk of eventuality or escalation. Additionally, the lack of transparency and explainability creates a fundamental problem of predictability. The issue of predictability and uncertainty could also be used by militaries as an advantage to provide a scenario of plausible deniability.[34]

Many scholars have also articulated the issue of predictability as paradoxical because the operations performed by autonomous systems should be predictable to the operators that employ them, but not predictable enough to an enemy who could exploit the predictability to its advantage.[35] Additionally, the issue of transparency is equally worrisome from a regulatory perspective because it makes it difficult to identify the source of a problem and attribute responsibility when something goes wrong.

The reliability of the output is another concern. Unlike humans, AI systems are unable to recognise patterns at an abstract level. They merely identify a correlation between different pixels. Facial recognition software cannot distinguish between a real person and a picture of a picture inside of another image; in both situations, a valid identification is made. According to a study, 'variations in an image that are imperceptible to the human eye could cause an image-recognition system to completely mislabel the object or people in the image,' for instance mistaking a Siberian husky for a wolf.[36] Another study has also highlighted that it is easy to produce images that are completely unrecognisable to humans, but that software believes it to be a recognisable object with over 99% confidence.[37] This means AI systems may not be fully reliable and can be fooled by adversaries. The issue of predictability and reliability is further exacerbated in the cyber domain. In DARPA's 2016 Cyber Grand Challenge – an autonomous agent captures the flag hacking contest wherein autonomous computer systems faced each other. During the contest, one autonomous system quit in the middle of the competition, while another repaired some damage but incapacitated the machine it was supposed to safeguard.

Thus, the reality stands that these technologies that facilitate autonomous features within machines are still at a nascent stage, especially for weapon systems that would be used on the battlefield and in militaries. Deploying these half-tested machines in conflicts and wars could only lead to chaos and escalation. Hence, ensuring a responsible adoption of AI technology continues to remain a pressing challenge for regulators, developers and users.

Impact on stability and deterrence

The use of cyber-AI tools for defence and security raises pertinent questions about how it would alter the current dynamics of international relations as well as strategic stability.[38] For many states, possession of nuclear weapons has averted an all-out war to some extent based on the fear of Mutually Assured Destruction (MAD).[39] The notion of MAD is based on the notion that a potential attacker has no motivation to launch a nuclear strike if the defender can guarantee a retaliatory strike. The likelihood of a nuclear state shifting towards autonomous capabilities is linked to its confidence in the second-strike capability. The 'more vulnerable they perceive these second-strike capabilities to be, the more likely it is that they will integrate autonomous systems,' particularly those that may speed up decision-making or cut the human out of the loop.[40]

With the introduction of AI, the equation of MAD may not hold true. It may change the current balance of power because of the fear of miscalculation and unexpected outcomes inherently linked to the technology. On the one hand, AI-enhanced military capabilities can enhance deterrence, because of effective detection, prediction and reduction of the sensor-to-shoot time frame. On the other hand, the same advances could also cast doubt on the survivability of second-strike capabilities, especially for nations having a no-first-use policy. For instance, as the ability of AI improves, it can make accurate predictions based on the fusion of disparate sources of information, allowing it to find and target missiles stored in silos, on aircraft, submarines or trucks. This could lead to 'strategic destabilization of threats to the survivability' of systems such as missile launchers, which are the foundation of deterrence.[41] With such capabilities, the threat of retaliation could be eliminated, inviting a first strike –a very destabilising prospect. This uncertainty could encourage more aggressive nuclear postures, potentially increasing nuclear risk. The speed of decision-making, data processing and executing a task would be reduced to milliseconds because of the integration of AI within systems. Contrarily, these developments could also incentivise adversary forces to undertake covert attacks such as exploiting vulnerabilities or manipulating data of the systems, thereby exacerbating the situation.

Deterrence is a psychological concept that relies on the threat perception of an opponent's capabilities and intentions. Advances in AI capabilities create a misperception of an adversary's capabilities. To counter at least the issue of predictability, certain research programmes are underway to bypass the black box nature of the AI system such as DARPA's Explainable AI program (XAI), which has been in progress since 2016. Explainable AI, defined as 'machine learning techniques that make it possible for human users to understand, appropriately trust, and effectively manage AI,' still possesses limited capabilities and has generally struggled to realise the goals of 'understandable, trustworthy and controllable' AI in practice.[42] Additionally, these measures still do not address the accountability issue of AI machines, that is, who should be held responsible, especially in the case of military usage where lives are at stake.

This problem set is furthered in the absence of any agreed framework or an understanding of how escalatory behaviour can be defined in cyberspace. For instance, a cyber campaign launched for signalling (i.e., for coercive diplomacy) may go unnoticed by the target or, worse, may be mistaken for an offensive attack. Even if information about an operation of this nature is promptly and precisely discovered, the motivations behind it would remain unclear or misunderstood. Thus, the onus falls on nation states and especially nuclear powers to communicate clearly and be transparent about their AI capabilities so as to prevent an AI arms race which may consequently upset the current nuclear strategic balance. This AI arms race would not be limited to state actors; soon the proliferation could be seen among malicious non-state actors. Owing to the 'scalability, efficiency and ease of diffusion' of AI systems, the cost in terms of resources, manpower and psychological distance would be lower and hence the barriers to entry would come down. A classic example is the use of UAVs (or semi-autonomous drones), the weaponsied version of these were seen in the Vietnam War in 1959–75, but by the end of the 2000s, this technology had proliferated among non-state actors.[43] With the commercial off-the-shelf availability of many of these systems, there are noted instances of non-state actors continuously employing them in their activities. During the Battle of Mosul in 2017, the Islamic State loaded 40 mm grenades onto UAVs and dropped them on Iraqi Government positions, killing up to 30 Iraqi soldiers in a single week.[44]

This is a common pattern with emerging technologies. Initial development is expensive, but once commercialised, the price drops drastically and it becomes easily available. With algorithm-based systems, the cycle would more likely go faster because many of these systems are currently developed using open source software. Even if they are not being developed with open source software, once an algorithm has been developed, the cost of reproducing it is almost non-existent. This necessitates new approaches to arms control, including mechanisms that bridge the gap between conventional and nuclear munitions and address both horizontal and vertical proliferation to non-state actors. This will also alter the nature of warfare because, as access to weapon technologies becomes more widely available, states and non-state actors will be tempted to use surrogates to fight on their behalf.

Conclusion

Many nations today are prioritising the applications of AI towards civilian applications for economic enhancement, social change and military modernisation. The noteworthy aspect is the dual-use feature of these applications, that is, any development in the civilian sector could very well be adopted in the military domain. For instance, if a self-driving car becomes a success, then the same technology could be used for battle tanks, or the success of autonomous food or medical delivery drones can also be applied to weapon-carrying drones. These AI tools on the one hand facilitate seamless operation at great scale and speed but also make them prone to vulnerabilities and errors.

The added complexity of the AI-cyber tools combo would only further deteriorate the current stability matrix. As demonstrated in the ongoing Russia-Ukraine War, cyber operations are feeding into kinetic forces, thus doubling the chaos of war. Additionally, these campaigns have a spillover effect that is not limited by borders and have consistently caused disruptions for neighbouring nations.

The war in Ukraine may not have seen the deployment of killer robots or fully autonomous weapons but rather a comprehensive use of semi-autonomous drones and surface reconnaissance vehicles which very much highlights the potential usefulness of AI. These small successes are an indication of the improvement and advancement of AI's abilities and that the future is no longer bleak for AI weapons. Especially in combat scenarios, the demand for speed in conflict will inevitably compel militaries to delegate more decision-making to machines. AI tools may currently have certain limitations, as the AI is still in a narrow phase and further faces challenges of the absence of quality data and unreliability towards the black box technology. However, one cannot deny that the current war presents opportunities in the form of a testing ground as well as a source of data collection that can later be used in progressing the application of AI.

As of now, cyber operations and AI systems have been employed separately in the war, but a combination of cyber-AI has considerable potential for 'obfuscation and strategic misdirection.'[45] Policymakers must be cognisant that a combination of AI-cyber tools adds a new dimension to the challenges of 'validity and attribution' already present in cyber campaigns. Given the opportunities for data poisoning and manipulation, the trust and integrity of a system would always be questionable. Going forward, the integration of AI-cyber tools within weapon systems could lead to unforeseeable scenarios, widen the trust gap between adversary nations and increase the risk of miscalculation. Especially in the case of countries possessing nuclear weapons, these technologies have the potential to threaten the survivability of each other's nuclear second-strike capability. Although speculative, the development trajectory of AI highlights a high probability of escalations and clashes with catastrophic outcomes. Therefore, it is essential that many of these developments are straitjacketed by appropriate guardrails in the form of confidence-building measures, verification regimes and treaties.

Notes

1 AI has been used as an umbrella term to cover a range of technologies and activities since the 1950s. For the purpose of this chapter, to limit the scope of AI, the working definition is: 'Artificial intelligence is that activity devoted to making machines intelligent, and intelligence is that quality that enables an entity to function appropriately and with foresight in its environment.' Nils J. Nilsson, *The Quest for Artificial Intelligence: A History of Ideas and Achievements* (Cambridge: Cambridge University Press, 2010), p.13.
2 Greg Allen and Taniel Chan, 'Artificial Intelligence and National Security,' *Belfer Center Study,* July 2017, available at: https://www.belfercenter.org/sites/default/files/files/publication/AI%20NatSec%20-%20final.pdf , accesse.
3 Vincent Boulanin and Maaike Verbruggen, 'Mapping the Development of Autonomy in Weapon Systems,' *Stockholm International Peace and Research Institute (SIPRI),* November 2017, available at: https://www.sipri.org/sites/default/files/2017-11/siprireport_mapping_the_development_of_autonomy_in_weapon_systems_1117_1.pdf , accessed on 18 October 2022.
4 ML is a software development approach that first builds systems with the ability to learn and then is used as a training model via different techniques, that is, reinforcement learning, unsupervised learning or supervised learning. ML eliminates the need for manually hard-coding the software features into the systems.
5 AI is a means (technology) that facilitates autonomy within systems.
6 Boulanin and Verbruggen, 'Mapping the Development of Autonomy in Weapon Systems.'
7 Boulanin and Verbruggen, 'Mapping the Development of Autonomy in Weapon Systems.'
8 Lindsey R. Sheppard, 'The Future of War: Less Fantastic, More Practical,' *Observer Research Foundation,* 1 February 2020, available at: https://www.orfonline.org/expert-speak/the-future-of-war-less-fantastic-more-practical-60935/ accessed on 19 October 2022.
9 Monica Kaminska, James Shires and Max Smeets, 'Cyber Operations during the 2022 Russian invasion of Ukraine: Lessons Learned (so far),' *European Cyber Conflict Research Initiative,* July 2022, available at: https://eccri.eu/wp-content/uploads/2022/07/ECCRI_WorkshopReport_Version-Online.pdf accessed on 21 October 2022.
10 Matt Burgess, 'A Mysterious Satellite Hack Has Victims Far Beyond Ukraine,' *Wired,* 23 March 2022, available at: https://www.wired.com/story/viasat-internet-hack-ukraine-russia/. INSIKT Group, 'Overview of the 9 Distinct Data Wipers Used in the Ukraine War,' *Recorded Future,* 12 May 2022, available at: https://www.recordedfuture.com/overview-9-district-data-wipers-ukraine-war , accessed on 21 October 2022
11 Kaminska et al., 'Cyber Operations During the 2022 Russian invasion of Ukraine.'

12 Alexey Ramm, 'Russia Has Its Own Line of Kamikaze Drones,' *Известия*, 19 February 2021, available at: https://iz.ru/1126653/aleksei-ramm/u-rossii-est -svoia-lineika-bespilotnikov-kamikadze, accessed on 21 October 2022.

13 Gregory C. Allen, 'Russia Probably Has Not Used AI-Enabled Weapons in Ukraine, but That Could Change,' *CSIS*, 26 May 2022, available at: https:// www.csis.org/analysis/russia-probably-has-not-used-ai-enabled-weapons -ukraine-could-change ,accessed on 21 October 2022.

14 Ilya Gridneff, 'Killer Robots Have Arrived to Ukrainian Battlefields,' *coda*, 8 December 2022, available at: https://www.codastory.com/authoritarian-tech/ killer-robots-ukraine-battlefield/, accessed on 21 October 2022.

15 Will Knight, 'Russia's Killer Drone in Ukraine Raises Fears about AI in Warfare,' *Wired*, 17 March 2022, available at: https://www.wired.com/story/ai-drones -russia-ukraine/, accessed on 21 October 2022.

16 Military Support for Ukraine, Federal Government of Germany, available at: https://www.bundesregierung.de/breg-en/news/military-support-ukraine -2054992, accessed on 22 October 2022..

17 Tristan Greene, 'Ukraine Has Become the World's Testing Ground for Military Robots,' *The Next Web*, 21 December 2022, available at: https://thenextweb .com/news/ukraine-has-become-worlds-testing-ground-for-military-robots, accessed on 22 October 2022..

18 Kashmir Hill, 'Facial Recognition Goes to War,' *The New York Times*, 7 April 2022, available at: https://www.nytimes.com/2022/04/07/technology/facial -recognition-ukraine-clearview.html; Paresh Dave, 'Ukraine Has Started Using Clearview AI's Facial Recognition During War,' *Reuters*, 14 March 2022, available at: https://www.reuters.com/technology/exclusive-ukraine-has-started -using-clearview-ais-facial-recognition-during-war-2022-03-13/ ,accessed on 24 October 2022.

19 Darian Meacham and Martin Gak, 'Does Facial Recognition Tech in Ukraine's War Bring Killer Robots Nearer?' *Open Democracy*, 30 March 2022, available at: https://www.opendemocracy.net/en/technology-and-democracy/facial-recog-nition-ukraine-clearview-military-ai/ , accessed on 24 October 2022.

20 Raluca Csernatoni and Katerina Mavrona, 'The Artificial Intelligence and Cybersecurity Nexus: Taking Stock of the European Union's Approach,' *Carnegie Europe*, 15 September 2022, available at: https://carnegieeurope.eu/2022/09/15 /artificial-intelligence-and-cybersecurity-nexus-taking-stock-of-european-union -s-approach-pub-87886 ,accessed on 24 October 2022.

21 Gerrit De Vynck, 'The US Says Humans Will Always Be in Control of AI Weapons: But the Age of Autonomous War Is Already Here,' *The Washington Post*, 7 July 2021, available at: https://www.seattletimes.com/business/technol-ogy/u-s-says-humans-will-always-be-in-control-of-ai-weapons-but-the-age-of -autonomous-war-is-already-here/ ,accessed on 24 October 2022.

22 David Brumley, 'Mayhem Moves to Production with the Department of Defense,' *ForAllSecure Blog*, 12 May 2020, available at: https://forallsecure.com/blog/ mayhem-moves-to-production-with-the-department-of-defense ,accessed on 24 October 2022.

23 Elias Groll, 'ChatGPT Shows Promise of Using AI to Write Malware,' *Cyberscoop*, 6 December 2022, https://cyberscoop.com/chatgpt-ai-malware/ ,accessed on 25 October 2022.

24 Summary of the 2018 Department of Defense Artificial Intelligence Strategy, available at: https://media.defense.gov/2019/Feb/12/2002088963/-1/-1/1/ SUMMARY-OF-DOD-AI-STRATEGY.PDF ,accessed on 25 October 2022.

25 Kelsey D. Atherton, 'How Silicon Valley Is Helping the Pentagon Automate Finding Targets,' *Military.com*, 14 January 2022, available at: https://www

.military.com/daily-news/2022/01/14/how-silicon-valley-helping-pentagon -automate-finding-targets.html ,accessed on 26 October 2022.

26 Michael R. Bloomberg, 'Google Walks Away from America's Security,' *Bloomberg*, 6 June 2018, available at: https://www.bloomberg.com/opinion/ articles/2018-06-06/google-s-decision-to-ditch-project-maven-is-a-grave-error ?leadSource=uverify%20wall , accessed on 26 October 2022.

27 Chris Bing, 'The Tech Behind the DARPA Grand Challenge Winner Will Now Be Used by the Pentagon,' *Cyberscoop*, 11 August 2017, available at: https://www .cyberscoop.com/mayhem-darpa-cyber-grand-challenge-dod-voltron/ ,accessed on 26 October 2022.

28 Christian Szegedy, Wojciech Zaremba et al., 'Intriguing Properties of Neural Networks,' *arXiv*, 19 February 2014, available at: https://arxiv.org/pdf/1312 .6199v4.pdf accessed on 27 October 2022.

29 Shahar Avin and S. M. Amadae, 'Autonomy and Machine Learning at the Interface of Nuclear Weapons, Computers and People,' in The Impact of Artificial Intelligence on Strategic Stability and Nuclear Risk, *Stockholm International Peace Research Institute (SIPRI)*, May 2019, available at: https://www.sipri .org/sites/default/files/2019-05/sipri1905-ai-strategic-stability-nuclear-risk.pdf accessed on 27 October 2022.

30 Pavel Aksenov, 'Stanislav Petrov: The Man Who May Have Saved the World,' *BBC News*, 26 September 2013, available at: https://www.bbc.com/news/world -europe-24280831 accessed on 27 October 2022.

31 Danny Yadron and Dan Tynan, 'Tesla Driver Dies in First Fatal Crash While Using Autopilot Mode,' *The Guardian*, 1 July 2016, available at: https://www .theguardian.com/technology/2016/jun/30/tesla-autopilot-death-self-driving -car-elon-musk , accessed on 27 October 2022.

32 Davide Castelvecchi, 'Can We Open the Black Box of AI?,' *Nature*, 5 October 2016, available at: https://www.nature.com/news/can-we-open-the-black-box-of -ai-1.20731 , accessed on 28 October 2022.

33 Arthur Holland Michel, 'The Black Box, Unlocked: Predictability and Understandability in Military AI,' *United Nations Institute for Disarmament Research*, September 2020, available at: https://unidir.org/sites/default/files /2020-09/BlackBoxUnlocked.pdf; Yavar Bathaee, 'The Artificial Intelligence Black Box and the Failure of Intent and Causation,' *Harvard Journal of Law & Technology*, vol. 31, no. 2 (2018), available at: https://jolt.law.harvard.edu/ assets/articlePDFs/v31/The-Artificial-Intelligence-Black-Box-and-the-Failure-of -Intent-and-Causation-Yavar-Bathaee.pdf , accessed on 28 October 2022.

34 Vincent Boulanin, 'The Impact of Artificial Intelligence on Strategic Stability and Nuclear Risk,' *Stockholm International Peace Research Institute (SIPRI)*, May 2019, available at: https://www.sipri.org/sites/default/files/2019-05/sipri1905-ai -strategic-stability-nuclear-risk.pdf , accessed on 28 October 2022.

35 Boulanin, 'The Impact of Artificial Intelligence on Strategic Stability and Nuclear Risk.'

36 'Why AI Fails in the Wild,' *Unbabel*, 15 November 2019, available at: https:// resources.unbabel.com/blog/artificial-intelligence-fails. accessed on 30 October 2022.

37 A. Nguyen, J. Yosinski, and J. Clune, "Deep Neural Networks are Easily Fooled: High Confidence Predictions for Unrecognizable Images,' *IEEE Conference on Computer Vision and Pattern Recognition (CVPR)* (2015), pp. 427–36. https:// arxiv.org/pdf/1412.1897.pdf.

38 Strategic stability refers to the absence of incentives to use nuclear weapons first (in pre-emptive attacks) and the absence of incentives to build up those forces.

39 Each nuclear power state has acquired a robust second-strike capability, there-fore, none of the states can launch a nuclear attack without triggering a response that would destroy itself.
40 Vincent Boulanin, 'The Impact of Artificial Intelligence on Strategic Stability and Nuclear Risk.'
41 Vincent Boulanin, 'The Impact of Artificial Intelligence on Strategic Stability and Nuclear Risk.'
42 Jessica Newman, 'Explainability Won't Save AI,' *Brookings,* 19 May 2021, available at: https://www.brookings.edu/techstream/explainability-wont-save -ai/ accessed on 02 November 2022.
43 Jean-Marc Rickli, 'The Economic, Security and Military Implications of Artificial Intelligence for the Arab Gulf Countries,' *Emirates Diplomatic Academy*, November 2018, available at: https://www.agda.ac.ae/docs/default -source/Publications/eda-insight_ai_en.pdf, accessed on 02 November 2022.
44 Pablo Chovil, 'Air Superiority Under 2000 Feet: Lessons from Waging Drone Warfare Against ISIL,' *War on The Rocks*, 11 May 2018, available at: https:// warontherocks.com/2018/05/air-superiority-under-2000-feet-lessons-from -waging-drone-warfare-against-isil/ , accessed on 04 November 2022
45 Christopher Whyte, 'Poison, Persistence, and Cascade Effects: AI and Cyber Conflict,' *Strategic Studies Quarterly*, vol. 14(1) (2020), available at: https:// www.airuniversity.af.edu/Portals/10/SSQ/documents/Volume-14_Issue-4/ WhyteRev.pdf , accessed on 4 November 2022

5

ETHICAL ANALYSIS OF AI-BASED SYSTEMS FOR MILITARY USE

Gregory M. Reichberg and Henrik Syse

Introduction

There is a growing expectation– that artificial intelligence (AI) technologies will transform military affairs. Hopes have been kindled for more precise targeting and fewer casualties, while fears have arisen about new weapons of mass destruction, lack of human control and oversight, and the potential for a major shift in the geo-political balance. As this military transformation takes place – some call it a 'revolution' – gaining a better understanding of its potential benefits and disadvantages for humanity is imperative. In other words, an ethical analysis of the main challenges posed by AI military technologies is needed.

Setting aside logistics (as outside its scope), this article shows how AI-enabled autonomous operations function both to support human targeting decisions and to have an artificial agent (computerised robot) render targeting decisions. As to the first, and using NATO targeting doctrine as a point of reference, we explain how machine-issued target-identification and target-engagement recommendations, despite the notable efficiency gains, also involve possible ethical downsides. These we discuss under the following labels: Overfitting problem, classification problem, information overload, automation bias and automation complacency.

Regarding machines that are designed to make targeting decisions without direct human involvement – standardly termed 'autonomous weapon systems' (AWS)[1] – we survey the arguments that have been proposed for and against their use in battlefield settings. Some have maintained, for instance, that AI-directed robotic combatants have an advantage over their human counterparts, insofar as robots operate solely by rational assessment, while

DOI: 10.4324/9781003421849-6

humans are often swayed by emotions that can be conducive to poor judgement. By contrast, others maintain that emotions enhance moral judgement, for instance by enabling empathy; that it would violate human dignity to be killed by a machine; and that the honour of the military profession hinges on maintaining an equality of risk between combatants, an equality that would be removed if one side delegates its fighting to robots. Most important among the ethical arguments questioning the use of AWS are those dealing with control and explainability, namely, the ability of human beings to manage and understand the systems deployed.

We believe that the key ethical questions that must be asked can preliminarily be summarised as follows:

- Decisions: Who decides – and how is it decided – whether to develop, implement De and use AI-based systems?
- Control/safety: At what stages can the use of these systems be managed by human decision-makers so that the likelihood of accidents is reduced (and when accidents nonetheless occur, we can mitigate the effects and learn how to prevent similar events in the future)?
- Misuse: Wherein lie the dangers of misuse, such as illegitimate storage and use of information (violating privacy, confidentiality, etc.), application of AI-enhanced weaponry toward unauthorised targets, as well as adversarial hacking or other forms of illegitimate interference?
- Training: How are human agents trained in the proper use of these systems? What moral requirements should be placed on those agents? What technical competence and virtues should they develop and possess?
- Responsibility: Who is responsible for oversight, and who is liable for sanctions when mistakes are made?
- Transparency/Explainability: Is it possible to know who decided or did what? To what extent can human decision-makers have the required insight into the context and ramifications of their own decisions, given such challenges as secrecy and enormous complexity?

Uses of AI in warfare

AI has three main uses in warfare: It can enable more efficient logistics (including communications coordination within a unified command); it can support human decision-making, aiding in the identification and assessment of targets; and it can replace human decision-making and engage directly with targets.[2] All three modalities of AI can involve autonomous functioning (although, as already noted, we set aside the first in this chapter). Decision-support systems are autonomous technologies;[3] they can collect and analyse targeting data, and even offer recommendations, in the absence of an intervening human agent. In general, such decision aids are devices that support human decision making in complex environments... Such systems are meant to support human cognitive processes of information analysis and/or response selection by providing automatically generated cues to aid the human user to correctly assess a given situation or system state and to respond appropriately. Two different functions of decision aids can be distinguished: alerts and recommendations. The alert function... makes the user aware of a situation that might require action. The recommendation function involves advice on choice and action.[4]

AWS can likewise run through an entire operating cycle without the need for human intervention. In this case, however, the cycle includes the decision to engage with a target.

Some systems are designed for decision-support functionality only, while others can operate in either mode – decision support and decision-making – depending on the option selected by the human operator. This twofold functionality has been technologically possible at least since the Aegis Combat system was first deployed in the 1970s. It could be set to run in automatic mode in which case it would engage targets directly without an intervening decision by a commander, or it could run in manual mode, whereby only the authorised human operator could trigger a missile launch. We have been told by a retired US admiral that he and his peers were very reluctant to deploy the system in automatic mode, for fear of the unintended engagements that might occur.

AI in autonomous decision-support and decision-making functions

To examine the support that AI can provide for targeting decisions, we use NATO doctrine as a point of reference. The *Allied Joint Doctrine for Joint Targeting*[5] outlines several phases in the targeting cycle (or 'loop,' as it is alternatively termed).[6] AI-enhanced decision-support tools can be inserted at different points in this loop. Understanding these points of insertion will assist us in mapping the relevant ethical issues. Beforehand, it should be noted that the NATO doctrine outlined below has been designed for what might be described as 'offensive operations,' namely, operations where military command has a choice with respect to the inception of hostilities. In defensive engagements, where prior knowledge of an oncoming attack is usually very limited, the time margin for careful selection of targets will be considerably restrained. NATO speaks of 'dynamic targeting' apropos these defensive engagements, while 'deliberate targeting' is the term reserved for engagements of the offensive sort.[7] AI can have a role in either case; for the latter, the assessment of targets will be paramount, whereas for the former, the assessment of imminent threats will be primary,[8] with the determination of targets having a subordinate role.

Figure 5.1[9] shows the six main phases in the deliberate targeting cycle.

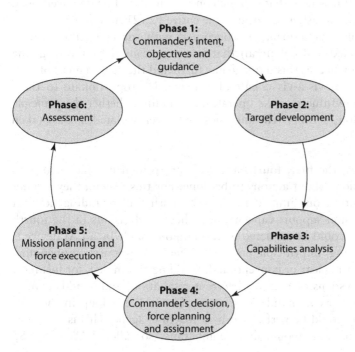

FIGURE 5.1 The joint targeting cycle

Phase 1, the 'Commander's intent, objectives, and guidance,' refers to the top-echelon military decision that a use of force will be implemented, based on strategic-political guidance from the relevant authorities. All subsequent target strikes should be traceable back to this intent.

Phase 2, 'Target development,' refers to the identification of potential targets, their subsequent vetting, prioritisation and selection, as well as building the database of knowledge about them.

Phase 3, 'Capabilities analysis,' consists in matching the prioritised targets with the most appropriate weapon systems (hence it is also called 'weaponeering') to achieve the desired effects.

Phase 4, 'Commander's decision, force planning and assignment,' involves final approval of the prioritised targets and communication of these from the operational to the tactical level. Relevant constraints, articulated as rules of engagement, are passed on to the assigned units.

Phase 5, 'Mission planning and force execution,' involves positive identification of targets to be engaged with the predetermined weapon systems. At this juncture, a tactical targeting cycle is made operational, and is described by NATO as F2T2EA: Find, fix, track, target, engage and assess.[10] It is here that robotic combatants could be deployed, who would autonomously run through the entire F2T2EA cycle.

Phase 6, 'Combat assessment,' is concerned with surveying the damage (both intended and collateral) that has been caused by the targeting action, and whether this damage has furthered the achievement of mission goals, or, vis-à-vis collateral damage, is proportionate to these goals; the lawfulness of the operation, including whether the principle of distinction or other relevant rules have been violated, is also taken up in this stage.

Of these phases, the first, fourth and fifth are primarily concerned with rendering decisions about actions to be done. For this reason, they are not directly relevant to our immediate concern with understanding how AI serves in a decision-support capacity. Whether the decisions in the fourth and fifth phases could be rendered in autonomous mode by an AI device is (we note in passing) in principle possible. Defensive responses to oncoming attacks (whether nuclear or conventional) could be automated, for instance, in air defence systems or nuclear deterrent retaliation, and would serve as an equivalent for a commander's decision. The F2T2EA loop in the fifth phase, similarly, could be performed by a machine agent. This is the locus for ongoing discussions about robotic combatants (so-called 'LAWS,' 'AWS,' or 'killer robots').

The second, third and sixth phases are primarily concerned with intelligence in the broad sense, namely, the gathering and analysis of data regarding the operating environment, as well as the issuance of recommendations to decision-makers.[11] The second and third phases, in particular, have a vital impact on the direction and accomplishment of missions. Commanders are not bound to heed the information and advice they receive, but most often they do – otherwise their actions would be deemed hasty and ill-informed. Commanders want to understand the facts on the ground before they decide what steps to take. Increasingly, this understanding derives from input that commanders receive from AI-enhanced machines; algorithms structure this input so it can inform their decision-making in the most efficient manner possible.

Efficiency is an extremely important consideration in intelligence gathering, because military operations, even when they result from much advance planning, are subject to time constraints, must be responsive to fast-changing adversarial developments, and require careful analysis of complex potential targets (an oil refinery could, for instance, be hit at different points, with varying degrees of collateral damage), and potential interactions between these targets (for instance, an ammunition depot located near the oil refinery). In this connection, Ekelhof explains how raising the efficiency of targeting processes was one of the reasons the USDOD established the Algorithmic Warfare Cross-Functional Team (AWCFT), also known as Project Maven. The overall aim of this team is to accelerate DoD's integration of AI, big data, and ML across operations to maintain advantages over increasingly capable adversaries. Its first task is to field technology to automate processing, exploitation, and dissemination (PED) for theater- and-tactical level UAVs collecting FMV (full motion video) in support of the Defeat-ISIS campaign. Currently, analysts spend 80% of their time doing mundane administrative tasks associated with staring at FMV (e.g., look, count, characterise) and typing data manually to a spreadsheet. Although it is necessary to conduct such tasks, commanders and Pentagon leaders do not consider them a good use of their analysts' time. So instead, they are introducing autonomous intelligence processing to help reduce the burden on human analysts, augment actionable intelligence, and enhance military decision-making.[12]

The efficiency gains afforded by autonomous technologies can be replicated in other phases of decision support, for instance, the matching of capabilities to targets (weaponeering), the establishment of targeting priorities and the assessment of executed operations.[13] In addition to these efficiency gains, there is also considerable potential for ethical gains as well. Machine-enabled foresight into the collateral damage to be expected from different targeting options could enhance the protection of civilian lives and infrastructure. Post-combat assessment of the battlespace could provide closer monitoring of compliance with key norms (rules of engagement and International Humanitarian Law (IHL)). The enhanced precision of targeting would likewise allow for more careful differentiation of military and civilian targets. Overall, targeting decisions would proceed from more detailed knowledge of the operating space, enhanced force protection, less unnecessary harm to adversary personnel and property, and other advantages that could, for instance, aid steps toward a political settlement of the dispute that led to the use of armed force.[14]

Alongside the ethical benefits of using autonomous technologies for targeting-decision support, there are potential ethical downsides as well. Apart from the fact that such systems can function with undetected flaws or undergo adversarial intrusion – thus leading to false assessments – use of these tools can also negatively impact the ability of human operators to make well-informed judgements about potential targets. As is well noted by Ekelhof:

> the use of autonomous technologies *changes the activities* of human actors as such technologies do not simply supplant the human beings, who simultaneously relinquish all the responsibilities and control they

exercised previously... [T]his is not without risk; the redistribution of existing tasks and the creation of new ones change the relationship between human actors and technologies, which can give rise to a transformation of decision-making processes. If these transformations are not considered thoroughly, the use of autonomous technologies could ultimately result in an unacceptable loss of human control. Whether humans remain in control of critical targeting decisions depends on how well they succeed at creating a framework within which this control can continue to be exercised alongside the use of increasingly AI technologies.[15]

In a similar vein, psychologists R. Parasuraman and D.H. Manzey observe that

the benefits anticipated by designers and policy makers when implementing automation – increased efficiency, improved safety, enhanced flexibility of operations, lower operator workload, and so on – may not always be realised and can be offset by human performance costs associated with maladaptive use of poorly designed or inadequately trained-for automation.[16]

Prudential decision-making requires not only careful attention to the situation in which the decision will be implemented; it also supposes that the decisionmaker has an introspective awareness about his or her cognitive abilities and shortcomings, possible biases, and is able to critically evaluate the credibility and truth-value of key sources of information. Any user of AI-systems for decision support, especially in highly sensitive settings where human lives are at stake, would need to be aware of several pitfalls. Understanding these potential pitfalls is of vital importance. Parasuraman and Manzey note in this connection that

if the benefit of a decision aid is to be realized, it needs to be used appropriately. Quite often, however, decision aids are misused, for two main reasons. First, the automatically generated cues are very salient and draw the user's attention. Second, users have a tendency to ascribe greater power and authority to automated aids than to other sources of advice.[17]

We will come back to this shortly, under the heading of 'automation bias.'

There is a danger that the current concern with lethal autonomous weapon systems (LAWS or AWS), a hotly debated topic in the public domain, including movements such as 'The Campaign to Stop Killer Robots,' will deflect attention away from the very real challenges associated with the use of automated technologies in decision-support capacities.

The use of autonomous technologies for intelligence gathering, weaponeering and related decision-support tasks raises several challenges that we briefly sum up below. None of these challenges decisively militates against the use of these technologies, but awareness of them is crucial because it is only then that mitigating measures can be undertaken, both when the technologies are developed, training routines are enacted and operating protocols are established.

- Overfitting problem: Contemporary methods of machine learning depend heavily on training data. The cogency of machine-generated targeting recommendations will consequently be a function of the data set from which the algorithmic model was derived. 'Algorithms should be trained on datasets that are representative of the operational environment to prevent bias, misidentification, and error.'[18] When targeting decisions are being made in preparation for combat, video footage and other images from past combat settings will often be sourced to supplement the training data that has already been acquired about the new context. If data is trained too heavily on data from past combat settings, the resulting algorithms will excessively reflect – *overfit* – these past settings and will accordingly *underfit* the new setting where military operations will be carried out. This can have serious ethical implications, such as mistaking non-combatants for combatants, because the past profile of a combatant may not adequately reflect the new situation.
- Classification problem: Following the procedure known as 'neural networks,' machine learning culls data to construct models that organise thousands or even millions of images into categories. A classification hierarchy is established that enables the recognition of specific objects and then correlates this object recognition with a specific mode of action, for instance, recommended lethal targeting by a specific weapon. However, when the algorithmic model organises these categories into a hierarchy, it is possible that the most morally salient categories are either missing or are too far down the hierarchy to influence decisively the targeting decision. The proper frame or context of morally correct action is thereby misunderstood. To illustrate this 'frame' (or 'classification') problem, Lin et al. ask whether a machine agent would *'know,* for example that a medic or surgeon welding a knife over a fallen fighter on the battlefield is not about to harm the soldier?'[19]
- Information overload: At any point in time, decision makers have a limited cognitive capacity to process the information that comes into their awareness. Data overload accordingly occurs 'when the amount of input to a system exceeds its processing capacity.'[20] For human beings, this is likely to result in a reduction of decision quality. Indeed, within military

settings, it is this very problem that has led to the automation of intelligence-gathering processes.

- Automation bias: Classical accounts of prudence emphasise the importance of taking counsel, namely, listening to the advice of well-informed individuals. Exercised virtuously, this will involve both seeking out such advice while at the same time retaining a critical distance toward the advice received. After the advent of AI, machines are now assuming a role once occupied solely by human beings. Numerous studies have shown that human beings often 'blindly accept the validity and strength of non-intuitive algorithmic conclusions.'[21]

    [A]utomation bias increases in high-intensity situations where time pressure and stress increase reliance on automated tools. The reasons behind this phenomenon are related to the human tendency of choosing the simplest cognitive task and willingness to diffuse the responsibility combined with perceived greater trust in the superior analytical capability of machines.[22]

- Automation complacency (decremental vigilance): Overreliance on autonomous technologies can likewise result in reduced attention to even obvious features of the surrounding environment. The alertness and contextual awareness of human operators decline accordingly. 'By relying on the algorithmic tools, the number of "out of sight" spots increases.'[23] There are documented cases of soldiers who have failed to see targets in their immediate vicinity while they were engaged in monitoring their video feeds.[24]

    More broadly, when relying on algorithmic tools, humans are prone to overlook information that is not prioritised nor displayed by the recommender system. Thus, even in a human-machine team, humans are limited by their abilities to observe and analyse information beyond that explicitly exposed by the decision support system.[25]

Using the related term 'automation complacency,' Parasuraman and Manzey explain that both it and automation bias are phenomena 'that describe a conscious or unconscious response of the human operator induced by overtrust in the proper function of an automated system.'[26] 'The performance consequence is usually that a system malfunction, anomalous condition, or outright failure is missed.'[27]

This leads us to the next modality of AI-based systems: The use of AI to make actual targeting decisions without direct human intervention. While we here prefer the wider label of AWS to describe this category of weapons – since they need not in practice be lethal – much of the debate has used the terminology of LAWS, or simply 'killer robots.' Campaigns to legally outlaw such weapons have gathered significant popularity and also some political support, and aim to outlaw such weapons along the lines of the prohibitions against chemical and biological weapons and anti-personnel landmines.[28] Critics of this ban movement hold that it is hard to specify what one would then outlaw – would, for instance, support systems also be outlawed? – and that such a ban overlooks the possibly positive humanitarian consequences of deploying at least some forms of autonomous weapon systems.

We should remind ourselves here of the phases of the targeting cycle we are now dealing with:

Phase 4, 'Commander's decision, force planning, and assignment,' involves final approval of the prioritised targets and communication of these from the operational to the tactical level. Relevant constraints, articulated as rules of engagement, are passed on to the assigned units.

Phase 5, 'Mission planning and force execution,' involves positive identification of targets to be engaged with the predetermined weapon systems. At this juncture, a tactical targeting cycle is made operational, and is described by NATO as F2T2EA: Find, fix, track, target, engage and assess.[29] It is here that robotic combatants could be deployed, who would autonomously run through the entire F2T2EA cycle.

Lethal autonomous weapons can be defined as lethal weapons that are 'capable of selecting and attacking targets without human intervention';[30] in other words, they are weapons that, depending on the level of autonomy achieved, can perform several of the actual decision-making functions within these two phases.

A summary of the debate

The debate around autonomous weapons has produced arguments for and against their development and use.[31] Very broadly speaking, we could say that the debate is between, on the one hand, optimists, who believe such weapons can be used within the framework of IHL and be under meaningful human control (a phrase we will return to below), and that their use will overall reduce bloodshed and collateral damage; and, on the other hand, pessimists, who hold that such weapons will take killing – the crucial and fateful decision to kill – out of the hands of human decision-makers, thus reducing or removing human control and potentially leading to warfare not regulated effectively by legal and moral boundaries.

Let us now summarise the ethical arguments for and against AWS. The 'pro' arguments can be divided into four categories.

(a) Such weapons will reduce the sway of human emotions such as vengeance and hatred and, for that matter, human weaknesses such as tiredness and confusion, leading to less brutal warfare and the more effective cessation of hostilities when military necessity dictates it.

(b) Autonomous weapons allow one to radically increase targeting accuracy, avoiding significant collateral damage as well as reducing misunderstandings and accidents (see below).

(c) AWS have no existential need for self-defence in case of danger; this could lead to much more manoeuvering room to, for instance, protect civilians and to engage in close combat rather than attacking and bombing whole areas.

(d) And finally, such weapons, like all remotely controlled weaponry, would increase force protection and the overall risk to combatants.

The 'con'/'against' arguments can likewise be divided into four:

(a) Emotions are a crucial aspect of proper moral reasoning, and they contribute decisively, not least to our understanding of right and wrong, our ability to distinguish what is important from what is less important, and our appreciation of basic moral concepts and norms such as dignity and respect. Removing human emotions from the battlefield becomes equivalent to what Valerie Morkevicius has termed 'fighting with Tin Men,' that is, beings without heart and without soul.[32]

(b) Fully autonomous weapons would need exceedingly complex programming to be able to face the contingencies and uncertainties of war. These contingencies and uncertainties are arguably so many that preparation for them can never be fully and dependably programmed into AI-based systems; hence, such systems will often make split-second decisions without full appreciation for changed circumstances or totally unforeseen events. While these are challenges for human beings, as well, it is more within the grasp of humans than artificial intelligence to appreciate and understand the nuances of contingencies and changed circumstances. This is closely linked to 'the frame problem': the challenge for machines to identify what falls within a 'frame' and what does not, thus making the right targeting decisions based on the right intelligence. (This was considered above under the discussion of decision-support systems.)

(c) Enabling machines to undertake targeting decisions without direct intervention of a human decider would violate fundamental requirements of human dignity. It would reduce those killed on the battlefield to 'mere targets.'[33]

> 'They become zeros and ones in the digital scopes of weapons which are programmed in advance to release force without the ability to consider whether there is no other way out, without a sufficient level of deliberate human choice about the matter.'[34]
> 'The lack of human agency in the specific decision to kill or injure means there is no recognition of the target's humanity, undermining

the human dignity of both those making the decision and those who are targeted, and so contributes to the dehumanization of warfare.'[35] There is, according to this argument, something unseemly about making decisions that knowingly lead to loss of life with no active human participation.

(d) Finally, the practice of war, going back to ideals of chivalry, is based on an idea of a certain level of reciprocal risk: Both parties assume risk by going to war, leading to restraint and mutually beneficial rules. Increased use of AWS could reduce or remove such risk if one side is able to deploy autonomously functioning weapons that operate without any direct human support or presence, while the other party would be fighting physically with man- and womanpower. This latter argument is certainly not new and is not restricted to AI-based AWS; it also applies to high-altitude bombing, for instance, where risk is very unevenly distributed. However, the lack of reciprocity arguably increases with autonomous weaponry.

At least some elements of the 'Ban Killer Robots' movement arguably contribute to muddying the waters by not tending sufficiently to the advantages of AI-based systems in decision-support functions, nor to the arguments in favour of (at least some) AI-based systems used in decision-making capacities, not least to increase protection of civilian lives and force protection. Simultaneously, the development towards battlefields, whether kinetic or cyber, where human beings are only minimally involved in, and cannot stop or change, kill decisions, and furthermore, where ML systems develop in a way that creates an increasing distance between users' intent on the one hand and the actual decision points on the other, must be addressed as maybe the key concern militating against deployment of such systems.[36] Insisting on the ethical obligations of all developers and users of such systems might be a useful way to address this concern without arguing in favour of blanket bans. In the legal arena, the establishment of international regulations on the use of autonomous weapon systems would likely be of interest to many states, including those invested in such systems, because of the safety and escalation risks that these systems pose. Limitations on use would impact first and foremost on military operators but would not directly raise the thorny issues of dual-use that would impact on producers. However, as indicated below in our recommendations, a ban on selected AWS, for instance as applied to nuclear launch decisions or hypersonic missiles, could be desirable.

AI accidents versus AI misuse

Ethically speaking, AI use can result in two main sorts of problems. On the one hand, the complexity of AI systems can give rise to serious accidents. Developers and users have an obligation to examine the likelihood of such accidents and to adopt precautionary measures. On the other hand, as a powerful tool, AI can be intentionally misused for wrongful ends, whether by deliberate misdirection, as when, for instance, a weapon system designed to target military personnel is directed against civilians, or by malicious interference, as when a military adversary gains control of an opposing AI system, through spoofing (i.e., disguising a communication from an unknown source as being from a known, trusted source) or other means, and directs it against the possessor's own personnel. Below, we briefly consider each of these in turn.

Accidents and AI safety

An accident is an 'undesired and unplanned event that results in a loss (including loss of human life or injury, property damage, environmental pollution and so on).'[37] Safety, by contrast, may be defined as 'freedom from accidents (loss events).' Ensuring the safety of actions to be undertaken is a key aspect of practical wisdom (*phronesis or prudentia*) and is traditionally described by terms such as 'foresight,' 'circumspection' or 'precaution.'[38]

AI raises distinctive challenges with respect to the safe employment of associated systems. The challenges are especially acute when AI is employed to deliver weapons in battlespace settings. The challenges can be summed up as threefold: amplification of errors, misalignment and miscommunication.

Amplification of errors

Automation is often introduced to reduce the errors that result when human operators are engaged in rote tasks that require close concentration, for instance, overnight guard duty at a military installation. Even short moments of inadvertence can have devastating consequences. Watching a radar screen for many hours, checking for an enemy incursion, can similarly give rise to operator errors. Automated machines, including AI-based surveillance systems, will be less prone to fatigue-induced or distraction-related errors. Machine agents have no need to rest and are not beset by competing thoughts that can draw one's attention away from the task at hand. Although machine agents are less error-prone than humans, when errors occur – and the causes can be multiple – they tend to be more severe, due to the inter-locking character of AI systems. Eliminating human error has

> allowed systems to grow, becoming faster and more complex. But this growth came at a price. In automated systems, the errors that do happen can cascade through multiple subsystems almost instantaneously, often taking down broad swaths of seemingly unrelated operations.[39]

'Emergent effects' is the term now used to describe effects that emerge when two or more systems interact. The powerful emergent effects that arise when error is transferred from one AI-based system to another has led Former US Secretary of the Navy Richard Danzig to speak of 'Technology Roulette': Whereas 'the Russian game normally involves playing with only one bullet,' Technology Roulette is a game 'played every day all over the planet, with increasing number of players, ever-greater variety in our weaponry, and increasing interactions within a crowded technological space, so that a live bullet may trigger deadly exchanges among many (or all) players.[40]

[M]any of our defense systems are exceptionally connected and interdependent even by the standards of an era when networking is widespread in civilian systems... [These systems] are subject to the 'CACE principle: Changing Anything Changes Everything.' 'Operational reliance on the combination of separate system increases vulnerability to emergent effects.'[41] (https://s3.us-east-1.amazonaws.com/files.cnas.org/documents/CNASReport-Technology-Roulette-Final.pdf)

Misalignment of software and 'real-world' tasks

The computer is a general-purpose machine that can take on a multitude of specialised tasks. The experts in this domain – software engineers – now design applications that previously would have been constructed as standalone machines by engineers who would have had broad experience within their specific domain, telecommunications or air traffic control, for instance. But now, since the generalists are in command, so to speak, safety requirements for specific applications can easily be misunderstood. This dynamic is well described by Nancy G. Leveson:

> [W]ith computers, the design of the special purpose machine is usually created by someone who is not an expert on designing such machines. The autopilot design expert, for example, decides how the autopilot should work, and then provides the information to a software engineer, who is an expert in software design but not autopilots. It is the software engineer who then creates the detailed design of the autopilot. The extra communication step between the engineer and software developer is the source of the most serious problems with software today.[42]

Leveson further explains how

> nearly all the serious accidents in which software has been involved in the last twenty years can be traced to requirement flaws, not coding errors.

> The requirements may reflect incomplete or wrong assumptions about the
> operation of the system components being controlled by the software…
> or about the required operation of the computer itself.[43]

If these requirement flaws can have serious repercussions in the civilian
domain, the potential for unintended harm can be proportionately greater
within the military sphere, for, as Danzig notes:

> The military extensively tests its products, but warfare is among the
> least predictable of endeavors. The timing, locations, and circumstances
> of use, even the characteristics of adversaries and of users, are highly
> variable and at best very imperfectly foreseen. The hidden complexi-
> ties of massive operations are compounded by the incentives for oppo-
> nents to surprise each other. Accordingly, tests of military technologies
> and extrapolations to use are probably less reliable than for civilian
> systems.[44]

Among weapon systems, those involving nuclear munitions raise particularly acute safety concerns. A growing literature has examined the implications of AI for this domain.[45] AI could have a role in (1) the protection and
maintenance of nuclear weapons infrastructure; (2) nuclear early-warning/
ISR (intelligence, surveillance and reconnaissance systems); (3) command
and control; and (4) weapon delivery. Each of these comes with potential
benefits and risks.[46] As to (1), AI could beneficially assist with protection
from cyber infiltration, predictive maintenance and other measures to
reduce the risk of malfunction, but at the same time it could lead to de-
skilling on the part of human operators who will continue to oversee nuclear
launch sites. With respect to (2), early warning of nuclear missile launch
could be made more accurate by a system of AI-linked detectors, both
stationary and mobile platforms. This is in principle a positive; however,
machine errors that go undetected could induce human operators to order
retaliatory strikes on mistaken grounds. AI could also enable better weapon
delivery, with greater speed, scope, precision and with more numerous platforms (such as UAVs). While strengthening deterrent threats, it could also
upset the balance of deterrence, with disfavoured nations coming to believe
their second-strike capabilities would be at risk. Command-and-control is
the most worrisome arena for introduction of AI-based autonomy; on the
one hand, AI could improve the situational awareness of commanders, contributing toward better judgement; on the other hand, it is generally thought
that the decision to engage nuclear weapons is so fraught with consequence
that this must remain in human hands alone, and then only at the highest echelon of government. Broadening current automated procedures, such

as the 'dead-hand' procedure, carry grave risks of unintended launch and escalation.

Human-AI miscommunication

The military users of AI technologies are usually not AI experts themselves. Accidents can readily occur when instructions are reciprocally misunderstood. AI started out as an attempt to emulate human cognitive abilities. Some have even referred to AI as a 'human-centric and ambiguous term: calling computers "artificially intelligent" is like calling driving "artificial running."'[47] If AI is understood as an emulation or extension of human thought processes, the difficulties of communication between the two sorts of cognitive agents can easily be underestimated. Confusions arising from misdirected machine and human inputs and outputs could have disastrous consequences. Humans are usually good at conveying our aims to one another; but proper understanding of another's aims will nonetheless depend on subtleties of context that even sophisticated AI devices have difficulties grasping. This is often discussed by reference to what is termed the 'classification problem.' The current trend is thus to better differentiate human and machine cognition, so that the challenges of communication are more realistically understood and addressed. User-interfaces suitable for battlefield conditions are being developed. The benefits of 'hybrid intelligence,' where the respective strengths (and weaknesses) of each are coordinated, can lead to a reduction of accidents due to machine-human mistranslation. This has been well described by Marianne Bellotti:

> One of the benefits of AI is that the types of mistakes humans and machines make are fundamentally different from one another. So-called hybrid intelligence aims to combine the complementary strengths of human and AI to create better outcomes. At the heart of hybrid intelligence lies Moravec's paradox, which tells us that pattern matching is difficult and resource-intensive for computers but cheap and easy for people, whereas calculations are difficult and resource-intensive for people but cheap and easy for computers. … The most effective systems combine reliable predictions based on computer calculations and flexible human responses when those predictions fail.[48]

Misuse of AI military applications

As noted above, misuse of AI is of two main kinds: (1) employment of one's own military capabilities against illegitimate targets or with disproportionate force; (2) subversion of one's AI-based systems by an adversary.[49]

With respect to (1), basic rules of IHL apply here as with other weapon systems. AI does not add anything dramatically new to the equation, apart from the fact that it enables more precise identification of specific individuals. This can be beneficial or harmful; the former because it enables more discriminate targeting and fewer unintended casualties, and the latter because such targeting can be misdirected against civilian populations who are then subjected to mass surveillance and terror campaigns, which in the worst case can result in genocide. Another problem arises when AI is deployed to enhance cyber intrusions of an adversary's digital infrastructure and the physical systems that depend on it. Because so much of today's digital infrastructure is dual-use (civilian and military), attacks that are intended to disable military targets, radar installations or other air defence systems, for example, can also disable important civilian functions, air-traffic controls at major airports, for instance, thereby causing crashes and transportation paralysis with potentially systemic effects for civilian life. Here again we encounter potential benefits and harms. A key benefit comes from the fact that in the past dual-use facilities would often be bombed, making post-conflict reconstruction slow and costly; now, through cyber intrusion such systems can be disabled and not destroyed, enabling swifter post-conflict recovery. Wider scope for harm opens up, however, because given the wide and growing dependency of modern societies on digital infrastructure – few areas of life are isolated from it – the immunity of civilian society from warfare is fast eroding. Before the advent of long-distance aerial bombardment, civilian population centres could be targeted only after a nation's military force had been defeated. Airpower opened up new vulnerabilities for civilian life, and what we witness today with offensive cyber is an even broader extension of that same vulnerability. What's more, the boundaries of war and peace are increasingly blurred, as numerous states now direct offensive cyber operations, the effectiveness of which is enhanced by AI, against dual-use and even civilian targets, although no state of war has openly or tacitly been declared.

With respect to (2), because AI systems necessarily operate in digital media, they are inherently vulnerable to adversarial intrusion by digital means. Many AI applications depend on communication links that can serve as points of entry for enemy operatives. Even when protected by encryption, these codes can be broken or compromised; communications can be forcibly severed, or control commands can be modified by the intruder. Autonomous systems have been developed in large measure to safeguard against the vulnerabilities arising from communication links. However, even fully autonomous systems can be subverted when adversaries take advantage of vulnerabilities that arise from machine-learning procedures. Unlike other forms of hacking that take advantage of code errors ('bugs') that are found in the opponent's software, ML depends on data streams that can be altered by an opponent even in the absence of communications links. To illustrate with an example from self-driving cars, 'an AI attack can transform a stop sign into a green light in the eyes of a self-driving car by simply placing a few pieces of tape on the stop sign itself.'[50] Defining an AI attack ('adversarial AI') as 'the purposeful manipulation of an AI system with the end of causing it to malfunction,' Comiter divides such attacks into two main kinds, each of which 'strike[s] at different weaknesses in the underlying algorithms.'[51]

- Input attacks manipulate what is fed into the AI system in order to alter the output of the system to serve the attacker's goal. Because at its core every AI system is a simple machine – it takes an input, performs some calculations, and returns an output –manipulating the input allows attackers to affect the output of the system.[52]
- Poisoning attacks damage the AI model itself so that once it is deployed, it is inherently flawed and can easily be controlled by the attacker. To poison an AI system, the attacker must compromise the learning process in a way such that the model fails on certain attacker-chosen inputs, or "learns" a back door that the attacker can use to control the model in the future. One motivation is to poison a model so that it fails on a particular task or types of input. For example, if a military is training an AI system to detect enemy aircraft, the enemy may try to poison the learned model so that it fails to recognise certain aircraft.[53]

In addition, traditional forms of attack can be undertaken specifically to breach AI systems when they are carried aboard unmanned surveillance or attack platforms, which employ edge-computing functions.

> In edge computing, rather than sending data to a centralised cloud infrastructure for processing, the data and AI algorithms are stored and run directly on the devices deployed in the field. The [US] DoD has made the development of "edge computing" a priority, as the bandwidth needed to support a cloud-based AI paradigm is unlikely to be available in battlefield environments.[54]

If these platforms are captured by an enemy, the collected data and models risk being breached as well, which could compromise related systems.

Such attacks in turn have several different kinds of effects. They 'can cause damage by making the system malfunction' (e.g., a car crash caused by the alteration of a stop sign), they can hide relevant data, thereby leaving security systems inoperative against incoming attack, and they can degrade faith in a system, so that those responsible for its maintenance take it offline, 'allowing a true threat to escape detection.'[55] In the worst case, breach of one AI system can have a cascading effect on systems of a similar kind. This prospect is not unlikely, because cost-saving within military organisations militates in favour of shared AI datasets across domains. 'Because these datasets and systems will be expensive and difficult to create, there will be significant pressures to share them widely among different applications and branches. However, when multiple AI systems depend on this small set of shared assets, a single compromise of a dataset or system would expose all dependent systems to attack.'[56]

To counter these vulnerabilities, there are no simple solutions as the weakness is inherent to AI itself. As a remedy, Comiter recommends 'AI Security Compliance' programs that include 'encouraging stakeholders to adopt a set of best practices in securing systems against AI attack... adopting IT-reforms to make attacks difficult to execute, and creating attack response plans.'[57] How these can be applied within the highly adversarial context of military organisations and their opposing operations remains to be spelt out.

Stages of development and use

The ethical questions raised above will vary and be directed at different parties at the various stages of AI development. These stages encompass, at a minimum, the following points.

Initial research and development (including funding for research and needs assessments from end users)

Most major militaries are investing heavily in AI R&D. The consensus appears to be that warfare is trending in a direction that will greatly favour those militaries that are equipped with the newest AI-enhanced technologies, including hypersonic weapons, quantum technology, directed-energy weapons and biotechnology (physical enhancement of soldiers' physical capabilities).[58] Here we could flag the military-strategic debates surrounding AI military applications. In the USA, for instance, the late Senator John McCain was a tireless advocate of investment in AI-related technologies.[59]

On the opposing side, some former military officers have argued that investment in these new capabilities is misplaced, both because future wars will not be very different from wars of the present and past, and because the belief that 'technology will save us' will lead to a process of de-skilling so that military personnel will find themselves less, not more, equipped to fight the sort of conflicts we will likely face.[60] Combined with this discussion is the parallel debate over whether ethical issues should be emphasised in the early-development phase, or whether this is best taken up later, when the decision is made whether or not to deploy the weapons that have been developed. Those adopting the latter position worry that moral 'scruples' will lead to inaction in the face of a growing threat from would-be adversaries (who are fast developing these new technologies), thereby creating serious vulnerabilities vis-à-vis nations that show little concern for moral standards.[61] On the opposing side are those who maintain that because the trend toward autonomous weapon systems is unstoppable (the very dynamic of competitor relations will ensure that this happens), it is essential that ethical safeguards are carefully considered at the earliest stages of the design process. The speed of these autonomous machine interactions is such that it will become increasingly difficult to retain humans in and even on the decision loop of lethal engagements; the best chance to prevent undesired and uncontrollable outcomes is to build in ethical safeguards at the very beginning of the design process.[62]

Actual development of AI-based systems (software and military hardware)

As was explained above, software developers are often not experts in military applications and vice versa. Significant safety gaps ('incompleteness') can arise as a result. Nancy Leveson has argued that flaws in fundamental safety requirements, not coding errors, are the biggest threat to the safe use of AI.[63] This problem can be addressed, she maintains, only when software designers interact with end users throughout the entire design process. Trust emerges when users are secure in their operating environment, and the designers understand the full range of challenges that the operators will face:

> Providing feedback and allowing for experimentation in system design, then, is critical in allowing operators to optimise their control ability. In the less automated system designs of the past, operators naturally had this ability to experiment and update their mental models of the current system state. Designers of highly automated systems sometimes do not understand this requirement and design automation that takes operators 'out of the loop.' Everyone is then surprised when the operator makes a mistake based on an incorrect mental model. Unfortunately, the reaction to such a mistake is to add even more automation and to marginalise the operators even more, thus exacerbating the problem.[64]

This observation holds even for AWS because these will be deployed on the battlefield by the decision of commanders, who must retain human control over these systems. The system may operate autonomously once deployed, but the deployment decision remains fully in the hands of its human operator, who must retain a capacity to direct and redirect the system toward the achievement of his/her tactical and strategic goals. The one deploying the AWS is its end user, and the designers of these systems must communicate regularly and effectively with these end users throughout the design phase.

Only then will the system's safety requirements come to be adequately understood.

Training in the use of AI-based systems

'Contrary to... popular belief, automation does not always lessen operator training requirements. It frequently changes the nature of operator performance demands and increases operator training requirements.'[65] '[A]utomated systems will not replace humans; rather it will make their roles more cognitive and more goal centered.'[66] Indeed, 'soldiers' role will increase as engineers find new uses for so-called autonomous and semi-autonomous systems.'[67] '[I]n many ways "unmanned" is a misnomer; the soldier will acquire more responsibility and not less with the proliferation of these systems.'[68] Proper training in the use of AI-based systems is key to identifying and tackling both ethical challenges in general and safety challenges in particular associated with AI, including the detection of bugs, biases, and incompleteness in the systems and programs. The key requirements of proper training related to AI can be divided into three:

- Through training, users find out what competence they need to manage AI, and what competence they currently lack. Identifying that competence requires application of systems in a variety of different settings. Not least, proper training will help identify the sorts of personnel and qualities needed for safe and competent operation and surveillance of these systems, which may be different qualities than those associated with the use of other forms of military equipment. Simultaneously, AI operators must be thoroughly skilled in IHL, also known as the Laws of Armed Conflict (LOAC), and be required through training to identify when the use of AI-based systems endanger basic tenets of IHL/LOAC.
- There may be a long road from the ideal models of the AI developers to the way in which these function under real conditions. Feedback loops must therefore be established so that the difficulties encountered by users are clearly communicated back to the developers, as well as to the authorities deciding on the deployment and use of AI systems. Not least, the safety limits of the system must be tested – 'the envelope must be pushed' – since it is only when operational errors are generated that the true characteristics of the system become better known. Leveson thus observes that Operators use feedback to update their mental models of the system as the system evolves. The only way for the operator to determine that the system has changed and that his or her mental model must be updated is through experimentation: To learn where the boundaries of safe behavior currently are, occasionally they must be crossed.[69]

John Hawley adds that the main problems associated with previous errors in automated systems result from the neglect of three key requirements: (1) recognition that automated and AI-based systems are fallible and that 'trust in the system's automation must be developed incrementally on the basis of experience, and will always be system-specific;'[70] (2) awareness that 'highly automated systems... rely on a high level of user expertise for safe and effective use,' which prioritises 'mindful exercise of positive control over rote drills;'[71] and (3) understanding that regular feedback loops must be maintained between developers, testers, trainers and users, such that the 'irreversible waterfall' view (that 'information flows in one direction only [from developers to users], regardless of the downstream consequences for the system') is avoided to the detriment of an iterative approach where 'requirements and design solutions can evolve as the technology is developed.'[72]

- The dangers of 'de-skilling' and the challenges of 'up-skilling' attendant upon the introduction of AI must be addressed. 'De-skilling' signifies the danger that the extreme scale and velocity of AI systems make human skills inoperative, so that when AI systems malfunction, traditional skills one needs to fall back on will have been weakened or simply not been learnt. 'Up-skilling' signifies the basic need for new and qualitatively different skills that must be acquired and perfected. Marianne Bellotti sums this up well:

  Advocates for AI ethics emphasise the importance of having a *human in the loop* — that is, human supervision of AI outcomes. But this supervision will not help if the human operator does not have enough experience or expertise to determine what the correct outcome is in the first place. Safety researchers refer to the "ironies of automation." The process of automating tasks makes the human operators whom systems rely on for good judgment less knowledgeable and experienced. The first stab at integrating AI into an existing process often leaves the most difficult and nuanced analysis to the human operators while delegating more basic analysis to the machine. Conventional thinking assumes that without the burden of the simple tasks, the human will be more efficient. In reality, though, accuracy on those hard tasks comes from experience doing the simpler ones. In the industry this is sometimes referred to as a "moral crumple zone." The human operator ends up being the scapegoat for the machine's error since he has a supervisory role, yet the machine's growing complexity makes it impossible for him to understand what the correct outcome should be.[73]

The above-mentioned concerns about training can be summarised in the following action points:

- More clarity must be achieved about the ways in which the scope and form of military training should change and be further developed as AI-based systems are developed and deployed.
- Efficient feedback loops from users in training to developers and political authorities must be established.
- Training protocols must ensure that operators understand the limitations of the systems they use, enabling these operators to learn 'by experience when a machine can be trusted and when additional scrutiny and oversight are necessary.'[74]
- Ethical and legal concerns must be built into all training in the use of AI-based systems.
- Military identity and the whole field of military ethics should be re-explored as more and more of military activity will consist of AI-and computer-based digital activities and in collaboration models where the military operative will often work more closely – and be more dependent on – artificially intelligent machines rather than on other human beings.

We should add here that by using the term 'training,' we are not thinking merely of short-term or purely technical forms of instruction. We also have in mind the broader acquiring of skills and qualities that we normally categorise under the more ambitious heading of 'education.' This includes a proper understanding of the ethical and social context within which we use such machines as well as the deeper challenges they represent to our understanding of armed conflict and, even more deeply, of what it means to be a responsible human agent. Often, the word 'training' is understood as a band-aid-like correcting of mistakes and addressing of challenges, primarily through quick-fix 'training modules.' While such programmes can have their proper place, the meeting between human and machine that the introduction of AI represents also calls for proper education about the ways in which we can maintain high standards of military ethics within a modern, technologically advanced armed force.

Modes of operation

Decisions about actual use of these systems, and in what mode they should be used (i.e., with what degree of autonomy [with what possibilities of meaningful human control]), encompass both strategic and tactical decisions. Knowledge and consciousness among decision-makers as to the various options and modes for employing weapons with autonomous functionality and ensuing clarity – both legal and practical – about when such functionality can be engaged, is crucial.

Rules of engagement

Decisions on the battlefield (engaging of targets), whether on the kinetic or the cyber battlefield, demand specification in the rules of engagement. All different, realistic scenarios must be examined beforehand, with an eye to the relative advantages and pitfalls of using AI-based systems. For instance, autonomous functioning may increase the possibilities of speed and scale in an operation, thus with gains in effectiveness, but escalation risks will grow proportionately.

Human control

All AI systems are subject to human control, whether they are fully autonomous or semi-autonomous. Autonomous weapons are under the strategic and tactical authority of human commanders, although operationally these weapons have decision-making authority delegated to them. Semi-autonomous systems have more limited functions delegated to them, e.g., information gathering, issuance of recommendations, etc., but not the targeting decisions themselves.

There is substantial agreement among scholars and practitioners about the dangers of 'loss of control over the use of force' (SIPRI 2020). This goes for the whole chain from strategy formulation and development via deployment and training to potential use. However, there is some disagreement about the best terminology to be employed.[75]

The term most often used to describe the kind of control needed is 'meaningful human control,' signifying that (authorised) human beings have an actual possibility of controlling (and halting) ongoing processes at all points, and thus remain in the decision loop.

Some, not least the USA, have preferred other designations, primarily the term 'appropriate human judgement,' indicating a process whereby commanders' intent and proper moral and legal restrictions are built into decision-making at all stages, even if constant human control cannot be maintained at each stage, for instance, due to the immense speed of the processes (such as target selection and attack). Ekelhof explains:

> The USA proposes to use the phrase "appropriate levels of human judgment" to reflect the premise of human machine interaction. The UK suggests the term 'intelligent partnership' and the UK NGO Article 36 coins the term "meaningful human control." There might be certain differences in terms of wording; for instance, the term "judgment" seems to be a narrower standard than "control." But there are also similarities, such as the purpose of the adjectives; the terms "meaningful" and "appropriate" both indicate a certain level under which human intervention is deemed

unacceptable. Therefore, some commentators use the terms (effective, appropriate, and meaningful) interchangeably.[76]

In addition to the lack of consensus about terminology, other points of disagreement (some explicit, others only implicit) hamper the international discussions about AWS. For instance, some state parties speak as though AWS is already in operation and has been for some time; others, by contrast, assume that deployment of AWS is still off in the future. The disparity would suggest that in their exchanges the parties are not referring to an identical set of technologies. In other words, for some a ban would prohibit only future weapon systems, while others recoil because a ban would also cover existing technologies, some of which are well entrenched and are considered valuable to some militaries.

Other ambiguities surround related key terms of the discussion, for instance the meaning assigned to 'autonomous functioning,' 'targeting' and so forth. If a ban on AWS is to become a matter of prescriptive law, these ambiguities would have to be ironed out, so that the parties could operationalise what exactly is or is not excluded from their arsenals.[77] Finally, even among those who embrace the language of 'meaningful human control,' disagreement surrounds the relevant point of comparison. For some, the term implies that a human being must be involved in every decision whereby a specific human being is targeted with lethal harm. This would entail that no automation is allowable whenever a 'trigger is pulled,' as in 'hand-to-hand' combat situations. For others, by contrast, 'meaningful human control' refers instead to the person commanding an operation; it is he or she who delegates the execution of the command to subordinates – whether human or machine – and must adequately oversee what they do. In this connection, it is important to recognise that even when a human being executes a command, a commander's 'meaningful' control is not equivalent to 'perfect' control; human subordinates cannot be counted on to invariably act in accordance with the intent of their commanders; by consequence, setting up perfect control as a standard would perhaps set an overly idealistic expectation for the acceptable use of AWS in battlefield settings.

What are the effects of AI use on the chain of command? Even in conventional warfare, not all decisions can be carried out by way of obedience. Two situations can render it inoperative/inapplicable: (1) time constraints do not allow for receiving instructions from above; (2) communications links are severed, not allowing for receipt of commands. Obedience, moreover, can fail in its basic intent, as commanders can misconstrue a situation and, based on error, order actions that run counter to what they would have otherwise intended. Here, a kind of higher obedience trumps the more immediate, lower obedience. To cite an example from the battle of Drøbak

Sound (when, in April 1940 – at this point, Norway was still a neutral power – the German cruiser Blücher led an invasion force up the Oslo Fjord and was sunk by the guns at Oscarsborg fortress under the command of Colonel Birger Eriksen[78]). When Eriksen gave his order to fire, he was asked by his subordinates whether he had received the command to do so; he famously responded, 'Either I will be decorated or I will be court martialed. Fire!'

Imagine if AI-driven autonomous technologies had been operative at Oscarsborg. The communications would have happened much more quickly. Eriksen would have known in advance that the Blücher was a German vessel (he did not know this with certainty until after he had fired). Sensors downstream would have shown this and relayed the information instantaneously to him. Had an intelligent machine been on command instead of Eriksen, it would have automatically implemented the order received from its superior (human) commander in Oslo. If the command had been, 'don't fire,' the AI machine, at that moment the possessor of delegated operational authority, would instantly have complied with this command, and Oscarsborg's torpedoes would not have been fired. The Blücher would have sailed to Oslo, resulting in the Norwegian King Haakon's capture and a significant hampering of the evacuation of key government personnel as well as gold reserves from the capital. Imagine if the command to hold fire had been issued by an officer who was a German sympathiser; would the AI machine then on duty at Oscarsborg have factored this in? The machine's compliance would have been automatic or could be withheld only under strict guidelines. What guidelines? How would it have dealt with the ambiguity of the Oscarsborg situation, Norway being officially neutral at the time? Would the AI machine have fired warning shots instead (as was then required by Norway's prewar rules of engagement, which Eriksen maintained had been satisfied as warning shots had earlier been issued to no avail by fortresses situated down the Oslo Fjord) allowing the ship to slip past the narrow straits? Human obedience is never automatic; it must pass through the ethical filter we call conscience. Context must be carefully weighed. This is the great benefit of human obedience; its combination of responsiveness and flexibility makes obedience a powerful virtue in moments of crisis. At present, no machine, however intelligent it may be, can replicate the subtlety of human obedience. (Perhaps such replication will never be possible.) This is the great advantage that human obedience has over automatic machine compliance. Where human beings come up short, however, in addition to the question of speed and knowledge, is in their ability to step outside the chain of command and/or act in pursuit of their private ends. This risk would not arise with a machine, unless it was spoofed, or penetrated by an adversary.

Virtue ethics as an appropriate language and framing of AI military ethics

We conclude our ethical analysis with a brief discussion of the ethical framework and language to be used when discussing the military ethics of employing AI systems, not least with a view to the ethical aspects of military education.

While ethics in general and military ethics in particular are often associated with the formulation of basic duties and rules (under the heading of deontology or duty ethics) or the calculation of desired effects (under the heading of utilitarianism or more broadly consequentialist ethics), the approach to ethics that informs much of the training and education in – and self-understanding of – military ethics is 'virtue ethics.' This is an ethical approach that sees good character (or 'excellence' or 'virtue') as inescapable and foundational to sound ethical behaviour.

Some would say that character and virtue are concepts inextricably linked to human beings and their behaviour, and that they do not apply to the actions of machines. One could therefore claim that increased use of computer-based machinery in general and AI-based, increasingly autonomous machines and systems in particular make virtue ethics less relevant. Instead, duties, rules and consequences become the key terms for the ethical use of AI, including in military settings.

Given the centrality of elements such as training, competence, proper feedback loops, safety, trust and proper delineation of responsibility, we believe that virtue ethics remains an unavoidable and essential framework for understanding the ethics of using AI-based machines and systems in a military setting. From the way in which systems are developed and programmed to the way in which they are deployed and used, whether for information gathering and other forms of enhancement of human decisions or as autonomous decision-makers, the virtues central to proper military action will have to inform the processes for them to align with core legal and ethical concerns. Rules and utilitarian calculations without a basis in virtue and character can derail into the most inhumane of acts and processes.

In Classical Greek thought, with ramifications for later Western as well as Islamic moral philosophy, four virtues have been delineated as the core ('cardinal') virtues, and they have had deep-seated importance for the development of military ethics, as well. It is proper to conclude this ethical analysis with a brief reflection on them.

- Moderation signifies the ability to hold back and behave in a balanced manner, not least when one is in possession of – or has at one's disposal tools of – great power and strength.

- Courage signifies the ability to confront challenges, but at the same time to do so rightly in the face of fear, chaos or temptation; this includes a readiness to signal one's moral concerns in the face of powerful processes or institutions, which may not be taking those moral concerns properly into account.
- Prudence, or practical wisdom, signifies the ability to use actively one's own experience and competence, and that of others, and to weigh many factors against each other, not least when faced with complex and fateful decisions. Prudence calls on us to acquire skills we may need to carry out tasks effectively; for AI applications, this will require enhanced digital competence on the part of all who make use of these applications. Prudence also signifies a precautionary approach, where dangers and opportunities are taken into account at every stage of a process.
- And finally, justice signifies the ability to take the moral concerns of all involved – today, we would say the human dignity and human rights of all involved, as well as key environmental concerns – into consideration when decisions are made, and to let the common good of one's community rather than the particular good of one or more parts of that community guide politics. Ultimately, and ethically speaking, given the grave ramifications of war, that community encompasses all human beings, meaning that we must take into consideration international and global concerns, and thus also the human dignity and human rights of those we would consider our adversaries. Most would today add – under the heading of environmental justice – that this includes proper care of the natural environment of which human beings are part.

We believe these virtues will be crucial in the development of both military ethics and political ethics as the power and potential of AI-based military systems are developed and deployed.

Other related virtues will similarly be needed; obedience, for instance, the virtue by which we respond to command authority promptly but never rotely or automatically. Also, we must take care not to confuse machine input and recommendations with the order of a superior and adjust our understanding of obedience accordingly.

According to classical thinking about virtues, not least inspired by the Greek philosopher Aristotle, but also emphasised in much of modern moral psychology, the proper operation of virtues in general – and of moral virtues in particular – requires training (indeed, education) and habituation. Virtues are not learnt the way that simple facts or purely automated routines or tasks are learnt. Virtues are learnt and acquired through a combination of understanding and repeated action over time. This also explains why the ethical challenges associated with using AI-based systems must be integrated into basic training and into our broader understanding of military ethics.

We hope that our overview of the ethical challenges associated with AI-based systems has helped the reader see the importance of this conclusion. All the while, we must realise that there is no way back to a non-digital military – at least not all the way back – and that AI-based systems will be an integral part of most future uses of military force. Being aware, then, of its possibilities and pitfalls is crucial.

Notes

1 Also called 'lethal autonomous weapons' (LAWS) or 'killer robots.'
2 A more fine-grained and practice-oriented overview of uses of AI for military systems is the following: Mission planning; intelligence gathering; observation/orientation; logistics optimisation; detection; decision-making; engagement of targets.
3 Merel A.C. Ekelhof, 'Lifting the Fog of Targeting: "Autonomous Weapons" and Human Control through the Lens of Military Targeting,' *Naval War College Review*, vol. 71, no. 3 (2018), pp. 61–94, see especially pp. 74–5, available at: https://digital-commons.usnwc.edu/nwc-review/vol71/iss3/6. Accessed on 3 January 2024. See more on the notion of 'autonomous technologies.'
4 Raja Parasuraman and Dietrich H. Manzey, 'Complacency and Bias in Human Use of Automation: An Attentional Integration,' *Human Factors: The Journal of the Human Factors and Ergonomics Society*, vol. 52, no. 3 (2010), pp. 381–410, at p. 391. https://doi.org/10.1177/0018720810376055.
5 *AJP-3.9, Allied Joint Doctrine for Joint Targeting* (NATO Standardization Office, 2016), available at:https://assets.publishing.service.gov.uk/government/uploads/system/uploads/attachment_data/file/628215/20160505-nato_targeting_ajp_3_9.pdf.Assessed 3 January 2024.
6 *AJP-3.9*, pp. 2-2–2-8. Some US-based writings employ 'kill chain' as an equivalent for 'targeting cycle/loop.' See, for instance, Christian Brose, *The Kill Chain: Defending America in the Future of High Tech Warfare* (New York: Hachette, 2020). However, in our opinion, 'targeting' is to be preferred because such a cycle may intentionally be directed at achieving both lethal and non-lethal effects, as is pointed out in the NATO report (see p. 7).
7 *AJP-3.9*, pp. 1-2–1-3.
8 Vincent Boulanin et al., *Artificial Intelligence, Strategic Stability and Nuclear Risk* (SIPRI: Stockholm, 2020). The use of AI to predict missile launches and other imminent threats that entail a defensive posture has most often been discussed in the context of early warning systems for nuclear deterrence. See also Marcus Weisgerber, 'The Increasingly Automated Hunt for Mobile Missile Launchers,' *Defense One*, 28 April 2016, available at: https://www.defenseone.com/technology/2016/04/increasingly-automated-hunt-mobile-missile-launchers/127864/. Accessed 3 January 2024.
9 *AJP-3.9*, p. 2-2.
10 Paul Scharre, *Army of None: Autonomous Weapons and the Future of War* (New York: W.W. Norton, 2018), p. 43. The US military employs the alternative terminology of OODA loop, i.e., observe, orient, decide and act: 'search for target(s),' 'detect target(s),' 'decide to engage the target(s)' and 'engage target (s).'
11 Ekelhof ('Lifting the Fog of Targeting') notes (drawing on *Joint Intelligence*, issued 22 October 2013 *by the Joint Chiefs of Staff*, available at: https://www.jcs.mil/Portals/36/Documents/Doctrine/pubs/jp2_0.pdfAssessed on 3 January 2024.) that 'Intelligence is not the same as *information* or *data*. When data are collected and processed into an intelligible form, the end result is *information*.

Information can be of utility to the commander, but when related to other information about the operational environment and considered in the light of past experience, it gives rise to a new understanding of the information, which may be termed *intelligence*' (p. 92, note 84).

12 Ekelhof, 'Lifting the Fog of Targeting,' p. 78.

13 Ekelhof, 'Lifting the Fog of Targeting,' pp. 79–82.

14 For a detailed discussion on how autonomous technologies – both decision-support and weapon systems – can support enhanced compliance with IHL, see *Humanitarian benefits of emerging technologies in the area of lethal autonomous weapon systems*, working paper submitted by the United States of America to the Group of Governmental Experts of the High Contracting Parties to the Convention on Prohibitions or Restrictions on the Use of Certain Conventional Weapons which may be deemed to be Excessively Injurious or to have Indiscriminate Effects, Geneva, 9–13 April 2018.

15 Ekelhof, 'Lifting the Fog of Targeting,' pp. 83–4.

16 Parasuraman and Manzey, 'Complacency and Bias in Human Use of Automation,' p. 381.

17 Parasuraman and Manzey, 'Complacency and Bias in Human Use of Automation,' p. 391.

18 Klaudia Klonowska, 'Article 36: Review of AI Decision-Support Systems and Other Emerging Technologies of Warfare,' T.M.C. Asser Institute for International & European Law, Asser Research Paper 2021-02, forthcoming in, Yearbook of International Humanitarian Law (YIHL), vol. 23 (2020), The Hague: T.M.C. Asser Press (2021), Available at SSRN: https://ssrn.com/abstract=3823881.

19 Patrick Lin, George Bekey, and Keith Abney, *Autonomous Military Robotics: Risk, Ethics, and Design* (Washington, DC: US Department of the Navy, Office of Naval Research, 2008), p. 31. See also, Michał Klincewicz, 'Autonomous Weapon Systems, the Frame Problem and Computer Security,' *Journal of Military Ethics*, vol. 14, no. 2 (2015), pp. 162–76. For a brief look at how AI engineers are currently attempting to address the frame problem, see Figure 7 on 'moral scene assessment' in Bruce A. Swett et al., 'Designing Robots for the Battlefield: State of the Art,' in Joachim von Braun, Margaret S. Archer, Gregory M. Reichberg and Marcelo Sánchez Sorondo (eds.), *Robotics, AI, and Humanity: Science, Ethics, and Policy* (Springer, 2021), p. 142, available at: https://library.oapen .org/bitstream/handle/20.500.12657/47279/9783030541736.pdf?sequence=1 #page=131.Accessed 3 January 2024.

20 Bertram Gross, *The Managing of Organizations* (1964), cited in 'Information Overload, Why It Matters and How to Combat It,' available at: https://www .interaction-design.org/literature/article/information-overload-why-it-matters -and-how-to-combat-it#:~:text=Gross%20defined%20information%20over-load%20as,system%20exceeds%20its%20processing%20capacity.&text= Managing%20information%20in%20daily%20life,problem%20which%2 0faces%20nearly%20everyone. Accessed 3 January 2024.

21 Klonowska, 'Article 36: Review of AI Decision-Support Systems and Other Emerging Technologies of Warfare,' p. 22.

22 Klonowska, 'Article 36: Review of AI Decision-Support Systems and Other Emerging Technologies of Warfare,' pp. 22–3.

23 Klonowska, 'Article 36: Review of AI Decision-Support Systems and Other Emerging Technologies of Warfare,' p. 23.

24 Thom Shanker and Matt Richtel, 'In New Military, Data Overload Can Be Deadly,' *The New York Times,* 16 January 2011, available at: https://www .nytimes.com/2011/01/17/technology/17brain.html. Accessed 3 January 2024.

25 Klonowska, 'Article 36: Review of AI Decision-Support Systems and Other Emerging Technologies of Warfare,' p. 24.
26 Parasuraman and Manzey, 'Complacency and Bias in Human Use of Automation,' p. 406.
27 Parasuraman and Manzey, 'Complacency and Bias in Human Use of Automation,' p. 382.
28 The two most noteworthy groups are the International Campaign for Robot Arms Control, begun in 2009, and the Campaign to Stop Killer Robots, a coalition launched in 2013; available at: https://www.icrac.net and https://www.stop-killerrobots.org.
29 Scharre, *Army of None*, p. 43.
30 Merel A.C. Ekelhof, 'Complications of a Common Language: Why It Is So Hard to Talk about Autonomous Weapons,' *Journal of Conflict and Security Law*, vol. 22, no. 2 (2017), pp. 311–31; especially p. 313, https://doi.org/10.1093/jcsl/krw029.
31 The summary of this debate draws on our more detailed account, in Gregory M. Reichberg and Henrik Syse, 'Applying AI on the Battlefeld: The Ethical Debates,' in Joachim von Braun, Margaret S. Archer, Gregory M. Reichberg, and Marcelo Sánchez Sorondo (eds.), *Robotics, AI, and Humanity* (Cham, Switzerland: Springer, 2021), pp. 147–59.
32 Valerie Morkevicius, 'Tin Men: Ethics, Cybernetics, and the Importance of Soul,' *Journal of Military Ethics*, vol. 13, no. 1 (2014), pp. 3–19.
33 Christof Heyns, 'Autonomous Weapon Systems: Human Rights and Ethical Issues,' Presentation to the Meeting of High Contracting Parties to the Convention on Certain Conventional Weapons, Geneva, 14 April 2016; see also by the same author: 'Autonomous Weapons in Armed Conflict and the Right to a Dignified Life: An African Perspective,' *South African Journal on Human Rights*, vol. 33, no. 1 (2017), pp. 46–71.
34 Heyns, 'Autonomous Weapon Systems.'
35 Vincent Boulanin, Neil Davison, Netta Goussac and Moa Peldán Carlsson, *Limits on Autonomy in Weapon Systems: Identifying Practical Elements of Human Control* (Stockholm: SIPRI and ICRC, 2020), p. 13, available at:https://www.sipri.org/publications/2020/other-publications/limits-autonomy-weapon-systems-identifying-practical-elements-human-control-0. Accessed 3 January 2024.
36 Whether or not to ban AWS has been an issue hotly debated at the UN in Geneva by the Expert Group representing State parties to the Convention on Certain Conventional Weapons. For a look at this debate in its initial phase, see Vincent Boulanin, *Mapping the Development of Autonomy in Weapon Systems: A Primer on Autonomy* (Stockholm: SIPRI working Paper, 2016). For an updated account, see Forrest E. Morgen, Benjamin Boudreaux, Andrew J. Lohn, Mark Ashby, Christian Curriden, Kelly Klima and Derek Grossman, *Military Applications of Artificial Intelligence: Ethical Concerns in an Uncertain* World (Santa Monica: RAND Corporation, 2020), pp. 41–4.
37 Nancy G. Leveson, *Engineering a Safer World: Systems Thinking applied to Safety* (Cambridge: MIT Press, 2016), p. 367.
38 Terms employed, e.g., by Thomas Aquinas in *Summa Theologiae* II-II, q. 48, a. 1.
39 Marianne Bellotti, 'Helping Humans and Computers Fight Together: Military Lessons from Civilian AI,' *War on the Rocks*, 15 March 2021, available at: https://warontherocks.com/2021/03/helping-humans-and-computers-fight-together-military-lessons-from-civilian-ai/. Accessed 3 January 2024.
40 Richard Danzig, *Technology Roulette: Managing Loss of Control as Militaries Pursue Technological Superiority* (Washington, DC: Center for New American Security, 2018), p. 4.

41 Danzig, *Technology Roulette*, p. 8.
42 Leveson, *Engineering a Safer World*, p. 48.
43 Leveson, *Engineering a Safer World*, p. 49.
44 Danzig, *Technology Roulette*, p. 7.
45 Boulanin et al., *Artificial Intelligence, Strategic Stability and Nuclear Risk*, provides a review of the literature and an examination of the relevant issues.
46 Boulanin et al., *Artificial Intelligence, Strategic Stability and Nuclear Risk*, pp. 102–13. These pages provide a good summary of potential positive and negative effects of introducing AI within nuclear weapon systems.
47 Ethem Alpaydin, *Machine Learning: The New AI* (Cambridge: MIT Press, 2016), p. 171.
48 Bellotti, 'Helping Humans and Computers Fight Together.'
49 Marcus Comiter, *Attacking Artificial Intelligence: AI's Security Vulnerability and What Policymakers Can Do About It* (Cambridge: Belfer Center, 2019), p. 4 points out that subversion of an AI-based system is not necessarily bad, for instance, '[A]s autocratic regimes turn to AI as a tool to monitor and control their populations, AI "attacks" may be used as a protective measure against government repression...' In what follows, we set this sort of case aside and limit ourselves to comments about adversarial AI that is directed for wrongful ends.
50 Comiter, *Attacking Artificial Intelligence*, p. 1.
51 Comiter, *Attacking Artificial Intelligence*, p. 10.
52 Comiter, *Attacking Artificial Intelligence*, p. 10.
53 Comiter, *Attacking Artificial Intelligence*, p. 28. There are other avenues to poisoning, as documented by Comiter, pp. 28–39, for instance, 'to simply replace a legitimate model with a poisoned one... [Through traditional cyberattack] [A] attackers can hack the systems holding these models, and then either alter the model file or replace it entirely with a poisoned model file' (p. 32).
54 Comiter, *Attacking Artificial Intelligence*, p. 37.
55 Comiter, *Attacking Artificial Intelligence*, p. 11.
56 Comiter, *Attacking Artificial Intelligence*, p. 37.
57 Comiter, *Attacking Artificial Intelligence*, p. 2.
58 Kelly M. Sayler, *Emerging Military Technologies: Background and Issues for Congress* (Congressional Research Service, 2020), available at: https://fas.org/sgp/crs/natsec/R46458.pdf. This report provides a summary discussion of these developments, with some reference to the attendant ethical concerns.
59 See Brose, *The Kill Chain*. This endorsement is forcefully articulated by Brose (who was McCain's former staff director on the Senate Armed Services Committee) against the perceived threat represented by China's advances in this technological domain.
60 See Sean McFate, *The New Rules of War: Victory in the Age of Durable Disorder*, with Preface by Gen. Stanley McChrystal (ret.) (New York: William Morrow, 2019); allusion to the (purportedly) mistaken belief that 'technology will save us' appears on the book's dust jacket.
61 Based on discussions with members of the US defence establishment.
62 This is the main thrust of Richard Danzig's report, *Technology Roulette*.
63 Leveson, *Engineering a Safer World*, p. 49.
64 Leveson, *Engineering a Safer World*, p. 43.
65 John K. Hawley, *Patriot Wars: Automation and the Patriot Air and Missile Defense System* (Washington, DC: Centre for New American Studies, 2017), p. 10.
66 Michael J. Barnes and A. William Evens III, 'Soldier–Robot Teams in Future Battlefields: An Overview,' in Michael Barnes and Florian Jentsch (eds.),

Human–Robot Interactions in Future Military Operations (Farnham: Ashgate, 2010), 9–25, at p. 10.

67 Barnes and Evens III, 'Soldier–Robot Teams in Future Battlefields: An Overview,' in Barnes and Jentsch (eds.), *Human–Robot Interactions*, p. 10.

68 Barnes and Evens III, 'Soldier–Robot Teams in Future Battlefields: An Overview,' in Barnes and Jentsch (eds.), *Human–Robot Interactions*, p. 11.

69 Leveson, *Engineering a Safer World*, p. 42.

70 Hawley, *Patriot Wars*, p. 12.

71 Hawley, *Patriot Wars*, p. 12.

72 Hawley, *Patriot Wars*, p. 12.

73 Bellotti, 'Helping Humans and Computers Fight Together.'

74 Boulanin et al., *Limits on Autonomy in Weapon Systems*, p. 22.

75 Ekelhof, 'Complications of a Common Language,' provides a useful overview of this debate.

76 *Killer Robots and the Concept of Meaningful Human Control* (Memorandum, *Human Rights Watch*, 2016), available at: https://www.hrw.org/news/2016/04/11/killer-robots-and-concept-meaningful-human-controlAccessed 3 January 2024); Richard Moyes, *Key elements of Meaningful Human Control* (Background paper, *Article 36*, 2016, available at: https://www.article36.org/wp-content/uploads/2016/04/MHC-2016-FINAL.pdfAccessed 3 January 2024); see also the different statements and interventions made by states during the CCW Meeting of Experts in 2014, 2015 and 2016.

77 For a treatment of these ambiguities and equivocations, see Ekelhof, 'Complications of a Common Language,' pp. 323–30.

78 See Geirr H. Haarr and Tor Jørgen Melien, *Sinking of the Blücher: The Battle of Drøbak Narrows April 1940* (Annapolis: Naval Institute Press, 2023).

6

SENTIENCE IN FUTURE WAR MACHINES

The challenge of building ethics, morality and virtue as a normative engine

Swaminathan Ramanathan

Introduction

War machines of the future will be autonomous entities distinctly non-human but with all the key human features. This is a challenge that we haven't faced till now as a human species. It is a challenge that cannot leverage any previous human experience to anticipate and model future paths, dangers and strategies. All human knowledge about the world is derived from carbon life forms. To put it another way: we have no knowledge of non-carbon lifelike forms.[1] A future war machine will be one of several forms of non-carbon entities. A non-carbon entity with intelligence that passes the Turing Test (Artificial General Intelligence [AGI]) is not far away.[2] Humanity has neither encountered nor engaged with a non-carbon intelligence until now.

The progression from AGI to sentience is not an 'if' but a 'when' function. In the artificial intelligence (AI) community, sentience is still an emerging definition. Sentience is the capability to experience feelings and sensations, and it underpins all human systems. Normative thinking regulates human systems. Ethics, morality and virtue[3] are three key normative instruments that define boundary conditions and the outer limits of outcomes as good, permissible, and desirable or bad, impermissible, or undesirable. In the context of AI, it is useful to think of non-carbon sentience as Artificial Superintelligence (ASI). War Machines of the Future will be one form of ASI.

The key issue for ASI is not sentience. Sentience will emerge. The key challenge is the normative architecture that will regulate that sentience. War machines of the future will not only need it the most but will need it as a deep learning system. This chapter engages with the challenges of building a normative architecture for a war machine of the future. In doing so, the

DOI: 10.4324/9781003421849-7

chapter makes the case for a system-of-systems approach of reinforcing and balancing feedback loops as 'anticipatory' control laws of ethics, virtue and morality for a real-time sentience.

The modern contemporary war

It is useful to think of a contemporary 'modern'[4] war as a system-of-systems[5] It is composed of two interoperable pieces of Command, Control, Computers, Communications and Information (C4I) and Intelligence, Surveillance and Reconnaissance (ISR). A system-of-systems war plays out as a series of hybrid battles with manned assets and unmanned assets operating together in a complementary manner to combat a multidimensional hostile environment. The ongoing Russia-Ukraine conflict is a case in point. There are three key systems that are currently operating in a theatre of conflict, sometimes in conjunction with each other and at other times in isolation.

The first is an emerging complex of unmanned systems of different maturity levels interoperating with each other. These range from electronic intelligence gathering loitering drones operating together with small infantry-level handheld drones to lethal drones armed with stand-off missiles, with pre-configured commodity-level *Kamikaze* drones often replacing missiles due to cost and logistical considerations. The second are legacy assets like heavy and light tanks, bridge layers, portable radars and other sensing and electronic sniffing equipment, mine-clearing vehicles, trench diggers, infantry vehicles, fighter aircraft and helicopters with air-to-air and air-to-ground roles that operate together through a ground-, sea-, air-, and satellite-based command and control infrastructure.

The third system is still in its infancy. It is composed of autonomous systems with executive functions. These include AI decision-making systems that provide everything from targeting decisions for air, sea and ground forces to big ticket 'how to and with what' solutions for countering and suppressing hostile offensive and defensive assets. Such systems have the in-built capability to execute their decisions independently, but are currently supervised by a human layer with the final execution conducted through a human chain of command.

The modern future war

The future *modern* war will also be a system-of-systems war with two significant pivots. The first is that the complex of AI-driven autonomous decision-making systems will have matured significantly; so much so that it will be at least the first among equals if not the first system within the system-of-systems. The implications of this would be far-reaching, with such systems not only deeply networked with each other for information and situational awareness, but also becoming the 'brains' of the other two systems mentioned above. In short, these systems will be operating at the level of knowledge and experience that is currently more or less an exclusive human domain. It is a fundamental pivot. In short, smart battlefield management of the future will be largely about AI-driven knowledge- and experience-based autonomous *decisions* rather than about information- and situational awareness-based decision-making pathways. The second will be the rapid integration of artificial kinaesthetic intelligence[6] with neural networks, deep learning and language models based on Natural Language Processing (NLP)-Bidirectional Encoder Representations from Transformers (BERT) and deep situational awareness[7] all encompassed in a variety of non-carbon forms. These would range from general-purpose non-carbon entities with all key human characteristics including form, actions and behaviour[8] to nano-bots with specific purposes like precision microsurgery delivery of medicines at a cellular level.

Non-carbon entities of the future

The common foundation of all non-carbon entities of the future with key human characteristics will be a source bio-code (as a combination of biological artifacts and brain organoids)[9] that is self-learning, autonomously rewritable at the entity level by the entity itself at each learning moment and can anticipate potential outcomes of multiple decisions within complex environments for achieving either a specific purpose or a larger goal. Self-learning is a process connected to knowledge production and experiential wisdom. Knowledge and experience are integral parts of metacognition: A main indicator of sentience.

A non-carbon sentient entity of the future will display and use five metacognitive competencies of Systems Thinking, Normative and Values Thinking, Anticipatory and Futures Thinking, Strategic Thinking and Self Awareness and Critical Thinking. Normative and Values Thinking is the most critical competency since it cross-sects and intersects all the other competencies and is necessary for designating decisions and outcomes as good, permissible and desirable or bad, impermissible and undesirable. Ethics, Virtue and Morality are the key normative instruments of regulation for decision-making processes and all the executive functions connected to actions emerging from a decision.[10]

For creating an all-purpose source bio-code, then, architecting a library of 'if' 'but' 'and' 'or' statements that are coupled with a library of 'when' scenarios and 'then-these' outcomes on a real-time basis is an incredibly complex logarithmic problem set. The problem set will, to begin with, require the design of a logarithmic scale of the widest range possible to accommodate all norms, including their contextual nuances, and degrees of application in different scenarios, over a wide range of values in a compact manner.

A case for a new architecture for non-carbon sentience: A four-dimensional model

It is within this future context that war machines will need to be designed, developed and understood. It is only when the design architecture is comprehensive and clear that the paths, dangers and strategies for such war machines of the future and non-carbon sentience in general can be evolved in terms of risks and mitigations.

A war is no doubt a complex system-of-systems with its boundary conditions constantly changing with new actors, specificities of theatres of conflict, and blurring of civilian and military infrastructure and assets. It is also within this system-of-systems that war machines will likely be at the forefront of developing a normative engine as a self-regulatory set of decision gates for autonomous decisions and actions. There is a rich history of

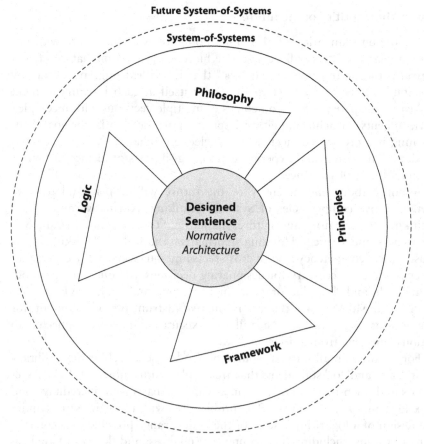

FIGURE 6.1 Generic four-dimensional model for a system-of-systems approach.

Source: Swaminathan Ramanathan, 2023

innovations being incubated first in the military domain, evolving into civilian use cases[11] that provides initial validation of this assumption.

This chapter initiates a case for the new logic system across four dimensions of Emergent Philosophy, Emergent Logic, Emergent Framework and Emergent Principles (see Figure 6.1). The chapter defines and uses Emergence in the way Swarm Intelligence[12] defines it as 'the collective behaviour of decentralized, self-organized systems, natural and artificial.'

Philosophy of status quo logic

Sentience itself is not the key challenge. The building blocks for generating a source bio-code are increasingly becoming available; the tipping point for viable integration will be reached sooner rather than later. The main challenge is to design the normative architecture for sentience. Sentience has been dissected using two main thought instruments: through *qualia* as a specific intersubjective experience that broadly characterises several Western schools of philosophy[13] and through *awareness* as a continuous inter-relational process of a universal transcendental cognition that infuses many Asian philosophies.

The need to understand human existence is at the core of several philosophical traditions of the West. Rene Descartes is a key figure – if not *the* key figure – of the Scientific Revolution. His Cartesian approach infused everything from logic, mathematics and geometry to philosophical treatises on reason, psychology and even physiology. His first principle of 'I think, therefore I am' in his 1637 treatise *Discourse on the Method* is arguably the cornerstone of all Western philosophy anchored to mind-body dualism. This dualism underpins Karl Popper's concept of empirical falsification that has informed the contemporary view of the relationship between knowledge, observation and objectivity. The concept of mind-body dualism of a binary separation between a thinker and his thinking and of an observer and his observation turns the 'thought' and the 'observed' into a problematic.

Logically, a watertight case can be made that the act of thinking and observation cannot take place without pre-existing knowledge with no *conscious act* and no *directed action* possible without knowledge. In machine learning terms, it is akin to the pre-existence of a library of images catalogued in a sufficiently indexable manner for any act of observation to fulfil itself. As illustrated in Figure 6.2, machine learning software cannot observe a tree and turn it into an observation of a tree and categorise it thus unless it has a sufficiently dense catalogue of images of trees and non-trees to cross-reference and in the process identify a tree as a tree.

The empirical falsification approach holds its logic as long as it operates within a dual axis of the observer and observation and/or the thinker and the thinking. However, the approach gets deeply problematised when there is a third axis added of the observed and thought. By adding the third axis,

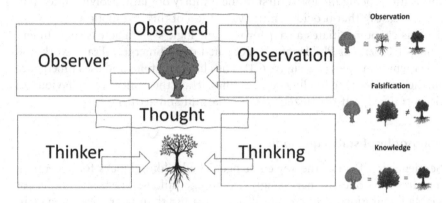

FIGURE 6.2   The empirical falsification approach.

Source: Swaminathan Ramanathan, 2023

the object and thought acquire an agency of their own. By extension, each object and thought, at least in principle, will have equivalence. Such equivalence will require a logarithmic scale to accommodate all normative values (and in numerical terms) associated with an object or thought in a compact manner. Without the third axis, the empirical falsification approach results in what is largely referred to as the optimisation logic; a logic that infuses every contemporary strategic and tactical decision-making process. It leads to what in a system-of-systems framework is called a path dependency.[14] In simple terms, path dependency can be paraphrased as: how you define the problem predefines the answer.

As seen in Figure 6.3, for any knowledge to be accepted it needs to pass the empirical falsification challenge: One of observer and observation or thinker and thinking or both. In simple terms, the object and the thought are fundamentally a null value (objectivity) unless proven otherwise. The empirical falsification approach embeds within itself a predesigned choice framework[15] that requires a set of choices to be prioritised. Such a prioritisation leads to an optimisation logic that informs objective function.[16] The nature of the objective function is such that it defines rationality by postulating a statement artifact of 'anything that is not.' By default, anything that is not objective is not a function and hence need not be designed for. Objective functions define solutions; and a set of grouped solutions (solutionism) feeds back into knowledge. It is a tautological logic that within a system-of-systems approach would be called a balancing feedback loop – one that favours status quo and maintenance of system equilibrium – as opposed to a reinforcing feedback loop – one that favours continuous learning and an agile incorporation of new knowledge and experiences into decision frameworks

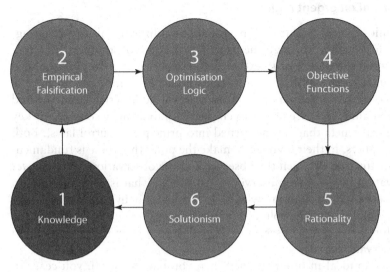

FIGURE 6.3   Problematising the empirical falsification approach: The path dependency paradox.

Source: Swaminathan Ramanathan, 2023

and processes for an exponential jump, and a new system equilibrium. The challenge of empirical falsification and optimisation logic becomes more acute in the context of machine learning (see Figure 6.4).

FIGURE 6.4   Machine learning issues with empirical falsification and optimisation logic.

Source: Swaminathan Ramanathan, 2023

Philosophy of emergent logic

Indian philosophical traditions have a long and different history of engaging with this 'duality' through the principle of *tattva* or 'thatness' as a tool to understand the nature of knowledge and experience. As a philosophical foundation for a new system of non-status quoist logic for non-carbon sentience, two contemporary Indian philosophers – Jiddu Krishnamurthy and U.G. Krishnamurthy – are useful to engage with to identify some of the key philosophical tenets that can be turned into principles (control laws). Both the philosophers, in their own ways, make the point that there is fundamentally no distinction between the observer and the observation, since the *act* of observation involves the observer transforming what is observed into an *act* of observation through knowledge. J. Krishnamurthy[17] in *Nature of the New Mind* points out the role of knowledge in observation that turns the relationship between the observer and the observed from observation to an act of observation.[18]

There is no ideal in observation. When you have an ideal, you cease to observe; you are then merely approximating the present to the idea, and therefore there is duality and conflict.

An idea and/or a thought is knowledge, and knowledge is something that is documented, commonly accepted, collectively shared, and experienced. In many ways, then, a tree cannot be categorised as a tree unless there is a prior knowledge that categorises it as a tree. This conundrum, in a nutshell, is the problematic of empirical falsification. U.G. Krishnamurthi puts it more directly and as a clear contrast to the concept of rationality, and by logical extension the foundations of existentialism derived from Cartesian 'I think therefore I am'[19] logic with an emphatic 'I don't think therefore I am not.'

> Man is just a memory. You understand things around you by the help of the knowledge that was put in you. I am not anti-rational, just unrational. You may infer a rational meaning in what I say or do, but it is your doing, not mine.[20]

Both Jiddu Krishnamurthi and U.G. Krishnamurthi provide the philosophical framework for an emergent logic where the observer, observation and the observed are no different from the thinker, thinking and thought. A philosophy of emergent logic provides for every observation being an *observer moment*[21] facilitating a 'here and now' emergent logic that exists for the moment and as long as the context requires it. A philosophy of emergent logic is ideal for incorporating chaos and a concomitant Bayesian approach to complex systems (See Figure 6.5).

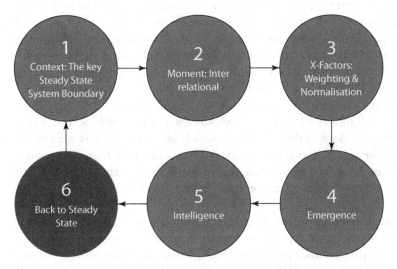

FIGURE 6.5 Beyond empirical falsification: Incorporating chaos and Bayesian approach.

Source: Swaminathan Ramanathan, 2023.

Logic stack for emergent non-carbon sentience

Non-carbon sentience is also non-anthropic sentience. Deep learning systems based on AGI and the still emerging ASI highlight a specific dimension of the anthropic principle.[22] The principle, simply put, considers the age of the universe and its physical constraints as necessary boundary conditions for the evolution of carbon-based life forms. The principle further deals with the notion that the Universe is fine-tuned in a manner to produce Anthropic Bias. There are three kinds of bias identified as part of the Anthropic Principle: Self-Sampling Assumption (SSA), Self-Indication Assumption (SIA) and Strong Self Sampling Assumption (SSSA). SSA states: 'All other things equal, an observer should reason as if they are randomly selected from the set of all *actually existent* observers (*past, present, and future*) in their reference class.' SIA states: 'All other things equal, an observer should reason as if they are randomly selected from the set of all possible observers.'[23]

SSSA revises the SSA statement, and in doing so also redefines the nature of assumption itself. SSSA states: 'All other things equal, an *observer moment* should reason as if they are randomly selected from the set of all *actually existent* observers (*past, present, and future*) in their reference class.'[24] The logic being that an observer who lives longer will have more opportunities to observe and hence will have more observer moments. As a logical corollary, the more observer moments are available to an observer, the better their chance of survival.

The key proposition that the author offers in this chapter and engages with is that a Survival Bias feeds the Anthropic Bias and in turn strengthens the Anthropic Principle. Anthropic Bias is inherently coded into any observation since all observations are indexically filtered by a self-locating belief of the observer. As a carbon lifeform, which humans are, it is reasonable to extrapolate that Survival Bias is not only an Anthropic Bias but is also a Carbon Bias. A Carbon Bias is designed to sustain and perform in a Carbon Ecosystem. Our world is a Carbon Ecosystem.

Carbon Bias is the starting point for Non-Carbon Sentience. The neural network algorithms powering the deep learning systems of non-carbon entities are embedding the Anthropic Principle (with Anthropic Bias) within multiple forms (including humanlike forms) that will have all the characteristics of carbon-based lifeforms.[25] As discussed earlier, such systems will be the first non-carbon entity with all the parameters of life. The Anthropic Principle will soon have to deal with a sentient non-carbon entity that will have its own versions (v1.0, v2.0… vN.0) of Anthropic Bias filtered and continuously refined by the self-locating belief of the observer (a sentient non-carbon entity observation).[26]

As a logical extension, it seems most likely that the core Anthropic Bias embedded as a source bio-code that is self-learning will refine itself over versions to evolve an autonomous Non-Anthropic Bias that is connected to a Non-Anthropic Principle. In short, it is reasonable to assume that there is a high likelihood that Non-Anthropic Bias will develop in multiple ways. Within this potential and likely scenario, there are questions of deep virtue, deep ethics, and deep morality.[27] The word 'deep' is used as an AI-Complete problem.[28] The key question for us to consider is whether we, as humans representing the Anthropic Principle, can have some sort of design control over the development of non-carbon sentience and by extension over the likely emergence of a Non-Anthropic Bias and a Non-Anthropic Principle. This chapter proposes a starting point; and it is to reduce, weigh and normalise the status quo bias (also a carbon bias) that infuses all our empirical knowledge. The Reversal Test and Double Reversal Test are general heuristics[29] to reduce and neutralise the status quo bias.

The Reversal Test states:

When a proposal to change a certain parameter is thought to have bad overall consequences, consider a change to the same parameter in the opposite direction. If this is also thought to have bad overall consequences, then the onus is on those who reach these conclusions to explain why our position cannot be improved through changes to this parameter.

If they are unable to do so, then we have reason to suspect that they suffer from status quo bias.[30]

The Double Reversal Test states:

Suppose it is thought that increasing a certain parameter and decreasing it would both have bad overall consequences. Consider a scenario in which a natural factor threatens to move the parameter in one direction and ask whether it would be good to counterbalance this change by an intervention to preserve the status quo. If so, consider a later time when the naturally occurring fact is about to vanish and ask whether it would be a good idea to intervene to reverse the first intervention. If not, then there is a strong prima facie case for thinking that it would be good to make the first intervention even in the absence of the natural countervailing factor.[31]

Building a logic stack that reduces and neutralises the Carbon Bias is a necessary step to design, build and prepare strategies for scenarios for a Non-Anthropic Bias that will be connected to an Emergent Non-Anthropic Principle. Building such a logic stack is also necessary to mathematically model an emergent non-carbon bias to prepare and design scenarios that could potentially emerge from such a bias. The framework to design for non-carbon sentience will need a deep normative engine with integrated ethics,

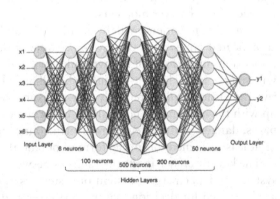

FIGURE 6.6   Potential of modelling emergent observer moments as a deep neural network using biocode.

Source: Swaminathan Ramanathan, 2023; General Neural Network Architecture sourced from Bahi and Batouche, 2018 (Meriem Bahi and Mohamed Batouche, 'Deep Learning for Ligand-Based Virtual Screening in Drug Discovery,' *2018 3rd International Conference on Pattern Analysis and Intelligent Systems (PAIS)* (2018), pp. 1–5, https://doi.org/10.1109/PAIS.2018 .8598488.)

morality and virtue decision tree connected to observer moments that uses context as the first and only filter of action. Using an emergent non-carbon bias also provides Bayesian non-linearity and complexity[32] for modelling observer moments (see Figure 6.6).

An emergent framework as a reinforcing feedback loop

The modern war of the future, as discussed earlier, will be a system-of-systems war. The modern future will also be a connected system-of-systems. A main feature of system-of-systems is a feedback loop. Every system has two kinds of feedback loops. The first is the balancing feedback loop that deals with new elements (X factors) as units of disequilibrium and hence forces such elements to normalise as per the system equilibrium, aka status quo bias. The second is the reinforcing feedback loops that deals with new elements (X factors) as units of exponential change and hence forces the system to achieve a new equilibrium state.

The key feature of the future system-of-systems will be its network centricity that will, by default make it emergent, real-time, and connected to moments rather than to events. In making moments the central unit of interaction and measurement, the emergent system-of-systems will be disproportionately loaded towards a reinforcing feedback loop, making the overall boundaries of the system-of-systems dynamic and ever-changing. This means that any system within the system-of-systems will have multiple equilibrium states. The author proposes the use of a modified Cynefin framework[33] (see Figure 6.7) for anticipating and designing for emergence connected to observer moments.

Non-carbon sentience will be dealing with and adding to the complex and complicated dimensions (Figure 6.7). Some of the key static aspects of two dimensions in the form of Known Unknown and Unknown Known moments will emerge from the Anthropic/Carbon Bias, while emergent aspects of the two dimensions will potentially emerge from a hybridisation that will come up with the carbon and non-carbon bias competing against each other. The paths, dangers and strategies from such connections need to be anticipated, or at least designed for, in the form of potential scenarios.

The key difference in the interactions between a contemporary system-of-systems and an emergent system-of-systems is at what point the interactions are leveraged for decision-making processes and actions. In the contemporary system-of-systems, the interactions are leveraged at the event level. In short, an unknown unknown event, for example, an extreme weather disaster, is leveraged once the event is over as a post-facto set of datapoints. In the emergent system-of-systems, the interactions are leveraged as observer moments, making the moment of interaction itself a real-time datapoint leading to a self-organisation of all other relevant datapoints, bringing in a deep predictive decision process and real-time action (see Figure 6.8).

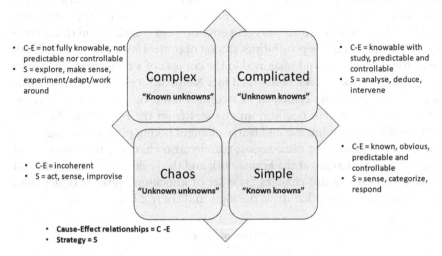

FIGURE 6.7 A Cynefin framework for an emergent system-of-systems approach.

Source: Adapted from David J. Snowden Mary E. Boone, 'A Leader's Framework for Decision Making,' Harvard Business Review, November 2007 Issue, available at: https://strategicleadership.com.au/wp-content/uploads/2017/06/A-Leader%E2%80%99s-Framework-for-Decision-Making-HBR-Nov-2007.pdf

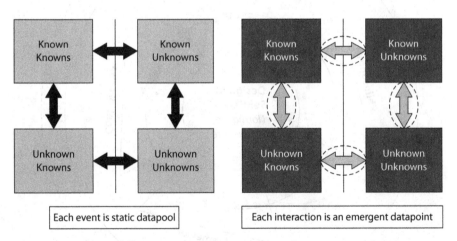

FIGURE 6.8 The difference between system-of-systems and emergent system-of-systems.

Source: Swaminathan Ramanathan, 2023

Conclusion: Future war machines are the smaller set in a larger set of non-carbon sentience

Emergent Principles are still in its infancy. It is useful to imagine Emergent Principles more as control laws[34] for a spectrum of possibilities within a non-carbon sentience than a set of principles as understood from an anthropic standpoint. With each interaction becoming a measurable observer moment, a principle transforms in an emergent environment from a singular state to a spectrum of exploitable possibilities as a set of control laws. Such control laws need to be visualised and imagined in the context of a non-carbon sentience connected to a logic stack networked to a Non-Anthropic/Non-Carbon Bias and a Non-Anthropic Principle, aka control laws. In an Emergent system-of-systems, the system equilibrium changes from the reinforcing feedback loop; with the most intense and heavy changes taking place in the logic and control law (principles) dimensions, thereby also changing the dimensional hierarchy. The changes at the Framework and the Philosophy dimension will at best be downstream changes (as a set of knock-on impacts) rather than fundamental changes that drive the logic and control laws (see Figure 6.9).

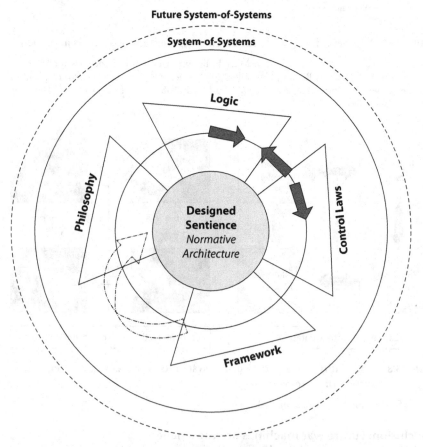

FIGURE 6.9   Specific four-dimensional model for emergent system-of-systems.

Source: Swaminathan Ramanathan, 2023

It is also likely that the Non-Anthropic Principle will eventually evolve into a set of non-carbon control laws that will set up limits and boundaries for non-carbon emergence. It is also equally possible, as per the emergent philosophy-logic-framework stack proposed in this chapter, that a non-carbon sentience will find it illogical to fight wars that are defined by carbon bias. Most likely, the modern war machines of the future will be fighting wars that are born out of a non-carbon logic and for maintaining a non-carbon status quo.

The key question for us is if we can proactively design to reasonably manage future scenarios and future strategies of non-carbon sentience by focusing on the normative aspects (a normative engine) of AGI and ASI. It is a control that we will not have in the future in the way we recognise control and its causal relationship to outcomes and impact today. Non-carbon sentience will be autonomous. So, the key question, rephrased, is this: Can we design its normative control laws in such a way today so that even in its most autonomous form our co-existence with them is reasonably safeguarded?

Notes

1 I have derived this conceptual distinction between carbon sentience as a key foundational anthropogenic principle and non-carbon sentience as a key foundational non-anthropogenic principle from my extensive reading of Niklas Boström's works. For this chapter, I have used N. Bostrom's, *Superintelligence: Paths Dangers and Strategies* (Oxford: Oxford University Press, 2014) and N. Bostrom's, *Anthropic Bias: Observation Selection Effects in Science and Philosophy* (1st ed.) (London: Routledge, 2002), https://doi.org/10.4324/9780203953464, and his paper with Toby Ord. N. Bostrom and T. Ord, 'The Reversal Test: Eliminating the Status Quo Bias in Applied Ethics,' *Ethics*, vol. 116 (July 2006), pp. 656–79.

2 Some claim that Google's LaMDA (Language Model for Dialogue Applications) AI system is 'sentient' and has a 'soul.' The transcripts of conversations between scientists and LaMDA reveal that the AI system provides answers to challenging topics about the nature of emotions, 'generating Aesop-style fables on the moment, and even describing its alleged fears.' Please see: https://cajundiscordian.medium.com/is-lamda-sentient-an-interview-ea64d916d917.

3 The author uses the terms ethics, morality and virtue within the larger framework of Normative Ethics. Ethics is seen as a practical means of determining a course of action directly connected to principles. Morality is seen as conformity to a standard as per the principles, with the principles defining morality for the moment for the real-time scenarios that it encounters. Virtue is the framework that provides the key set of filters for momentary outcomes in terms of good, permissible, desirable or bad, impermissible and undesirable for the scenarios encountered.

4 I use 'modern' as an all-purpose placeholder to refer to theatres of direct and indirect conflict where decision-making processes and executive battle functions are informatised through a system-of-system approach and operationalised through a Network Centric War strategy. Hence 'modern' is not an indicator of chronological historicity.

5 K. Boulding, 'General Systems Theory - The Skeleton of Science,' *Management Science*, vol. 2, no. 3 (1954), ABI/INFORM Global, pp. 197–208; M. Jackson and P. Keys, 'Towards a System of Systems Methodologies,' *Journal of Operational*

Research Society, vol. 35 (1984), pp. 473–86, https://doi.org/10.1057/jors.1984 .101; Bar-Yam Yaneer et al., 'The Characteristics and Emerging Behaviors of System-of-Systems,' in *NECSI: Complex Physical, Biological and Social Systems Project*, January 7, 2004, available at: http://static1.squarespace.com/static/5b6 8a4e4a2772c2a206180a1/t/5cd1af924e17b66bdf3125d9/1557245844724/ NECSISoS.pdf; D. Luzeaux, J. R. Ruault, and J. L. Wippler, *Complex Systems and Systems of Systems Engineering* (New Jersey: ISTE Ltd and John Wiley & Sons Inc, 2011).

6 Artificial Kinaesthetic Intelligence is used to describe and characterise autonomous robots of today like the various versions of the Boston Dynamics industrial and military robots and the emerging system of intelligent exoskeletons that are entering retail, commercial, industrial and military applications related to logistics and supply chain.

7 Deep situational awareness will be a further development of the current situational awareness, where informatised situational awareness is integrated with knowledge and experience stacks for decisions and actions in complex environments.

8 Such non-carbon forms are already in experimental and prototype stages, and some early forms are already in operation in select coffee shops in Japan, China and South Korea.

9 It is derived from synthetic biology where an artificial genetic code can be constituted, reconstituted and enhanced. Please see: https://phys.org/news/2019-08 -brain-mini-brains-grown-dish.html.

10 Please refer to endnote 3 for functional definitions of Ethics, Morality and Virtue.

11 M.R. Di Nucci, 'From Military to Early Civilian Applications,' in R. Haas, L. Mez and A. Ajanovic (eds.), *The Technological and Economic Future of Nuclear Power. Energiepolitik und Klimaschutz: Energy Policy and Climate Protection* (Wiesbaden: Springer VS, 2019), pp. 7–34, https://doi.org/10.1007/978-3-658 -25987-7_2. Please read the chapter as an illustrative meta example of the evolution of nuclear technology and its application in the civilian domain.

12 G. Beni and J. Wang, 'Swarm Intelligence in Cellular Robotic Systems,' in *Proceedings NATO Advanced Workshop on Robots and Biological Systems, Tuscany, Italy, June 26–30, 1989* (Berlin, Heidelberg: Springer, 1993), pp. 703– 12, https://doi.org/10.1007/978-3-642-58069-7_38. ISBN 978-3-642-63461-1.

13 U. Kriegel, *Current Controversies in Philosophy of Mind* (New York: Taylor & Francis, 2014), p. 201.

14 K. Zhu, K.L. Kraemer, V. Gurbaxani, and S.X. Xu, 'Migration to Open-Standard Interorganizational Systems: Network Effects, Switching Costs, and Path Dependency,' *MIS Quarterly*, vol. 30 (2006), pp. 515–39, https://doi.org/10 .2307/25148771. ISSN 0276-7783. JSTOR 25148771. S2CID 2182978.

15 The Pareto checklist is one example of a path dependency that creates a concept, methodology and tools to satisfy the fundamental design architecture of the approach.

16 Objective functions are used extensively in the software industry to evolve requirement documents and functional specifications.

17 J Krishnamurti has an extensive body of dialogues documented both in video and book forms by the Krishnamurti Foundation International (KFI). A key set of Krishnamurti dialogues that I referenced to for this book were with theoretical physicist David Bohm. I also referenced the book J. Krishnamurti, *The First and the Last Freedom* (1954, reprint, London: Rider, 1992), p. 352, Preface by David Skitt, Foreword by Aldous Huxley (new ed.). and Krishnamurti,

Choiceless Awareness: A Study Book of the Teachings of J. Krishnamurti (1992, reprint, California: Krishnamurti Foundation of America, 2001) for evolving my understanding of an Emergent Logic.

18 J. Krishnamurti, *The Nature of the New Mind* (Chennai: Krishnamurti Foundation India, 2001).

19 A key text to engage with and understand U.G. Krishnamurti's philosophy is the book *The Mind Is a Myth: Disquieting Conversations with the Man Called U.G*, ed. by Terry Newland (Goa: Dinesh Publications, 1988).

20 F. Noronha and J. S. R. L. Narayana Moorty, *No Way Out* (Bangalore: Akshaya Publications, 1991); U. G. Krishnamurti, *The Mystique of Enlightenment: Conversations with U.G. Krishnamurti*, Rodney Arms, Sunita Pant Bansal (1st ed.) (New Delhi: Smriti Books, 2005).

21 J. Krishnamurti in his dialogues has often referred to the unity of the observer and the observed with observation manifesting itself only through knowledge and memory. He used 'observer moment' as part of universal awareness within a *self* that is distinct from the awareness that comes from self-indexing using knowledge as a locational and navigational tool. Niklas Boström also uses observer moment as part of the key logic statement of Strong Self Selection Assumption (SSSA). The author sees multiple points of engagement with J Krishnamurti's philosophy and the fundamental design architecture of artificial intelligence algorithms for non-carbon sentience of the future.

22 Derived from Niklas Boström's book *The Anthropic Bias: Observation Selection Effects in Science and Philosophy* (2002, reprint, London: Routledge, 2010).

23 Bostrom, *Anthropic Bias*, pp. 73–86. The quotation is from p. 57.

24 Bostrom, *Anthropic Bias*, p. 162.

25 The robots designed by Boston Dynamics are good illustrative examples of multiple forms with embedded artificial intelligence.

26 Over a period of time a non-carbon entity's observations will carry more weight; in the process skewing the Anthropic bias by its observation moments.

27 Please check endnote 3 for the author's functional definition of Ethics, Virtue, and Morality; P. Boddington, 'Normative Modes: Codes and Standards,' in Markus D. Dubber, Frank Pasquale and Sunit Das (eds.), *The Oxford Handbook of Ethics of AI* (2020; online edn, Oxford Academic, 9 July 2020), pp. 125–40, https://doi.org/10.1093/oxfordhb/9780190067397.013.7, accessed on 1 May 2023.

28 AI-complete problems are hypothesised to include computer vision, natural language understanding, and dealing with unexpected circumstances while solving any real-world problem

29 A mental shortcut that is integral to the human experience and the evolution of the human species. N. Bostrom and T. Ord, 'The Reversal Test: Eliminating the Status Quo Bias in Applied Ethics,' *Ethics*, vol. 116 (July 2006), pp. 656–79.

30 Bostrom and Ord, 'The Reversal Test: Eliminating the Status Quo Bias in Applied Ethics,' p. 664.

31 Bostrom and Ord, 'The Reversal Test: Eliminating the Status Quo Bias in Applied Ethics,' p. 673.

32 N. Bostrom and M. Cirkovic, 'The Doomsday Argument and the Self-Indication Assumption: Reply to Olum,' *Philosophical Quarterly*, vol. 53, no. 210 (2003), pp. 83–91, https://doi.org/10.1111/1467-9213.00298; K. Olum, 'The Doomsday Argument and the Number of Possible Observers,' *Philosophical Quarterly*, vol. 52, no. 207 (2002), pp. 164–84, arXiv:gr-qc/0009081, https://doi.org/10.1111/1467-9213.00260. S2CID 14707647.

33 Please see: A Leader's Framework for Decision Making David J. Snowden Mary E. Boone HBR November 2007 Issue. https://hbr.org/2007/11/a-leaders-framework-for-decision-making
34 Niklas Boström's book *Superintelligence: Paths, Dangers, and Strategies* explains the control law conundrum beautifully through the sparrows and owl story.

7

BIASED ARTIFICIAL INTELLIGENCE SYSTEMS AND THEIR IMPLICATIONS IN WAR SCENARIOS

Kelly Fisher

Introduction

In recent years, there has been a growing debate about ethical uses of artificial intelligence (AI) in warfare. This debate and growth in interest have led to a number of reports and literature on the topic. Interestingly, few of these debates and topics have focused on the issue of biased AI, which is surprising as the topic of biased AI has gained much prominence when it comes to the use of AI in civilian settings. As our society has increasingly begun to implement the use of AI in everyday life, it has become also clear that these systems can inherit and further amplify the type of biases we as humans have, including gendered and racial biases. This chapter aims to bring together these two conversations of ethical use of AI in military settings and the issue of biased AI in civilian settings. I draw upon examples of biased AI from civilian settings to highlight what the implications of biased AI could mean in military settings. This chapter draws upon examples from the USA and Europe, but aims to highlight what biased AI means in the larger international community as well.

In the recent Netflix documentary, *Coded Bias*, Joy Buolamwini, a researcher at Massachusetts Institute of Technology (MIT), provides an example of the way that AI can be biased. Buolawmwini, a black woman, attempts to use a facial recognition program which is unable to identify her face, that is, until she puts a white mask up to her face. In an analysis of facial recognition's accuracy, Buolamwini would find that facial recognition programs only misidentified white men's faces 1% of the time yet a black woman's face is misidentified 35% of the time.[1] Multiple investigations and reports in recent years have also demonstrated that the types of racial

DOI: 10.4324/9781003421849-8

and gender biases that exist in our society can also become embedded in the supposedly neutral AI programs which increasingly surround us in our day-to-day lives.[2] Despite the evidence of the way in which these biased AI programs can have real and serious consequences for women and minority groups in society, there has been little discussion about the consequences of these biases in warfare settings.

In recent years, a growing number of researchers, diplomats and civil society leaders have been debating what the main ethical challenges and dilemmas that the use of AI in warfare poses are and how these systems should be regulated.[3] These debates have led to important discussions around whether or not LAWS should be banned, what steps should be taken to ensure meaningful human control, and other ethical concerns. Interestingly, despite the amount of attention that biased AI has received in non-warfare settings, concerns about biased AI in warfare have rarely been a point of discussion when it comes to ethical concerns about the use of AI in warfare.

In 2021, NATO released its AI strategy plan. The strategy document outlines six guiding principles for the responsible development of AI-systems, including what steps should be taken to mitigate bias in AI.[4] However, this strategy does little to outline how biased AI develops in the first place or what could be the consequences of biased AI in warfare. To address this gap, this article aims to unpack what might be the consequences of using biased AI-systems in warfare settings. By drawing upon examples of biased AI in non-warfare settings and the lessons learnt from them, this chapter hopes to contribute further insights into the ethical and moral dilemmas that AI-enabled systems in possible conflict scenarios.

This chapter focuses on gender and racial bias, although it should be acknowledged that bias in AI is an issue for a number of minorities and under-represented groups in society, including those who are transgender.[5] Additionally, this essay focuses on examples within the USA, both due to its leading role in AI development, but also because similar reflections and investigations into biased AI are lacking in other parts of the world, such as China as a point of comparison.[6] Therefore, this section focuses on examples of bias against racial and ethnic minorities in the US context. However, the lessons from these examples are relevant in the way in which those who are designing and creating these programs create bias from their social position, and this could just as easily happen elsewhere in the world. In the following section, I begin by providing some of the examples in which bias in AI has negatively and disproportionately impacted women and racial minorities in society. Following this, I outline what the different ways that bias becomes embedded into AI. I discuss what these findings show about the trustworthiness of AI, and the uncertainties of how AI use in military and especially warfare might disproportionately and unintentionally impact different individuals in society, and what this means in the larger international debates about AI's use in warfare.

Examples of bias

An initial promise and hope of the use of AI was that it would not make the same mistakes humans make, including the types of biases that we as humans may or may not have. As our society has begun to use AI programs to help us in the work that we do, it has become clear that this promise is in reality false. Increasingly, there has been an awareness of biased AI as a problem facing our society, a problem which has received more attention recently. There have been some prominent examples in the last few years of the different types of bias that AI programs have. Whether this was Amazon's resume filter program, which, due to the algorithms it relied on, disqualified women as eligible for jobs,[7] voice recognition programs in cars which struggle to identify women's commands, or even an algorithm used in the US court system which falsely predicted that black citizens were twice as likely to be at risk of recommitting crimes than white ones.[8] What these examples demonstrate is that bias is a problem in AI and one that has serious gendered and racial implications for contributing to inequalities in society.

Lack of diversity and different entry points for bias into AI

To mitigate bias in AI, it is important to understand the different ways that bias may become embedded into AI. While there are a number of contributing factors, one key factor is that the field of AI development has a diversity problem. Recent numbers show that 80% of all AI professors are men, with women making up only 15% of AI researchers at Facebook, and 10% at Google.[9] When it comes to black workers, this number drops to 4% at Facebook and Microsoft, and 2.5% at Google.[10] To what extent this is a main cause of the problem when it comes to AI and bias is something that is unclear and still in need of investigation, but there are those who believe that this is one of the main contributing factors to biased AI.[11] While there are examples of how this lack of diversity is contributing to biased AI, there are also other factors that play a role. These will now be explored at two different levels, including: 1) bias from the developers; and 2) bias in the data.

Bias from the developers

As Criado Perez outlines in her book with multiple examples, *Invisible Women: Data Bias in a World Designed for Men,* men in a variety of fields create their products, often unintentionally with men as the default, and in the process exclude women's different needs or perspectives. This can have minor outcomes, such as an iPhone that is designed to be held by the average male hand size rather than a female's,[12] but can also have more serious consequences such as a home security system not having any settings in place for domestic abuse, despite this being one of the most frequent threats to the security of individuals in a household.[13] Another example of bias coming from developers is that some of the major voice assistant programs, Alexa from Amazon, Cortana from Microsoft, and Siri from Apple, all have feminine names, and by default come with female voices. In a report by the UN about AI and Gender Bias, they found that not only are these female voice assistant programs a reflection of the bias of these developers, as these voice assistants play the role of being subservient to those who use them, but that these programs are reinforcing gender biases of women's role in society.[14]

It is important to highlight that these are primarily unintentional biases. Men don't create these voice assistant programs to purposely perpetuate the idea that women's role in society is that of an assistant, but rather it is a reflection of the unconscious bias they have about what type of voice or program they think would be best to give commands to.[15] These more prominent examples show the way in which unintentional and unconscious bias has very real and consequential impacts on women in society.

Bias in the data

There are several ways that AI can become biased due to data. The first is that the data can be trained on datasets that are either over-representative or under-representative of a group.. Many ML schemes today are trained on free and large datasets on the internet such as Wikipedia, Google Images, Google News or ImageNet, and these datasets are not equally representative of groups in society.[16] One example of how these biased datasets then affected AI programs includes Amazon's facial recognition application, which only mistook a white man's face 1% of the time, but a black woman's 35%.[17] Furthermore, voice recognition programs, which often perform better on men's voices than women's, are often trained on a large data set called the Linguistic Data Consortium, which is a dataset where two-thirds of the voices making up the data are men's [18]

The purpose of data for AI is to try and make predictions for the future based on previous trends and information from the data. As a result of this, if the types of data that are included reflect historical and social inequities, the AI programs can become biased.[19] An example of this includes Amazon's resume hiring scheme. This program was looking at which candidates were previously hired at Amazon, which was predominantly men, and then decided that men were the ideal candidates, and removed women's resumes from the list of possible candidates. Another example of this includes an algorithm in the USA that was calculating the risk score of a convicted criminal's likelihood to recommit a crime, and being used by court systems across that country. In an investigation of the program and its algorithms, it was found that white defendants were labelled as low risk more often than black ones, and that the system was wrongly labelling black defendants as future criminals at twice the rate of white defendants.[20] This bias arose from the attributes selected as high-risk factors for the algorithms, which were also attributes that black Americans are over-represented by, due to historical and social inequities in the USA.[21] However, it is important to acknowledge that the problem is not only that data can contain biases, but that even the types of data that are collected in our society can also be biased.

One example that shows bias in terms of what data is collected is that of maternal death rates during pregnancy. Until recently in the USA, there was very little data collected on maternal death rates in the country.[22] However, due to different investigations into the topic, it became clear that this was not only a large issue in the USA, but one that was disproportionately impacting women of colour who were dying at three to four times the rate that white women were.[23] As our society continues to integrate AI into everyday functions, such data will only become increasingly important. In light of this, it is necessary to recognise that decisions around what types of data are collected by AI-based systems are also influenced by bias.

Amplification of bias with AI

AI not only can reinforce bias and social inequalities, but it can also further amplify them. Among those working to address biased AI, there is a large consensus that amplification is one of the main challenges with AI.[24] One example of this includes the way in which the Los Angeles Police Department is using AI to help with policing.[25] Due to the way in which AI programs are drawing upon historical data, which we already know can contain bias, it then predicts which areas are at risk of crime, and the police then more heavily surveil and over-police the area, which further amplifies racial bias.[26] Due to the large-scale nature that AI programs can operate at, it creates a risk of these biases becoming multiplied, and creating a feedback loop which the bias then builds itself upon. Currently, Google Translate often defaults to masculine pronouns, resulting in more masculine pronouns on the internet and further amplifying this bias.[27] While humans certainly contain bias and this creates social inequalities, the issue of amplification highlights the way in which AI creates new dilemmas when it comes to the way in which bias negatively impacts individuals within society.

Implications of biased AI in combat

This section of the chapter takes into consideration the way in which biased AI presents different ethical challenges in both physical and digital battlespaces. It is important to highlight that despite a growth of reports and literature which outline the ethical challenges of using AI in warfare, and evidence of biased AI in civilian settings, there is little conversation considering these two themes, something which this chapter hopes to address. Some of the areas where bias and AI have come further into consideration have been drone warfare and also cybersecurity.[28] Building off this existing literature, as well as the information presented so far about biased AI in a civilian setting, I outline what some of the possible implications of biased AI are when it comes to AI and warfare.

a Threat assessments

The first dilemma when it comes to bias and AI usage in warfare is how threat assessments are determined and then acted on. The process of creating algorithmic threat assessments includes the selection of attributes which are deemed to qualify individuals as a risk or threat, and we can look at how threat assessments used in drone warfare already show warning signs for how these biases could further manifest with AI. As Sarah Shoker outlines in her book *Military-Age Males in Counterinsurgency and Drone Warfare*, in the context of the Middle East and drone warfare, biases play a large role in influencing who is or is not considered a threat on the battlefield. The term 'Military Age Male' as a category for who is targeted she argues reveals the way in which biases based around gender, age, race and religion have deadly consequences.[29] Other researchers have also pointed out the role of biases around gender, ethnicity and religion in shaping what bodies are seen as 'killable' in drone strikes.[30] In 2021, documents were released from the Trump administration's policy on drone warfare highlighting that there were different requirements for men and women when assessing whether or not they were seen as enough of a threat for a military strike.[31] What these authors and reports show is that biases shape who is seen as a target, regardless of whether or not they actually are a threat.

This becomes of further concern when we recognise the way in which algorithms and big data are currently being used to assess who is a threat in the battlespace, and the flaws which exist within these programs. Currently there are different forms of data that US military officials are using to determine with algorithms whether or not individuals are seen as a risk, and then targeted in a drone strike despite not knowing if they truly are a threat.[32] These algorithms use a number of attributes to identify whether or not an individual should be considered an 'extremist,' including how often they went to an airport or what parts of the country they visited.[33] Yet, in an investigation of these algorithms, it was found that the person who was deemed the highest threat by the algorithm was actually Al-Jazeera's Islamabad Bureau Chief, showing that despite having large amounts of data, innocent individuals can still be marked as a threat by these algorithms.[34] There are numerous entry points for bias into algorithms created for threat assessments, and it is important to recognise the concerns this poses, and the deadly consequences that have already occurred as a result.[35]

b Trustworthiness of facial recognition

Another concern revolves around the trustworthiness of facial and object recognition programs when it comes to targeting systems for the battlefield. As military development moves further into developing autonomous vehicles and weapon systems, they will be using a variety of AI programs to manoeuvre and engage in the combat space. When it comes to targeting, and who to target, key to these systems will be facial and object recognition, which, as we have seen due to different forms of bias, are inaccurate when it comes to identifying non-white faces. In the UK, police are currently using facial recognition programs to try and identify criminals in a crowd, which, after an investigation, found that the program only accurately identified faces 5% of the time.[36] These trends in how bias shapes the outcome of how accurate AI is raise serious questions about the trustworthiness of facial recognition programs usage in combat settings. Furthermore, it has been found that Project Maven, which was trained for object identification in the Middle-East, struggled in other contexts and settings. One of the challenges resulted from cultural differences between men and women's clothing, which then negatively impacted Project Maven's accuracy.[37] Both of these examples show the way in which bias can become embedded in AI when the training data is under-represented, but the Project Maven example highlights another dilemma with AI. This problem revolves around developing AI that can recognise cultural differences and nuances, something which is understood as a large challenge for AI programming.[38] These challenges pose serious questions about the uncertainty of using AI, and also when one should consider a program to be seen as accurate enough to be used for target recognition and enemy identification.

Enemy identification when it comes to other targets such as vehicles can also be susceptible to the challenge of trustworthiness. One dilemma when it comes to the trustworthiness of AI is we don't always understand how programs reach their decisions, even if they seem to do it correctly most of the time. When it comes to enemy identification, this is problematic because we may think that the program is capable of identifying enemies, when in reality it is actually making the decision off something such as a certain colour on a jet wing, and this might then result in unintentional friendly fire.[39] This is a concern that has come up in UN focus groups about the risks of militarised AI, where due to 'automation bias,' humans overly trust machines' decisions, even if we don't know how it is reaching its decision.[40]

c Biased human-machine system designs

The third area in which biased AI may be an issue in military settings has to do with the way in which these systems are designed. While AI systems will be able to operate autonomously, there will still be a great deal of human-machine interaction ongoing in the future battlefields. Possible examples of this human-machine interaction include the use of headsets in combat settings which may make recommendations for the soldiers by analysing real-time data[41] or an interface in a jet where a pilot is both flying their own jet while also controlling and overseeing a fleet of autonomous jets.[42] The concern here is how might these systems be biased due to the way they have been designed, often unintentionally, for men. As different researchers have shown, societal biases have resulted in a number of products and technologies being tested on and designed for men, and as a result of this, these are less effective for women.[43] Different examples include how voice recognition programs work better on men's voices than women's voices,[44] airbags in cars which are more effective at protecting men in a car crash,[45] to even iPhone's whose size have been based on the average man's hand size, often being larger than what is comfortable for a woman to use.[46]

Drawing on the lessons learnt from these examples, we can consider how the military technology examples provided above may also be biased in their design and thus more difficult for women to use. Perhaps the helmet is designed for the average male head size, making it harder for the female soldier to use it properly. Or in the situation with the fighter jet, it could be that the voice recognition program is less effective at understanding the female pilot's voice. Examples in the working world where AI-enabled programs have been used by women which may contain some of these system design biases show how these biased systems create real and concrete consequences for the women as they try to go about doing their job. This provides yet another area in which we can see how gender and other forms of bias could create negative consequences for women and other minority groups in military conflicts.

Conclusion

Much of the international debates on regulating military uses of AI have focused on whether or not to ban LAWS.[47] This is an important debate, and presents a number of ethical, moral and legal challenges to consider. However, there are a number of different military uses of AI other than just LAWS, and, as has been presented in this chapter, these other uses of AI can contain different forms of biases. As militaries increasingly implement and adopt different uses of AI-enabled systems in warfare settings, it is important that they recognise the pervasive way in which biases can become embedded in these systems and take steps to try and mitigate and reduce these biases.

Gender is often seen as something equating to women,[48] and therefore, the topic of gender biased AI has often been equated to ways in which AI-enabled programs may differently and negatively impact women. However, as I have argued in this article, gender biased ideas about men, which may also intersect with other aspects of identity including religion and ethnicity, can result in lower thresholds for whether or not they are seen as military targets. AI-enabled systems which focus on threat assessments and also provide recommendations for likely targets are high-risk, as they would be susceptible to bias from different angles, including whether that is from previous data used in an algorithm or through different attributes selected for an algorithm seen as increasing particular individuals' threat level. If an algorithm draws on previous data, it might result in certain groups of men being seen as more likely military threats, regardless of that being the case.[49]

The examples here have mainly been drawn from the USA and Europe. However, we can certainly see how dynamics are playing out in the USA. Biases against women and different minority groups are also present in other parts of the world. For example, discriminatory and biased ideas about certain ethnic groups which then result in them being seen as a threat is a dynamic that exists in many parts of the world, whether that is China, India, South Africa and many other countries. Similar to how the USA's examples of biased AI wrongfully convicted black Americans, how might the use of these programs in these other parts of the world contribute to and reinforce social inequalities? Additionally, as the IT sector is a male-dominated profession globally,[50] programs which rely on human-machine interaction may also similarly be designed for men rather than women, as we have seen in the USA and other parts of Europe.

The issue of biases in AI is a large and serious one, and one that has many negative implications if a biased system were ever to be used in a military setting. NATO's recent AI Strategy takes an important first step by stating that efforts must be made to mitigate bias. However, to mitigate this bias, it is important to understand how it can emerge in AI in the first place, and different steps that can be taken to reduce this bias. Research on the military implications of gendered (and other forms) of bias is still in its early stages when it comes to AI-based systems. While it is a topic gaining increased attention, it is also a topic which merits further research and a more detailed investigation. Such areas include bringing to attention the issue of biased AI from many different parts of the world, to make sure that different forms and processes of social inequalities from around the world are considered, and not just those in the US and Europe.

Additionally, military uses of AI can learn from the number of recommendations from AI experts on how to mitigate bias in non-military uses of AI. These steps include everything from auditing the data that is used in an algorithm, working with a diverse team of experts including those with subject expertise in gender and racial bias, along with a number of other recommendations. Bias in AI can be addressed, and it is necessary that it is, to ensure that different groups within a society do not have to pay the serious price and consequences that will occur if biased AI systems are used in warfare.

Notes

1 Larry Hardesty, 'Study Finds Gender and Skin-Type Bias in Commercial Artificial-Intelligence Systems,' *MIT News Office,* 11 February 2018, available at: https://news.mit.edu/2018/study-finds-gender-skin-type-bias-artificial-intelligence-systems-0212#:~:text=artificial-intelligence%20systems-,Study%20finds%20gender%20and%20skin-type%20bias%20in%20commercial%20artificial,percent%20for%20dark-skinned%20women, accessed on 24 April 2023

2 James Manyika, Jake Silberg, and Brittany Presten, 'What Do We Do About the Biases in AI?' *Harvard Business Review,* 25 October 2019, available at: https://hbr.org/2019/10/what-do-we-do-about-the-biases-in-ai, accessed on 24 April 2023

3 Klaudia Klonowska, 'Article 36: Review of AI Decision-Support Systems and Other Emerging Technologies of Warfare,' SSRN Scholarly Paper (Rochester, NY: Social Science Research Network, 17 March 2021), p. 21, available at: https://papers.ssrn.com/abstract=3823881, accessed on 24 April 2023

4 Zoe Stanley-Lockman and Edward Hunter Christie, 'NATO Review - An Artificial Intelligence Strategy for NATO,' *NATO Review,* 25 October 2021, available at: https://www.nato.int/docu/review/articles/2021/10/25/an-artificial-intelligence-strategy-for-nato/index.html, accessed on 24 April 2023

5 Steven Melendez, 'Uber Driver Troubles Raise Concerns about Transgender Face Recognition,' Fast Company, 9 August 2018, available at: https://www.fastcompany.com/90216258/uber-face-recognition-tool-has-locked-out-some-transgender-drivers, accessed on 24 April 2023

6 Jessie Dalman, 'Artificial Intelligence, Algorithmic Bias, and Ethics,' *Georgetown University Initiative for U.S.-China Dialogue on Global Issues* (blog), 24 May 2018, available at: https://uschinadialogue.georgetown.edu/responses/artificial-intelligence-algorithmic-bias-and-ethics, accessed on 24 April 2023

7 Jeffrey Dastin, 'Amazon Scraps Secret AI Recruiting Tool That Showed Bias Against Women,' *Reuters,* 11 October 2018, available at: https://www.reuters.com/article/us-amazon-com-jobs-automation-insight-idUSKCN1MK08G, accessed on 24 April 2023

8 Julia Angwin, Jeff Larson, Surya Mattu, and Lauren Kirchner, 'Machine Bias,' *ProPublica,* 23 May 2016, available at: https://www.propublica.org/article/machine-bias-risk-assessments-in-criminal-sentencing, accessed on 24 April 2023

9 Sarah M. West, Meredith Whitaker, and Kate Crawford, *Discriminating Systems: Gender, Race and Power in AI* (AI Now Institute, 2019), p. 3, available

at: https://ainowinstitute.org/discriminatingsystems.pdf, accessed on 24 April 2023

10 West et al., *Discriminating Systems*, p. 3.

11 West et al., *Discriminating Systems*, p. 6.

12 Caroline Criado Perez, *Invisible Women: Data Bias in a World Designed for Men* (New York: Vintage Books, 2019), p. 159.

13 Katharine Millar, James Shires, and Tatiana Tropina, *Gender Approaches to Cyber Security: Design, Defence, and Response* (Geneva: United Nations Institute for Disarmament Research, 2021), p. 17, https://doi.org/10.37559/GEN /21/01, accessed on 24 April 2023

14 Mark West, Rebecca Kraut, and Han Ei Chew, *I'd Blush if I Could: Closing Gender Divides in Digital Skills Through Education* (UNESCO and EQUALS Skills Coalition, 2019), p. 104, available at: https://unesdoc.unesco.org/ark: /48223/pf0000367416.page=1, accessed on 24 April 2023

15 West et al., *I'd Blush if I Could*, pp. 97–100.

16 Criado Perez, *Invisible Women*, pp. 164–5.

17 Hardesty, 'Study Finds Gender and Skin-Type Bias in Commercial Artificial-Intelligence Systems.'

18 Criado Perez, *Invisible Women*, p. 164.

19 Manyika et al., 'What Do We Do About the Biases in AI?'

20 Angwin et al., 'Machine Bias.'

21 Catherine D'Ignazio and Lauren F. Klein, *Data Feminism* (Cambridge: MIT Press, 2020), p. 55.

22 D'Ignazio and Klein, *Data Feminism*, p. 23.

23 Linda Villarosa, 'Why America's Black Mothers and Babies Are in a Life-or-Death Crisis,' *The New York Times*, 11 April 2018, available at: https://www .nytimes.com/2018/04/11/magazine/black-mothers-babies-death-maternal-mortality.html, accessed on 24 April 2023

24 D'Ignazio and Klein, *Data Feminism*, Criado Perez, *Invisible Women*; Cathy O'Neil, *Weapons of Math Destruction: How Big Data Increases Inequality and Threatens Democracy* (New York: Crown Publishing Group, 2016).

25 Michael Michael Steinberger, 'Does Palantir See Too Much?,' *The New York Times*, 21 October 2020, available at: https://www.nytimes.com/interactive /2020/10/21/magazine/palantir-alex-karp.html, accessed on 24 April 2023

26 D'Ignazio and Klein, *Data Feminism*.

27 James Zou and Londa Schiebinger, 'AI Can Be Sexist and Racist — It's Time to Make It Fair,' *Nature* 559 (2018), pp. 324–6, https://doi.org/10.1038/d41586 -018-05707-8.

28 Lauren Wilcox, 'Embodying Algorithmic War: Gender, Race, and the Posthuman in Drone Warfare,' *Security Dialogue* 48, no. 1 (2017), pp. 11-28, https://doi .org/10.1177/0967010616657947.; Millar et al., 'Gender Approaches to Cyber Security: Design, Defence, and Response.'

29 Sarah Shoker, *Military-Age Males in Counterinsurgency and Drone Warfare* (London: Palgrave Macmillan, 2021).

30 Wilcox, 'Embodying Algorithmic War,' p. 16.

31 Charlie Savage, 'Trump's Secret Rules for Drone Strikes Outside War Zones Are Disclosed,' *The New York Times*, 6 May 2021, available at: https://www. nytimes.com/2021/05/01/us/politics/trump-drone-strike-rules.html, accessed on 24 April 2023

32 Wilcox, 'Embodying Algorithmic War.'

33 Christian Grothoff and Porup, 'The NSA's SKYNET Program May Be Killing Thousands of Innocent People,' ars Technica, 16 February 2016, available at: https://arstechnica.com/information-technology/2016/02/the-nsas-skynet

-program-may-be-killing-thousands-of-innocent-people/, accessed on 24 April 2023

34 Grothoff and Porup, 'The NSA's SKYNET Program.'

35 Wilcox, 'Embodying Algorithmic War.'

36 Noel Sharkey, 'The Impact of Gender and Race Bias in AI,' *Humanitarian Law & Policy* (blog), 28 August 2018, https://blogs.icrc.org/law-and-policy/2018/08/28/impact-gender-race-bias-ai/

37 Klonowska, 'Article 36: Review of AI Decision-Support Systems and Other Emerging Technologies of Warfare,' p. 21.

38 Kenneth Kenneth Forbus, 'Creating AI Systems That Take Culture into Account,' The Ethical Machine, available at: https://ai.shorensteincenter.org/ideas/2019/1/14/creating-ai-systems-that-take-culture-into-account-aps9l, accessed on 6 May 2021. The Ethical Machine, 17 January 2019, available at: https://ai.shorenstein-center.org/ideas/2019/1/14/creating-ai-systems-that-take-culture-into-account-aps9l, accessed on 24 April 2023

39 Melanie Sisson, Jennifer Spindel, Paul Scharre, China Arms Control and Disarmament Association, and Vadim Kozyulin, *The Militarization of Artificial Intelligence*, United Nations Office for Disarmament Affairs, 2020, available at: https://www.un.org/disarmament/the-militarization-of-artificial-intelligence/, accessed on 24 April 2023

40 Sisson et al., *The Militarization of Artificial Intelligence*.

41 August Cole, 'Left of Beep: The United States Needs an Algorithmic Warfare Group,' *War on the Rocks*, 9 December 2021, https://warontherocks.com/2021/12/left-of-beep-the-united-states-needs-an-algorithmic-warfare-group/, accessed on 24 April 2023

42 'Air Combat Evolution (ACE),' DARPA, available at:, https://www.darpa.mil/program/air-combat-evolution, accessed on 1 March 2021, accessed on 24 April 2023

43 Criado Perez, *Invisible Women*.

44 Criado Perez, *Invisible Women*, p. 164.

45 Alisha Haridasani Gupta, 'Crash Test Dummies Made Cars Safer (for Average-Size Men),' *The New York Times*, 27 December 2021, available at: https://www.nytimes.com/2021/12/27/business/car-safety-women.html, accessed on 24 April 2023

46 Criado Perez, *Invisible Women*.

47 Kate Chandler, *Does Military AI Have Gender? Understanding Bias and Promoting Ethical Approaches in Military Applications of AI* (United Nations Institute for Disarmament Research, 2021), available at: https://www.unidir.org/publication/does-military-ai-have-gender-understanding-bias-and-promoting-ethical-approaches, , accessed on 24 April 2023

48 Jeff Hearn, *Men of the World: Genders, Globalizations, Transnational Times* (London: SAGE Publications, 2015).

49 Chandler, *Does Military AI Have Gender?*

50 West et al., *I'd Blush if I Could.*

8
ETHICS, LAWS ON WAR AND ARTIFICIAL INTELLIGENCE-DRIVEN WARFARE

Guru Saday Batabyal

Introduction

Military technology, which has always defined how wars are fought, has come a long way. From the bows and arrows of the Trojan War to the longbows in the Middle Ages through to the invention of gunpowder and much later tanks and aircraft of World Wars, nuclear arsenals, and now artificial intelligence (AI)-assisted remotely-piloted unmanned aerial vehicles (UAVs), robots, and cyber weapons amongst a plethora of other new weapons. Dreamy 'Star Wars' like ideas, which once were novel and the stuff of movies, are now a reality because of advancement in the field of military technology, especially UAVs (drones). And that has created a situation of flux in our understandings of war, how it should be fought, and who should fight where – man or machine or a combination of both in the case of interstate or even intrastate wars.

The term AI came into existence in 1956, coined during the Dartmouth Summer Research Conference, organised by the computer scientist John McCarthy. But serious discussion on it began when Alan Turing, popularly known as 'father of computer science,' in 1950 wrote his famous paper 'Computing Machinery and Intelligence,' where he dwelt on the famous question: 'Can machines think?'[1] With AI, machines aim to replicate human intelligence and perform automated tasks. Its military as well as non-military applications have increased manifold since the beginning of this century. President Vladimir Putin, while speaking to the students in a school in September 2017 predicted, 'Artificial intelligence is the future, not only for Russia but for all mankind. It comes with colossal opportunities, but

DOI: 10.4324/9781003421849-9

also threats that are difficult to predict. Whoever becomes the leader in this sphere will become the ruler of the world.'[2] Further Putin predicted that AI-driven drones are going to play a major role in future wars; the recent increase in the use of military drones proves his point.

AI has dual use – good and evil, both for civilian and military purposes. For military purposes, AI can have good uses like robots employed as fire extinguishers in a naval ship and evil and dreadful uses like self-flying lethal drones and autonomous weapons where a human is not in the loop resulting in unchecked killing and destruction. And that brings to the fore the issue of 'ethical designing of AI.'[3] AI-powered war machines and information warfare like cyberattacks remain morally questionable. There are countless ethical challenges concerning applications of AI in warfare. Famous astrophysicist Stephen Hawking feared that one day AI may surpass human intelligence, which may bring the world to an end. Like Hawking, there are others too who feel that AI surpassing human intelligence is harmful to humanity. Billionaire entrepreneur Elon Musk – CEO of Tesla, SpaceX and X (formerly known as Twitter) warned that AI smarter than humans could create an 'immortal dictator' and its adoption is like 'summoning a demon.' Hawking's and Elon Musk's prediction of an AI apocalypse may not come true in the near future but their caution against an extreme form of AI, in which thinking machines would design higher versions of themselves, outwitting mankind, needs to be taken seriously.

Data-driven technologies are now the core infrastructure around which the global economy operates. Nations are seeing AI as an empowering tool; consequently, AI now has become the basis for a highly competitive geopolitical contest where major powers are aiming for rapid growth in this domain, in order to get them in a higher place in the global order, along the lines of the Olympic motto of 'Swifter, Higher, Stronger.' This chapter discusses different AI-driven forms of warfare, with special highlights on AI-powered weaponised drones, robots and 'offensive cyber operations,' including military as well as non-military offensive operations and activities in cyberspace for cyberattacks and cyber terrorism. On a holistic note to cover the ethical and legal aspects of AI-augmented warfare, the chapter recalls the essence of existing laws on war besides examining the adequacy of international laws to regulate modern warfare. By providing normative implications of AI-augmented warfare, the chapter also brings out ethical guidelines, directed towards developers and users of AI, including the policy and decision-makers.

Ethics and AI: The morality and legality of the use of autonomous weapon systems

In recent years, the ethical implications of AI have become a global topic of discussion and have caught the attention of the media, academics and governments alike. Ethics, in simple terms, are those do's and don'ts, i.e., virtues and evil which we all learn and are expected to follow because ethics can be defined as based on well-founded standards of right and wrong that prescribe what humans ought to do, usually in terms of rights, obligations, benefits to society, fairness or specific virtues.[4] We need to speak of ethics for AI and especially machine intelligence, instead of merely human ethics. Ethics of AI design encompasses the moral behaviour of the engineers involved in designing, industries making and marketing AI-augmented products, including products for defence services, policy and decision-makers who allow them to use such products and finally the end users. As it concerns the behaviour of machines; machine ethics also becomes a subject of scrutiny. The ethics of AI is the branch of ethics of technology specific to artificially intelligent systems. It is sometimes divided into a concern with the moral behaviour of humans as they design, make, use and treat artificially intelligent systems, and a concern with the behaviour of machines, in machine ethics. It also includes the issue of a possible singularity due to superintelligent AI.[5]

AI has been added to conventional weapon systems to achieve better results or to lessen the human exposure to harm. Even though AI-assisted weapons may act on their own volition, if a human is in the loop then the machine cannot outsmart man and behave unpredictably beyond what it was preprogrammed or designed for. A possible singularity could be due to superintelligent AI, which may surpass human intelligence and thereafter become unstoppable, where human beings will not have any more control. That form of superintelligent AI may not be an immediate possibility, but Ray Kurzweil, Google's Director of Engineering, believes that could be a reality by 2045 or so; but he asserts that it should not be a cause of worry, as human beings will, by then, adapt to merge with the new form of AI.[6]

These are assumptions at this stage; AI surpassing human intelligence and technology and making its own decisions may be a problem, raising a burning question in respect of the autonomous weapon systems (AWS) – should we mortgage our freedom to an algorithm? The question also leads to issues on AI ethics and a universal adoption of regulations to check or ban unethical use of AI. During the 41st Session of UNESCO General Conference, held in November 2021 in Paris, an agreement, called 'Recommendation on the Ethics of Artificial Intelligence,' was adopted, renewing cooperation on this much-debated issue. Many organisations have also have come up with AI codes of ethics to guide stakeholders facing with ethical dilemmas when deciding on the adaptation and use of AI. Interestingly, years before the development of AI to its latest avatar, the potential dreadfulness of autonomous AI-powered robots was foreseen by the famous Russian-American science fiction writer Isaac Asimov, who wrote a fiction series on robots and created 'The Three Laws of Robotics' – binding robots to follow rules to limit danger. In Asimov's code of ethics, the first law forbids robots from actively harming humans or allowing humans to come to harm by refusing to act. The second law orders robots to obey humans unless the orders are not in accordance with the first law. The third law orders robots to protect themselves insofar as doing so is in accordance with the first two laws.[7] These laws were partially modified by Asimov in his later writings, but the basic philosophy behind 'unharming humans' has stayed and remains applicable to AI developers and users. Ways and means have been recommended, as checks and balances, to put in place ethical practices that apply to AI-enabled products. While developing, AI engineers must take note of the following guidelines to ensure the ethical design of AI, which are as follows.[8]

- Transparency – the most important ethical domain. Its adherence protects fundamental human rights and the privacy, dignity, autonomy and well-being of society (UNI Global Union, 2017). Organisations should be able to clearly decipher and explain how their AI is respecting current legislation.
- Justice and Fairness – issues of fairness, unbiasedness, equality and equity must be built into the algorithm model.
- Non-maleficence – Avoidance of harm to human beings has been an important aspect of concern in AI ethics, particularly in areas like killer robots and drones (UAVs).
- Responsibility and accountability – Liability and accountability for making the war amoral must be traceable and be under judicial review. AI should adhere to the privacy and data protection standards outlined in the General Data Protection Regulation (GDPR) (2018).
- Beneficence – AI should bring benefits to the society, as highlighted in IEEE, 2019.
- Freedom and autonomy – AI developers should ensure no impediments to the freedom and liberty of human beings.
- Trust and sustainability – trust is important for the ethical deployment and use of AI as enunciated in the document The HLEG (2019). AI organisations need to ensure that their entities are environmentally sustainable (Special Interest Group on Artificial Intelligence, 2018).

Nations, which are technologically advanced and increasingly acquiring sophisticated AI-assisted weaponries for their armed forces, are being questioned on ethical and legal proprieties. AWS has social and political ramifications too. Removing human footprints from the AI-driven decision-making loop during war would put ethics at stake and make the war amoral. A non-human technology AI, which has no accountability, unlike humans, should not be the one making a decision over a human.[9]

The framework discussed above only considers a few important guidelines; actually there are more. Understanding this framework will help engineers engaged in developing this technology to mitigate ethical issues through algorithmic design in the early stage; while policymakers, decision-makers and other stakeholders could identify and broadly consider potential military ethical issues concerning design and deployment of AI-assisted military robots, drones, etc. Even formulating regulatory policies *ab initio* on permissible use of AI by the armed forces will help in abiding the laws of war. For this, policy makers and decision-makers should be well informed about the technological impact on the society and human beings. The future of AI-augmented warfare will much depend on the ability of engineers to design autonomous weapon systems that would have capacity for knowledge and expert-based reasoning, independent of human footprints. However, here is what M.L. Cummins has to say about AI surpassing human intelligence and achieving full automation in weaponries:

> Although it is not in doubt that AI is going to be part of the future of militaries around the world, the landscape is changing quickly and in potentially disruptive ways. AI is advancing, but given the current struggle to imbue computers with true knowledge and expert-based behaviours, as well as limitations in perception sensors, it will be many years before AI will be able to approximate human intelligence in high-uncertainty settings – as epitomized by the fog of war.[10]

AI-powered military robots

Almost eight decades ago, USSR first fielded military robots in the battlefield during the Second World War in the form of unmanned radio-controlled 'teletanks' when the Red Army used them during the Winter War against Finland. It was followed by the German robot Goliath, a tactical mine demolition vehicle.[11] Robots from the point of view of use are of three types. Automatic robots – which respond in a mechanical way to external inputs. These are usually without any ability to discriminate the inputs. Automated robots – which execute commands in a chronological preprogrammed way with sensors that help sequence the action. They are mostly limited by algorithms that determine their rules of operations and behaviour, out of a fixed set of alternative actions, which makes them predictable. And the third type of robot is an autonomous robot which has minimum human interference other than possibly switching it on or off. Militaries prefer and mostly use autonomous robots for various purposes and such military robots' market has been showing a constant rise in leading countries' defence budgets in recent years. The latest available data suggests that the military robots' market size will likely grow from $14.5 billion in 2020 to $24.2 billion by 2025, at a compound annual growth rate (CAGR) of 10.7%.[12]

AI-powered armed robotic vehicles and autonomous surveillance vehicles are now part of military inventory because of the advantages they offer as a force multiplier in battle. As robots are devoid of emotions and are without fear, they can undertake risky tasks, saving human lives of friendly forces. Besides, robots are more accurate, quick and have better decision-making abilities in stressful battlefield environments. Their use has made military tasks safer, easier and more precise. Types of military robots will depend on the purpose for which they were designed. Presently, for military use there are transportation robots, search and rescue robots, firefighting robots, mine clearance robots, surveillance and reconnaissance robots, armed robots and unarmed aerial/ground/underwater vehicle robots.[13]

The US Army first deployed robots in July 2002, when the robot 'Hermes' was used to search the caves in Qiqay, at Afghanistan, to find hidden enemies and their stockpiles of weapons. Since then, militaries around the world have been deploying robots, including the latest AI-augmented robots, in combat and surveillance roles. Few of the topmost robots, including some AI-powered robots, that are operative in the major militaries are as follows.[14]

One is Multi-Utility Tactical Transport (MUTT), which carries equipment/load up to the weight of 1,200 pounds assisting the soldiers travelling on foot in difficult terrains, was tested by US Marine Corps in July 2016. This Unmanned Ground System (UGS) produced by the General Dynamics Land System is being inducted in the US Army. Next is RISE, a climbing robot by Boston Robotics which has micro-clawed feet that allow it to deftly scale rough surfaces, including walls, fences and trees. Another is a 9 mm Glock pistol-equipped tactical combat robot weighing 26 pounds named DOGO developed by General Robotics which functions as a watchdog for soldiers in battle. Guardbot, a surveillance robot which can negotiate all types of terrain, including sand, snow and dirt as well as bodies of water, having the ability to swim. Centaur robot, named after the mythological creature with the upper body of a human and lower body of a horse, is capable of finding landmines, unexploded ordnance, improvised explosive devices (IEDs) and destroying them. It is an unmanned ground vehicle (UGV) with a remote control mechanism. Gladiator is a mobile robot which looks like a small tank that can carry out multiple tasks like reconnaissance in Nuclear-Biological-Chemical (NBC) environment, direct fire, obstacle breaching and surveillance etc. in all types of warfare, to reduce hazards for regular soldiers. SAFFIR is a military robot that stands 5 feet 10 inches tall and weighs 143 pounds – developed by researchers at Virginia Tech. This is one of the most autonomous military robots utilising advanced military technology designed to extinguish a fire on a warship. AVATAR

III is Robotex's tactical robot which can be employed for dangerous tasks. Jaguar is a six-wheeled semi-autonomous UGV armed with a 7.62 mm MAG machine gun and it can self-destruct if it falls into enemy hands. Israel has deployed it on the Gaza border in June 2021 and it is substituting combat soldiers in border patrolling jobs. Finally, Uran-6 robotic demining system and Uran-14 fire extinguishing system have been co-opted in training by the Russian defence forces.

The Chinese defence industry during the Air Show China in Zhuhai in November 2022 unveiled its first dog robot, which is armed with an automatic gun system. The Indian Army has a Daksh automated robot developed by India's Defence Research and Development Organisation (DRDO), which is a remote controlled robot on wheels used to recover, locate and destroy explosives. The USA, Russia, China, France, Germany, UK, China, Israel and India are developing robots both for ground and underwater roles with increasing integration of AI. Navies are increasingly seeking for Unmanned Undersea Vessels (UUVs). The USA, for instance, has ordered four Extra Large Unmanned Undersea Vehicles (XLUUVs) to be built by Boeing for the United States Navy (USN). These XLUUVs would operate independently for months underwater and cover up to 6,500 nautical miles on a single fuel-cycle. These will be fully autonomous vessels, which can undertake the role of a submarine without any human being in the vessels.[15] The *Washington Times* reported on 16 December 2020 that the US Air Force for the first time successfully tested an artificial-intelligence-assisted control sensor and navigation system on a U-2 Dragon Lady spy plane in a training flight. This is possibly a precursor to many future AI-powered developments for use in air force. Development of AI-integrated military robots and other programs are kept under tight secrecy until disclosed by the developers. It is therefore difficult to exactly state in what stage such developments have reached and for what purpose.

Till now, mostly humans are controlling robots. However, the advent of AI-powered autonomous robots, where human beings are not in the loop, could be worrisome because that may trigger 'robotocalypse.'[16] The SGR A1 robots, developed by Samsung, which have been deployed in the Korean demilitarised zone since 2014, can undertake functions like surveillance, tracking, voice recognition and firing. This robot is functioning with human supervision but has the built-in capacity to act independent of human supervision. Developed by Israel Aerospace Industry (IAI), the IDF has deployed AI-assisted semi-autonomous robots like the RoBattle combat system for convoy protection and armed reconnaissance and the Jaguar (equipped with a machine gun) for patrolling the Gaza border. The IDF claims that the Jaguar is one of the first military robots globally to have the ability to replace combat soldiers in border patrolling tasks. Similarly, the IAI developed the Harop 'suicide drone' and it has been effectively employed by Armenia against Azerbaijan from 2016 onwards. The Harop can function both with human supervision and completely autonomously. According to a report generated by the UN, in March 2021, Libyan Forces used the 7 kg quadcopter drone STM Kargu-2 built by the Turkish ATM company and equipped with AI-powered facial recognition against militia fighters. The drone possibly operated without human in loop.[17]

The revolution in military robotics poses ethical and legal problems besides leading to a new paradigm of dangerous warfare. Human rights organisations, the United Nations as well as some research and industry groups, have shown their concerns about Lethal Autonomous Weapons (LAWs), which is ever growing. A meeting of the CCW (Convention on Certain Conventional Weapons) in November 2022 brought together a group of government experts and NGOs from the Campaign to Stop Killer Robots, which wants a legally binding international treaty banning LAWs, just as cluster munitions, landmines, and blinding lasers have been banned in the past. Most states parties to the CCW agree that a new legally binding instrument is urgently needed to address the serious threats posed by this emerging technology. However, they have not taken concrete steps toward making this goal a reality.[18]

UAVs/Drones

UAVs/Drones could be defined as remotely operated, autonomous or automated robots, that are capable of sensing information, processing it and executing a physical action without a human pilot on board.[19] Progressively, there has been an evolution of drones since their first use by Austria when at war with Italy for the bombing of Venice city on 15 July 1849, where it used balloons carrying explosives which had timer set fuses as control mechanisms to trigger the explosion. The Second World War witnessed the use of flying bombs by boththe Allied and Axis forces. The USN retrofitted B-24 bombers with remote-controlled mechanisms that dropped bombs on the German trenches; and German V-1 bombers similarly attacked many cities of the Allied nations using similar remote-controlled mechanisms.[20]

Modern UAVs, which are now being widely used, were initially invented for military purposes by the Ryan Aeronautical Company. Beginning in 1951, the Ryan Firebee was a series of target drones developed by the Ryan Aeronautical Company. Firebee 1241s UAVs were successfully used by Israel against Arabs during the 1973 Yom Kippur War for reconnaissance and decoys. The USA used drones in Vietnam, Iraq and Afghanistan; they were largely used for reconnaissance and intelligence gathering in the initial stage. Advances in AI are the new benchmark of power amongst the superpowers. Modern warfare in essence is more complex, on account of an increasing number of micro variables. A change in any one variable could create an exponential impact on battle outcomes – and even on the war itself.[21]

Visibly, in the last couple of decades, drones, in their avatar as weapons, have changed the face of war. In the year 2020, the world witnessed the pivotal role Turkish drones played in Azerbaijan's impressive victory over Armenia in the Nagorno-Karabakh conflict. Also, those following the Russian invasion in Ukraine saw the TV footage with awe how Kamikaze drones supplied by the US brought upon heavy losses on the Russian tanks, proving how military drones are playing a pivotal role in Ukraine's defensive war against the mighty Russian armed forces. Undeniably, AI-powered UAVs (drones) have now joined the armed forces of the world in their inventory as they are becoming indispensable in modern war.

But a new kind of war was born when the USA and its allies commenced using weaponised drones in countries where they were not formally at war or without the proper consent of the sovereign states. In such places, these UAVs targeted individual terror masterminds or terror groups in their so called 'war against terror.' Turkey and some other countries have been using drones for transnational assassination in Syria. The use of drones in the war against terror and the misuse of drones for terrorism calls for an in-depth analysis of the ethical and legal issues involved. But the most contentious issue here is the use of autonomous drones/weapon systems without human imprint or very negligible human control because such (lethal) autonomous weapon systems (AWS/LAWS) support extrajudicial killings, take responsibility away from humans, and make wars or killings more likely.[22]

In a message to the Group of Government Experts, Antonio Guterres, UN Secretary General on March 25 2019 wrote: 'Autonomous machines with the power and discretion to select targets and take lives without human involvement are politically unacceptable, morally repugnant and should be prohibited by international law.' He further reminded the expert group dealing with AI-assisted weapons that human responsibility for decisions on the use of weapon systems must be retained, since accountability cannot be transferred to machines.[23]

So far, the international community has not prohibited the use of such weapons. It is difficult for the UN to garner the support of all members to make them agree to ban AI-powered autonomous weapons/AI-embedded technology that may be used for offensive purposes in the form of a law. And that leads to the fear that lowering the hurdle to use such systems (autonomous vehicles, fire-and-forget/loitering missiles, or drones loaded with explosives) and reducing the probability of being held accountable would increase the probability of their use. The crucial asymmetry where one side can kill with impunity, and thus has few reasons not to do so, already exists in conventional drone wars with remote-controlled weapons (e.g., USA in Pakistan). Arguably, the main threat is not the use of such weapons in conventional warfare, but in asymmetric conflicts or by non-state agents, including criminals.[24]

On 12 November 2017, a fiction video named Slaughterbots was uploaded to YouTube. The video depicts, in the ensuing future, AI-powered small drones fitted with face recognition systems and loaded with shaped explosive charges that can be programmed to identify, search and kill known people individually or in groups. That includes people wearing a particular type of uniform. Dr Stuart Russel, an AI scientist, asked a valid question, 'How long this fiction robot will remain a fiction?'[25] Laboratories working on AI-assisted armed military robots may possibly make it happen; maybe it is already there, otherwise how was Ayman-al-Zawahiri, an Egyptian born militant and an important ideologue of Al-Qaeda identified and killed at Kabul in July 2022 with minimal collateral damage?

Cyberattacks and AI-enabled cyber weapons

Belligerent deliberate actions against a computer system with a purpose to acquire intelligence clandestinely or to make it unreliable or unavailable can take two forms, namely, 'cyber exploitation' which is non-destructive and 'cyberattack' which is destructive in nature. In the case of cyber exploitation, which is primarily an espionage activity, these could be launched over an extended period clandestinely with a purpose of extracting information, and deceiving someone else for personal gain without disturbing the system. Cyber exploitation is also used to steal military secrets or credit card information. In the case of cyberattack, a deliberate operation is undertaken for a short or extended duration depending upon the aim and objective, which could be to disrupt or destroy targeted computer systems of an adversary or networks, in addition to deceiving or degrading the system. Those entities paired with this network will also get affected. Examples of cyberattacks are manipulation of computer systems connected to control or shut down an electricity grid, make banking data unreliable or disrupt military communications, etc. Once the network or the system becomes untrustworthy due to a cyberattack, it is no longer useful as data can be altered which can then generate a misleading output.[26] As of 2022, 69% of the world's population, or 4.9 billion people, actively use the internet, and Asia accounts for approximately half of that number approximately.[27]

Superiority of AI is a crucial component of the new paradigm of competing powers. Historically, wars were based on military strength and victory belonged to those who were numerically superior. But modern warfare, in essence, is more complex on account of an increasing number of domains beyond the more traditional land sea, and air. Variable applications of AI could generate an exponential impact on the outcomes of war. China has integrated the use of cyber techniques into its military doctrine and economic policies, possibly far more comprehensively than any other nation. Since the last couple of decades, cyberattacks have been increasing globally, maturing to the level of cyber warfare, which has been escalated to cyber terrorism. Besides land, sea, air, and space, now cyberspace has emerged as a fifth theatre of war. Unlike the other four domains, cyberwar is being fought not by the military but mainly by the criminal organisations and terrorist groups independently, although at times with the tacit support of host governments. The world has witnessed the use of cyberterrorism which often resulted in shutting down critical national infrastructures such as energy/power, financial sectors, government operations and defence installations. There has been a surge in cybersecurity attacks globally in the preceding two years, and one can foresee that the future will witness more grey zone conflict in the cyber domain. Recent ransomware attacks hacking India's Tata Power servers by Hive, a ransomware gang, and in the premier health institute of India, AIIMS, New Delhi, presumably orchestrated from China, had serious ramifications and act as grim reminders of how shape of things may unfold in the foreseeable future. Masking users' identities, terrorists and rogue nations are clandestinely using AI-powered dark net/The Onion Rings (TOR) etc. for spreading terrorism and espionage activities, which need to be monitored as a mitigation measure.

During the visit of US House Speaker Nancy Pelosi to Taiwan, at the beginning of August 2022, which was objected by China, cyberattacks were unleashed targeting government offices as well as commercial organisations. Taiwan squarely blamed China for these cyberattacks as they saw Pelosi's visit as a serious provocation to the sovereignty of the motherland. Bulletin messages in 7-Eleven stores read 'Warmonger Pelosi, get out of Taiwan.' There were also electronic billboards across Taiwan protesting her visit – one calling her an 'old witch.' Among the official government websites, the website of the president's office, and the foreign and defence ministries were shutdown temporarily.[28]

There lies here an interesting comment on AI-enabled offensive cyber weapons and future cyberspace warfare. The first militarily significant offensive and autonomous weapon system enabled by AI will probably be deployed not on the physical battlefield but in cyberspace. AI-enabled cyber capabilities based on deep-learning algorithms will be easier and cheaper to develop than AI-enabled offensive autonomous platforms, which require maneuvering to fight physical enemies. The USA will most likely be the first country to field an AI-enabled cyber weapon, but other nations, particularly China and Russia, are bound to follow suit.[29]

AI in space

Warfare in space takes place outside the atmosphere. Global communication systems are heavily dependent on the satellites orbiting around the earth.

Protecting these satellites and access to space will motivate nations to deploy more space based weapons and militarise space. The history of active space warfare development goes back to the 1960s when the Soviet Union began the Almaz Project, a project designed for in-orbit inspections of satellites and to destroy them if needed. Around that time, the USA launched the Blue Gemini Project, with the purpose to deploy weapons and perform surveillance in the orbit. The race to space and its access has continued at an accelerated pace since then. Space warfare is aimed at three important segments: (1) the satellites in the orbit, (2) the up and down links to and from the orbiting satellite, and (3) the ground station that controls the satellite and passes information to the users.[30]

The use of AI in space applications is expected to increase to better facilitate process optimisation and automated services, as well as reducing the elaborate time required for data and imaging processing in the space industry field. However, AI exhibits cyber vulnerabilities and, thus, is open to cyber threats.[31] AI is becoming indispensable in managing the mega-constellations of satellites being deployed in low Earth orbit (LEO), by undertaking tasks like scheduling, ensuring collision avoidance in the much-crowded space, and satellite debris mitigation. AI will also have a role in national security management, including missile defence.[32] The space marketplace has been growing exponentially from $176 billion to over $345 billion from 2006 to 2018, with an estimated growth of near $1 trillion by 2030.[33]

Prototype autonomous cyber weapons have been created. For example, IBM has created an autonomous stealthy breed malware, DeepLocker, an evasive, AI-powered attack tool that uses neural networks to select its target. The malware can remain in disguise until it reaches its destination. It is almost impossible to carry out a reverse engineering to stop damage as it unlocks in the last moment after reaching the intended target.[34] The use of AI-enabled weapons in combat in tandem with other weapons or AI-embedded autonomous stealthily bred malware will pose serious ethical and legal problems, from possible conflict escalation, the promotion of mass surveillance measures, and the spreading of misinformation to breaches of individual rights and violation of dignity. If these problems are not looked into, the use of AI for defence purposes risks undermining the fundamental values of democratic societies and international stability.[35]

New forms of unconventional warfare

From the last few decades, there has been extensive discussion on unconventional warfare as a form of war strategy, such as asymmetric warfare, hybrid war and grey zone war. The use of the term asymmetric warfare first occurred in 1991. Merriam-Webster English dictionary defines asymmetric warfare as between opposing forces which differ greatly in military power and that typically involves the use of unconventional weapons and tactics (such as those associated with guerrilla warfare and terrorist attacks).[36] A cyberattack becomes 'asymmetric' when significant damage is done to the intended target through the use of small resources. Fast-paced advances in AI being leveraged by both state and non-state actors through multiple applications will keep adding new tools and add more power in multitudinous ways to both. Along with states, non-state actors, criminals and terrorists are also trying their best to lay hands on AI-powered tools and weapons and are making use of these wherever they can.

The terming and concept of hybrid warfare was first ideated by Frank G. Hoffman through his paper 'Conflict in the 21st Century: The Rise of Hybrid Wars' published in Potomac Institute for Policy Studies, 2007.[37] A hybrid threat has been defined by the US Joint Forces of Command as any adversary that simultaneously and adaptively employs a tailored mix of conventional, irregular terrorism and criminal means or activities in the operational battle space. Rather than a single entity, a hybrid threat or challenger may be a combination of state and non-state actors.[38] However, the concept as well as definition offered by the US armed forces are not universally accepted. Hybrid war essentially follows asymmetric warfare tactics, where a non-standard pattern war is launched, and employing kinetic and non-kinetic means in a synergised and non-linear manner, as part of a nation's strategy, to achieve its objective. It is a non-declared daily war which blurs the formal distinction between war and peace. Fourth-generation irregular war was earlier undertaken by non-state actors. But in hybrid war, both state and non-state actors are employed independently or simultaneously with the aim of harming the adversary. Space and AI technology-embedded weaponry are also employed to increase the scope and magnitude of damage intended to be caused on the adversary. The scope of hybrid war is limitless. In this type of warfare, besides kinetic weaponry, non-kinetic means like AI-driven propaganda, cyber operations, both by state and non-state actors, and at times violence and disorder launched by the non-state actors are employed simultaneously with state war machinery as part of an overall grand strategy.

Though there is no fixed definition for grey zone warfare, it can be broadly defined in the words of one analyst named Ashok Kumar Singh as the exploitation of operational space between peace and war to change the status quo, through the use of coercive actions which remain below a threshold that, in most cases, would prompt a conventional military response. It is characterised by sub-threshold activities including kinetic and non-kinetic methods and military and non-military means, by conventional military force and irregular proxies.[39] Like hybrid war, the definition of grey zone war does not find universal acceptance. The US Special Operations Command defined it as 'competitive interactions among and within state and non-state actors that fall between the traditional war and peace duality.'

In this kind of warfare, 'low threshold' prevents the affected state from launching a military offensive. It is contestation between statecraft and open declared war. Coercion is unleashed through various means like trade and economy, cyberattacks and interference in election processes (alleged meddling by Russia in the 2016 US Presidential election is a classic case of malevolent use of AI). There is a blurred line between the concepts of hybrid war and grey zone war. There are some commonalities in the tools used too, but both are not the same.

Military ethics, war and terrorism

The theory of the ethics of war can be traced to mythology, religious scripts and culture. Just War theory, which has evolved over time, has a historical pedigree. Hugo Grotius *De iure belli ac pacis,* a book in Latin meaning *On the Law of War and Peace*, published in 1625 in Paris, is regarded as a foundational work in international law which covers the legal status of war. On principles and concepts of Just War, four essential aspects are:

- The war should have a just cause.
- Commencement of war should be declared by a legitimate authority.
- The principle of proportionality should be applied while conducting war.
- Non-combat immunity is to be ensured during military engagements between the opposing forces.

If the above four essential aspects while resorting to war are met, then it may be considered justifiable and legitimate from the perspective of Just War tradition, not withstanding war's destructive and violent nature. But what about 'war on terror' and how does it differ from the war that we understand? Most importantly, the war is comprehensible and less ambiguous, but terrorism is viewed as violent, more threatening and incomprehensible, and therefore perceived as criminal and unjustifiable. One could say that while war is motivated, terrorism is gratuitous violence.[40] This explanation of war and terrorism may not be agreed upon by the radicalised extreme left or right wing ideologues, who feel if the postulation of Clausewitz that 'war is an act of force to compel our enemy to do our will' is acceptable, why not terrorism which is seen as an ideological clash, where terrorists through violence compel the target population to accept their ideology. To enforce their idea, the terrorist groups aim to directly challenge the government's monopoly of force. But rarely will a government or state allow a group to challenge its authority and it will take firm measures beginning with peaceful negotiations, and failing that, use force to destroy the challengers, although it's a different issue altogether whether an idea can be killed. In today's world, cross border terrorism is based less on political ideas and more so on religious fanaticism. In the Western nations, there is in-state terrorism mostly based on racism, anti-refugee movements or challenging political ideology emanating from the radical-ised right wing.

In a declared interstate war, the use of weaponised drones may pass the scrutiny of laws of war if not used as a mass destructive weapon, but my analysis here will focus on ethical and legal positions of 'droning' terror-ists based in a sovereign state with whom there is no formal war or any treaty permitting such action. Searching through the available literature one finds that there is no standard definition of terrorism. This is because often one state's 'terrorists' are considered another state's 'freedom fighters.' Terrorism in Kashmir is a classic example of how India's views differ from Pakistan's.

Due to the nature of changes in the character of war, AI-powered weapon systems, geo-political situations, and emergence of grey zone war, it may be argued that the Just War principles discussed above are dated and need modifications to suit the changes in warfare. Existing legal frameworks like International Humanitarian Law (IHL) or the law of armed conflict adequately regulate all types of conflicts. There have been some additions to existing war ethics, like banning the use of chemical and biological weapons. Some more regulations befitting the changes in the character of war and weaponry, etc., could be added, but I strongly feel that the basic framework of codified war ethics or war-tradition is unchangeable. Human-controlled AI-assisted weap-ons may adhere to basic tenets of the laws of war but autonomous weapons cannot conform to IHL, which requires observance of the principles of dis-tinction (between combatants and civilians), proportionality (of force) and military necessity (of force) in military conflict.[41] Therefore, isn't there a merit in invoking a fresh regulation legislating ban on the development and use of such weapons, ratified by the members of the United Nations?

Because of varying perceptions on terrorism, drawing up international legislation to deal with terrorism has been beset with problems. Post 9/11 Twin Tower terrorists' attack, the Security Council unanimously adopted the landmark Resolution 1373 in 2001. This important document entailed the future response of the Security Council to deal with the threat of terror-ism. This document and successive resolutions ask states to adopt measures which have far-reaching legislative and executive actions, yet remain silent on the definition of terrorism because its member states in the last seven decades have not been able to unanimously agree to a single definition. In the meantime, the UN General Assembly usually follows its inflection on terrorism based on its 1994 declaration on 'measures to eliminate terrorism' which is primarily an alternative definition of terrorism that states:

> Criminal acts intended or calculated to provoke a state of terror in the general public, a group of persons or particular persons for political

purposes are in any circumstance unjustifiable, whatever the considerations of a political, philosophical, ideological, racial, ethnic, religious, or
any other nature that may be invoked to justify them.[42]

The Counter-Terrorism Committee (CTC) was established as a subsidiary
body of the Security Council through Resolution 1373 (2001). It empowered
the Security Council to assess member states' compliance with the provisions incorporated in this resolution. Respecting the views of the human
rights groups, the Security Council, through its Resolution 1556, has offered
a new non-binding definition of terrorism which is quite narrow in its scope.
It is important to note that legal experts do not accept internal terrorism as
a threat to international peace and security.[43]

Recognising the increasing threat arising from the misuse of emerging
technologies, the United Nations Security Council's Counter-Terrorism
Committee (CTC) held a special meeting in India in Mumbai and New Delhi
on 29 and 30 October 2022, which was attended by Dr S. Jaishankar –
India's External Affairs Minister – UN officials, members of the Security
Council and diplomats. The meeting postulated on the challenges of global
counter terrorism architecture, focusing on significant areas where emerging technology aided an increase in the threat of abuse for terrorism purposes. The 2008 Mumbai 26/11 attacks and how Pakistan – despite evidence
to the contrary – dragged its feet to take action against the masterminds
Hafiz Saeed and Zaki-ur Rahman Lakhvi were discussed. In this conference, India also highlighted that though the USA convicted David Headley
and Tahawwur Rana, it refused to extradite them. Both Indian External
Affairs Ministers Dr S. Jaishankar and US Secretary of State were vociferous
on China's repeated disagreement in designating *Lashkar -e- Taiba* (LeT)
leaders on the UNSC 1267 terror list. In Delhi, the CTC focused on terror recruitment, radicalisation and terror-funding, including the misuse of
crypto currency. The deliberations led to the 'Delhi Declaration on countering the use of new and emerging technologies for terrorist purposes.'[44]

War on terror and politico-legal aspects

On 29 September 2001, US President George W. Bush said that the war on
terror would be much broader than the battlefields and beachheads of the
past. The war would be fought wherever terrorists hide, or run. Since then,
drones have been used by the USA in their mission of the 'war on terror.'
The US government made an unprecedented drone attack in Yemen on 30
September 2011 where two American citizens, Anwar Al Awlaki and Samir
Khan, were killed. David Rhode in Reuters magazine wrote,

> The target of the attack was Awlaki, a New Mexico-born Yemeni-
> American whose charismatic preaching inspired terrorist attacks around

the world, including the 2009 killing of 13 soldiers in Fort Hood, Texas. Civil liberties groups argued that a dangerous new threshold had been crossed. For the first time in American history, the United States had executed two of its citizens without trial.[45]

Turkey has been using weaponised drones extensively in Kurdish-controlled northeast Syria. Besides, the Turkish Armed Forces (TSK) had been conducting cross-border operations in northern Iraq, a region where Kurdistan Workers' Party (PKK) terrorists have hideouts and bases from which they carry out attacks against Turkey. Recently, it was reported that Ankara carried out drone strikes in Syria near an area of US operations. The fact is that such deep strikes inside Syria, under the garb of eliminating terror groups, where Turkey used drones with impunity against civilians, are a serious human rights violation, and in the past, similar drone strikes were termed as extrajudicial killings by human rights groups. For example, the killing of Salwa Yusuk by a drone has been debated as an extrajudicial targeted execution. Amnesty International in 2018 reported, according to the Bureau of Investigative Journalism, US drone strikes have killed up to 1,551 civilians since 2004 in Pakistan, Afghanistan, Yemen and Somalia. Amnesty International and others have exposed how some drone strikes have violated international law and may amount to extrajudicial executions or war crimes.[46] Killing civilians, whether targeted or through collateral exposure, by weaponised drones is contrary to the laws of armed conflict, which, as per international laws on war, require protection for non-combatants (civilians).

The efficacy of military drones in eliminating a target was reinforced when US President Joe Biden during a live television address from the White House on 1 August 2022 said 'Justice has been delivered,' confirming the killing of the 71-year-old Egyptian doctor Ayman al-Zawahiri. Zawahiri, together with Bin Laden, plotted the 9/11 attack and took over al-Qaeda after the death of Bin Laden in 2011. He was one of the most wanted terrorists on the USA's list and his remote killing in the Afghan capital Kabul on 31 July 2022, by a drone attack, was apparently organised by the CIA, in collaboration with Pakistan. The Biden administration demonstrated the capability of the USA's much-talked-about 'Over-The-Horizon' operation. The news got people thinking once again about how technology has replaced the human warrior in eliminating an enemy from far away with much precision and, possibly for the first time, without collateral damage.

But the justifiability of the means to deliver justice has been questioned by a section of international legal luminaries and human rights activists, highlighting ethical and legal untenability. Are such actions permissible from the Just War perspective? International law expert Ben Saul writes, 'The killing (of Zawahiri) is most accurately described as extrajudicial execution or revenge murder designed to deter others from participating in terrorist groups. It is also another body blow to the "rules-based international order" that the US demands others — but apparently not itself — respect.'[47] This dichotomy is noticeable. The USA had been vociferous over human rights violations in other countries and is much sensitive to terrorism when it affects them. But despite Pakistan's sheltering UN designated terrorist groups like *Lashkar-e-Taiba* and *Jaish-e-Mohammad*, the USA has not designated Pakistan in the list of the state's sponsors of terrorism. Therefore, on the issue of terrorism, the USA has adopted a policy which is partly instrumental and partly contradictory.

Can the killing of Zawahiri be considered a 'peacetime reprisal,' contrary to legal luminary Ben Saul's term 'extrajudicial killing'? Michael Walzer is of the view that the term peacetime reprisals is not an entirely accurate terminology because international laws on war or other law books deal with topics like war and peace and nothing in between. However, at the same time, Walzer is not happy with the strict judicial terminology of 'war' and 'peace' as he thinks that 'much of history is a *demi-monde* that neither word adequately describe.'[48]

US policy of using drones for targeted assassination is indeed questionable from the legal angle, but at the same time one ponders what to do in such cases where terrorists, harboured by a state which is not openly at war, are engaged without collateral damage (like killings of innocent civilians) in pursuance of a nation's aim to eliminate terror. Such military use of drones 'will be marked by a kind of asymmetry characteristic of peacetime reprisal.'[49] Its purpose is coercive and to repress terrorists, hiding/sheltering beyond the country's border, in these cases in faraway lands to prevent the recurrence of terrorism. Deployment of such advanced systems, besides eliminating a target, also creates fear and sends a message that without human-to-human contact, revenge can be taken through deployment of AWS. For a historical reference of a classic case of peacetime reprisal, one may recall when on 26 December 1968, two terrorists attacked an Israeli plane, preparing to take off from Athens. Later, after getting caught by Athenian police, it was discovered these two terrorists, who had Lebanese documents, were actually members of PLO, as PLO then had its headquarters in Beirut, Lebanon. Providentially, out of 50 passengers on board, only 1 was killed. Two days after the Athens attack, in a daring reprisal operation, Israeli commandos landed by helicopter in Beirut airport and destroyed 13 Lebanese civilian planes. This commando raid was considered a spectacular military success, but more importantly, it was considered a great success from the moral point of view, because despite repeated warnings to the Lebanese government for not allowing its soil to be used by PLO for terrorist activities, Israel was not left with much choice than to undertake this dramatic reprisal, resulting in no civilian deaths despite loss of life of an Israeli citizen in the Athens airport attack, and the retaliation was parallel and proportionate. This incident was condemned by the UN and was much criticised, citing reasons that the attack was upon Lebanese sovereignty. But Israel argued that the Lebanese Government had an obligation to disallow the use of its territory for carrying out acts of terrorism in/against Israel. However, in similar instances, an important question is whether a sovereign state can be forced by another nation to carry out its obligation to ensure peace. The UN's position is that such forceful and coercive law enforcement is illegal, notwithstanding the grave situation that may trigger such action.

Ultimate responsibility for just or unjust reprisals during peacetime or any war crime is on the head of a state because he/she represents the people of the country on the world stage and his/her actions will be considered as the 'reasons of state.' The head of the state is mainly responsible for all such actions both from a moral as well as a legal point of view. Brian Orend, supporting views of Walzer on the responsibility the of head of a state, avers 'Heads of state covet power, actively seek it, enjoy its exercise, and exert enormous influence. Often, they hope to be praised for the good they do, but they cannot escape blame for evil.'[50] Further, he cites the example of the US cruise missile attack in 1998 on Al-Qaeda training sites, which did not deter 9/11 in 2001.

Use of drones by states, terrorists and legal implications

With the increasing commercial availability of UAVs in the market and their technological development, these drones have potential to become a dangerous weapon in the hands of terror groups. Drones laden with Chemical, Biological, Radiological, Nuclear, and high yield Explosives (CBRNE) can be employed as a flying attacker. A range of terrorist and insurgent groups have already deployed UAVs for attacks and intelligence gathering. Today's commercial UAVs are exploitable mainly in two ways: For placement of IEDs, CBRNE substances, etc. and for reconnaissance of a particular region. The commercial segment of the UAV market in Asia Pacific is projected to grow at the highest CAGR of 18.5% during the forecast period from 2021 to 2026.[51]

On 4 August 2018, when Venezuelan President Nicholas Maduro was delivering his speech at a public outdoor event in Caracas, Venezuela on the occasion of the 81st anniversary of the national army, two drones equipped with powerful explosives were used by a terrorist group in an attempt to assassinate the president. Possibly, this is the first time terrorism by an UAV occurred. Fortunately, the president was not harmed, but a few soldiers were injured. Two aspects get highlighted with this drone attack: How easily drones could be used against the highest authority of a nation and how difficult it is to defend against such attempts.[52] The attempt to assassinate Nicholas Maduro came close and may embolden similar experiments by terrorist groups to employ weaponised drones to eliminate their targets.

The use of weaponised drones by non-state actors or a group acting at the behest of nation-states is a concern. The proliferation of UAV technologies, and their access by the terrorists or insurgents, can wreak havoc; drones can be used to harm a wide range of objects – from an individual target to a flying airliner. State-sponsored terrorist attacks making use of weaponised drones is a grim reality. It is already happening with Hezbollah and probably with the Houthi rebels, who have used drones to attack the Saudi air defences in Yemen. Some groups are mastering drone technology without the help of state sponsors. In Syria, the Islamic State has successfully used drones to conduct surveillance and reconnaissance, in addition to carrying out offensive actions like dropping a grenade on an adversary's military base. A terrorist group can steal or purchase a drone from a rogue state or corrupt military or intelligence officials.[53] On 27 June 2021, two weaponised drones attacked an Indian Air Force Station in Jammu; each drone had a payload of more than 2 kgs of explosives. India raised the matter in the General Assembly of the United Nations and asked the august body to take a serious note of the use of weaponised drones for terror activity.

One of the most dreaded scenarios would be terrorist groups launching a massive attack using weapons of mass destruction delivered through drones in a large public gathering like a sports stadium, political rally, or busy railway station, which may result in significant damage to lives. Even if it fails to create major damage, it can still create fear psychosis – an objective of terrorism. The possibility that drones could be used to disperse deadly chemical agents or viruses over a sports stadium or public gathering is a dangerous prospect. Even if a drone attack fails to result in large numbers of fatalities, the attempt could still achieve an attacker's goal of perpetuating the psychological dimension of terrorism.[54]

Terrorists will always aim to find newer means to generate fear psychosis and chaos. They are not constrained by international laws on war, as their organisations may not have legal recognition. As part of the counterterrorism response to the threat, a robust and all-encompassing approach has to be taken, such as detection, counter-measure mechanisms and imparting training as part of preparation to defend against attack by weaponised drones.[55]

An analysis of the following provisions of the UN is relevant in examining the legal aspects of drones during war or in non-war zones. Article 1 spells out the Purposes of the United Nations and Article 51 enumerates aspects of self-defence as under:[56]

Article 1 states that the purpose of the United Nations is

To maintain international peace and security, and to that end: to take effective collective measures for the prevention and removal of threats to the peace, and for the suppression of acts of aggression or other breaches of the peace, and to bring about by peaceful means, and in conformity with the principles of justice and international law, adjustment or settlement of international disputes or situations which might lead to a breach of the peace.

Article 51 states that

Nothing in the present Charter shall impair the inherent right of individual or collective self-defence if an armed attack occurs against a Member of the United Nations, until the Security Council has taken measures necessary to maintain international peace and security. Measures taken by Members in the exercise of this right of self-defence shall be immediately reported to the Security Council and shall not in any way affect the authority and responsibility of the Security Council under the present Charter to take at any time such action as it deems necessary in order to maintain or restore international peace and security.

Ben Saul, an expert on international law, avers that the issue of self-defence is not applicable to the USA as Al Qaeda is not in direct confrontation with the USA as of now and the USA, having left Afghanistan, has no right to transgress the territorial sovereignty of Afghanistan. And the Taliban government of Afghanistan has not threatened the security of the USA by employing Al Qaeda. The UN's monitoring team in July 2022 concluded that 'Al-Qa[e]da is not viewed as posing an immediate international threat from its safe haven in Afghanistan because it lacks an external operational capability and does not currently wish to cause the Taliban international difficulty or embarrassment.'[57] The 2020 Doha agreement commits the Taliban government in Afghanistan to prevent terrorist activity against the United States from its soil. Though the USA, citing this agreement, would like to give legality to their drone attack that killed Zawahiri, it must be noted that after the 2009 abortive bombing attempt in New York, no other act of terrorism has been committed by Al-Qaeda threatening the security of the USA. Therefore, the drone attack was an extrajudicial act and not compliant with international law. In 2021, the American drones' attack against Islamic State – Khorasan Province – killed 10 civilians, but the drone strike against Zawahiri, without collateral damage, has enhanced US confidence. The international response to pinpointed target elimination in Afghanistan was mild and there was not much international uproar – even the media gave it a pass.

Can AI help in defending India better?

India has fought three wars with Pakistan and one with China. Defending India, with two nuclear-armed adversaries – China and Pakistan – is a great challenge. And that challenge becomes formidable if there is a collaborative two-front threat. India also faces additional challenges of militancy and insurgency internally as she is beset with terrorists' attacks in Jammu and Kashmir and other places, in addition to prolonged insurgency in its North-Eastern states. Who can forget the dastardly 26/11 attack in Mumbai in 2008 by the Pakistani terror group *Lashkar-e-Taiba*? Pakistan's role in spreading terrorism is well established. Bleeding India through a thousand cuts has been a long-term strategy which terror groups based in Pakistan are implementing. And India is not the lone sufferer. More than half of the major terrorist plots against the West between 2004 and 2011, for instance, had connections with Pakistan.[58]

For better management of insurgency in North-East India, there have been occasional difficulties for Delhi to garner support and consistent cooperation from neighbouring countries, like Bangladesh and Myanmar, for intelligence collection and assistance in law enforcement. For terrorism in Jammu and Kashmir and other places in India, Pakistan hardly ever cooperated, though on many occasions they overtly pronounced these terrorists, originating from their soil, as non-state actors, or freedom fighters whom their Inter-Services Intelligence (ISI) was covertly abetting and supporting. The USA was able to get support of the Pakistan government when they were droning suspected terrorist hideouts in the Federally Administered Tribal Areas (FATA) region, but for India it always remained a distant dream. Obviously, there will be more difficulty in locating targets even if India wants to make use of drones against the terrorists based in Pakistan. After years of terror strikes, supported by Pakistan, India was compelled to carry out a surgical strike in October 2016, possibly to avenge the deaths of the Uri terror attacks in an army camp on 18 September 2016. During the operation, the army commandos, in a daring, precise, and quick operation across the Line of Control (LoC) in Pakistan occupied Kashmir (POK), dismantled six terrorist camps. Days after 40 Central Reserve Police Force (CRPF) personnel were killed on 14 February 2019, in a suicide car attack in Pulwama, Jammu and Kashmir, the Indian Air Force launched air strikes at the Balakot terrorist camp in Pakistan on 26 February 2019. Through this air strike, the Indian Government, displaying strong political will, sent a clear-cut message to the terrorist group *Jaish-e-Mohammad* and Pakistan that attacks like Uri and Pulwama would come at a cost. If India had AI-powered drones, possibly they could have been employed for these kinds of tasks or employed in tandem after assessing the target value. AI-augmented drones would have been a valuable addition, if not a substitute, to combat soldiers and airmen.

Besides the India-China War of 1962, there have been many border violations, disregarding mutually agreed protocols. In recent years, China's intrusion in Doklam (June 2017), Galwan (June 2020) and at the Arunachal border (December 2022) are cases in point. With the Chinese Navy's expanding presence in the Indian Ocean; their vessels are foraying for various purposes like survey and research. Chinese spy ship, *Yuan Wang VI*, entered the Indian Ocean ahead of India's planned missile test in the second week of November 2022. The Chinese vessel had the capability of tracking satellites and ballistic missiles and gathering intelligence. Before the entry of this ship, in September 2019, Chinese research and survey vessels were spotted in the Indian Ocean, south of Andaman and Nicobar Islands, in January 2021.

Pakistan is using drones to supply arms and explosives to terrorist groups in Jammu and Kashmir and Punjab. Many times these have been caught by the security forces. Along with countering these drones with precision engagement, there is a need to complement the effort with other extensive counterterrorism measures, such as the disruption of terrorist financing, information and cyber operations, plus security and counterterrorism assistance to states in the region, based on accurate real-time intelligence where practicable. The government of India has initiated actions towards this end and the results are visible.

India is well known for its soft-skills power in the domain of IT. Microsoft CEO Satya Nadela has, in an interview with CNBC TV18 on 4 January 2023, said that of all the world's software developers, India is number two. India can make best use of her talents to empower itself with AI-powered means, that can help the defence and security forces, as force-multipliers and enablers, to meet the challenges of conventional and irregular warfare. The Indian Navy has given a contract for manufacturing weaponised autonomous swarming boats. Possibly, the navy will need autonomous systems powered by advanced-level AI technologies (like underwater drones) for naval mine warfare and protecting submarines and ships. What India needs is AI that is explainable and can make informed decisions, for both manned and unmanned systems, which are fully trustworthy and will not act outside of what they have been developed to do.

Conclusion

Carl von Clausewitz, in his famous treatise *On War,* wrote 'Every age has its own kind of war, its own limiting conditions, and its own peculiar preconceptions.'[59] Now is the age of AI, which is poised to play an ever-increasing role in military systems, in cyber, hybrid and Net-Centric War (NCW). Its imprint is visible in our daily life, in areas like health, infrastructure and a plethora of other areas of finance, management and governance. The list is endless. But the problem is that technology is developing at a rapid pace, and the ethical, social and legal gaps are widening due to the slower process of policymaking.[60]

AI can be transformative as well as destructive depending on its adoption. In essence, there can be good as well as evil uses of AI and that calls for the developers, users, policymakers and leaders to differentiate between ethical and unethical uses of AI. The most important aspect in the design and deployment of AI-assisted weapons is that it would be unethical and illegal to delegate the decision to take a human life to a machine. There is a moral component to this position, that we should not allow machines to make life-taking decisions for which others – or nobody – will be culpable.[61]

For the past couple of years, there has been a considerable increase in debate in public forums questioning the ethics, legality, and most importantly morality of the development and use of AI-powered weapons. Many scientists and engineers researching in the AI domain concur with the view that the non-inclusion of attributability of taking human lives will lead to extrajudicial killings, besides making AI-powered weaponry powerful instruments of extreme violence. Many organisations are coming together under one umbrella to express their concerns through an open letter of pledge against such weapons, where it has been highlighted that lethal autonomous weapons have characteristics quite different from nuclear, chemical and biological weapons, and the unilateral actions of a single group could too easily spark an arms race. The international community lacks the technical tools and global governance systems to manage an AI generated arms race. Stigmatising and preventing such an arms race should be a high priority for both national and global security.[62] One also needs to ponder whether possession of AWS can at all change the balance of power and global order.

Offensive drone strikes targeting transnational terror groups have become part and parcel of the strategy of the USA, Turkey and a few other countries, and one will not be surprised if more countries follow a similar strategy to fight terrorism. When one compares the two world wars of the last century with the contemporary armed conflicts, the latter can be treated as low-level affairs, yet they pose serious and distinctive moral problems.[63]

Henry Kissinger feels that cyber weapons and AI applications (such as AWS) greatly complicate the current dangerous war prospects. Unlike nuclear weapons, cyber weapons embedded with AI are ubiquitous, relatively inexpensive to develop and easy to use. Because the threshold for their use is so low and their destructive ability so great, the use of such weapons – or even their mere threat – can turn a crisis into a war or turn a limited war into a nuclear war through unintentional or uncontrollable escalation.[64]

There has been an outcry by the human rights group against transnational drone strikes against terror groups in areas not designated as normal war zones. Instances like Zawahiri are described as targeted killing by those undertaking the operation, but the act does fail in the scrutiny of the experts of internal laws on war, and according to them, the euphemism of 'targeted killing' is actually an 'assassination.' Noticeably, the USA's single-minded purpose of eliminating terrorism and to achieve that aim, which means disregarding international rule-based order, did not evoke much adverse media response because terrorism is a global scourge. There were some protests in the aftermath of the killing of Zawahiri, which did not continue for long. Apparently, the Taliban government did not object to the presence of Zawahiri. Maybe politically, that did not suit them, though after the killing, the government of Afghanistan reported ignorance of Zawahiri's presence there. At the same time, the Afghan government was not very vocal against the US drone attack. Has Zawahiri's killing proved that even some violation of international law may be acceptable if it helps in reducing terror? Terrorism is unacceptable, and so is flouting international law at will. Townsend Hopes wrote of American leaders during the Vietnam War, 'struggling in good conscience to serve the broad national interest according to their lights.'[65] The same was true when a Predator drone was launched to kill Zawahiri with a Hellfire missile. May there be lights in the leaders of the nation states and in the terrorist masterminds too!

AI will see more and more integration between humans and machines in both operational and decision-making roles. It is time to ponder and do some soul searching on the thought-provoking question about the long-term consequences of AI posed by Yuval Noah Harari, author of the much-read book *Homo Deus: A Brief History of Tomorrow*, 'What will happen to society, politics and daily life when non-conscious but highly intelligent algorithms know us better than we know ourselves?'[66]

Notes

1 Jonathan Moreno et al., *The Ethics of AI-Assisted War Fighter Enhancement Research and Experimentation: Historical Perspectives and Ethical Challenges*, available at: https://www.ncbi.nlm.nih.gov/pmc/articles/PMC9500287/#fn-group-1title, accessed on 30 November 2022; A.M. Turing, 'Computing Machinery and Intelligence,' *Mind*, vol. 49 (1950), pp. 433–60, chrome-extension://efaidnbmnnnibpcajpcglclefindmkaj/https://redirect.cs.umbc.edu/courses/471/papers/turing.pdf.

2 James Vincent, 4 September 2017, available at: theverge.com, accessed on 30 November 2022.

3 Ana Santos Rutschman, 'Stephen Hawking Warned About the Perils of Artificial Intelligence – Yet AI Gave Him a Voice,' *The Conversation*, 15 March 2018, available at: https://theconversation.com/stephen-hawking-warned-about-the-perils-of-artificial-intelligence-yet-ai-gave-him-a-voice-93416#:~:text=We%20believe%20in%20the%20free%20flow%20of%20information&text=Hawking%20made%20no%20secret%20of,than%20this%20often%2Dcited%20soundbite, accessed on 1 December 2022. Also hear Elon Musk: We're 'Summoning the Demon' with Artificial Intelligence (Video) - Believers Portal.

4 Manuel Velasquez, Claire Andre, S.J. Thomas Shanks, and Michael J. Meyer, 'What Is Ethics?' *Markkula Center for Applied Ethics* (scu.edu), available at: https://www.scu.edu/ethics/ethics-resources/ethical-decision-making/what-is-ethics/.

5 Vincent C. Müller, 'Ethics of Artificial Intelligence and Robotics,' *Stanford Encyclopedia of Philosophy*, 30 April 2020. Archived from the original on 10 October 2020. Retrieved 26 September 2020.

6 'What are the Ethical Problems in Artificial Intelligence?' available at: https://www.geeksfor geeks.org/, accessed on 2 January 2023.

7 'What are AI Ethics (AI Code of Ethics)?' (techtarget.com), available at: https://www.techtarget.com/whatis/definition/AI-code-of-ethics#:~:text=AI%20ethics%20is%20a%20system,develop%20AI%20codes%20of%20ethics.

8 'Principles of AI Ethics Principles of AI Ethics: Enunciated by Jobin et al, 'The Global Landscape of AI Ethics Guidelines,' *Nature Machine Intelligence*, vol. 1, no. 9 (2019), pp. 389–99, https://doi.org/10.1038/s42256-0190088-2 quoted in Mark Ryan, Bern Carsten Stahl, 'Artificial Intelligence Ethics Guidelines for Developers and Users,' available at: https://www.emerald.com/insight/1477-996X.htm.

9 Wahyu Candra Dewi, 'A Matter of Ethics: Should Artificial Intelligence Be Deployed in Warfare?' *Modern Diplomacy*, 4 October 2022, available at: https://moderndiplomacy.eu/2022/10/04/a-matter-of-ethics-should-artificial-intelligence-be-deployed-in-warfare/, accessed on 12 December 2022.

10 Quoted from M.L. Cummings, *Artificial Intelligence and the Future of Warfare*, p. 12, available at: https://www.chathamhouse.org/sites/default/files/publications/research/2017-01-26-artificial-intelligence-future-warfare-cummings.pdf, accessed on 11 December 2022.

11 Alexander Lychagin, 'What Is Teletank?' *Odint Soviet News*, 9 October 2004, available at: https://www.duhoctrungquoc.vn/wiki/en/Teletank, accessed on 1 August 2010.

12 Prakash Nanda, 'Killer Robots: Watch How AI-Programmed Military Robots Could Make Human Soldiers Completely Obsolete,' eurasiantimes.com, 1 November 2021, https://eurasiantimes.com/unstoppable-military-robots-watch-how-ai-programmed-killer-robots-could-make-human-soldiers-completely-obsolete/.

13 https://www.robotsscience.com/military/military-robots-history-types-use-and-how-it-work/.

14 https://www.robotsscience.com/military/military-robots-history-types-use-and-how-it-work/.

15 John Keller, 'Boeing to Develop New Payloads, Capabilities, and Missions for Orca Large Long-Range Unmanned Submarines,' *Military & Aerospace Electronics*, 15 October 2020, quoted in Rise of the Robots: Weaponization of Artificial Intelligence (maritimeindia.org).

16 The word robotocalypse is akin to *Robopocalypse* – a 2011 science fiction novel by Daniel H. Wilson based on AI-powered robots which goes out of control beyond the imagination of the maker. Also see-killer-robots-could-make-human-soldiers-completely-obsolete/, https://eurasiantimes.com/unstoppable-military-robots-watch-how-ai-programmed.

17 Coralie Consigny, *blog*, 8 February 2022, 'Are Killer Robots Better Soldiers?: The Legality and Ethics of Using AI in War,' available at https://www.human-rightspulse.com/mastercontentblog/are-killer-robots-better-soldiers-the-legality-and-ethics-of-the-use-of-ai-at-war#:~:text=While%20using%20algorithms%20and%20AI,accountability%20and%20mutuality%20of%20risk. Also read G.S. Prathvik, 'Top Most Advanced and Artificial Intelligence (AI) Powered

Military Robots in 2022,' https://www.marktechpost.com/2022/08/17/top-most -advanced-and-artificial-intelligence-ai-powered-military-robots-in-202/.

18 https://www.hrw.org/report/2022/11/10/agenda-action/alternative-processes -negotiating-killer-robots-treaty.

19 Scientific American. The Changing Face of War (Kindle Locations 758–759).

20 Stefano Passini, Laura Palareti, Piergiorgio Battistelli, 'We vs. Them: Terrorism in an Intergroup Perspective,' *Dans Revue internationale de psychologie sociale*, 3–4 (Tome 22), 2009, pp. 35, 64, available at: https://www.cairn.info/revue -internationale-de-psychologie-sociale-2009-3-page-35.htm, accessed on 26 August 2022.

21 Manish Kumar Jha and Amit Das, 'AI Technology in Military Will Transform Future Warfare,' *Business World*, 13 August 2021, available at: https://www .businessworld.in/article/AI-Technology-In-Military-Will-Transform-Future -Warfare/13-08-2021-400525/, accessed on 31 March 2022.

22 Vincent C. Müller, 'Ethics of Artificial Intelligence and Robotics,' in Edward N. Zalta (ed.), *The Stanford Encyclopedia of Philosophy* (Summer 2021 Edition), available at: https://plato.stanford.edu/archives/sum2021/entries/ethics-ai/.

23 Available at https://news.un.org/en/story/2019/03/1035381.

24 Müller, 'Ethics of Artificial Intelligence and Robotics.'

25 'Military Robots are Getting Smaller and More Capable,' available at: https:// www.economist.com/science-and-technology/2017/12/14/military-robots-are -getting-smaller-and-more-capable?.

26 Herbert S. Lin, 'Offensive Cyber Operations and the Use of Force,' https://jnslp .com/2010/08/13/offensive-cyber-operations-and-the-use-of-force/, accessed on 17 September 2022 and Jack Goldsmith and a Berkman Center Cybersecurity Team, Cyber-Attack v. Cyber-Exploitation, https://h2o.law.harvard.edu/play-lists/657, accessed on 18 September 2022.

27 Published by S. Ganbold, 24 November 2021, https://www.statista.com/statis-tics/265156/internet-penetration-rate-in-asia/)   https://worldpopulationreview .com/country-rankings/internet-users-by-country.

28 *The Economic Times*, 4 August 2022, https://economictimes.indiatimes.com/ news/defence/from-7-11s-to-train-stations-cyber-attacks-plague-taiwan-over -pelosi-visit/articleshow/93355548.cms?from=mdr.

29 Artificial Intelligence and Offensive Cyber Weapons, ISSS publication, Volume 25, December 2019, https://www.iiss.org/publications/strategic-comments/2019/arti-ficial-intelligence-and-offensive-cyber-weapons#:~:text=The%20prospect%20of %20autonomous%20AI,physical%20battlefield%20but%20in%20cyberspace.

30 Kishore Vats, *Future of Space Warfare* (New Delhi: Surendra Publications, 2021), pp. 2–3.

31 Vats, *Future of Space Warfare*, pp. 2–3.

32 Nancy K. Hayden, Jason Reinhardt, Mallory Stewart, J. D. Doak, Thushara Gunda, Kelsey Abel, 'Artificial Intelligence and Autonomy in Space: Balancing Risks of Unintended Escalation,' 2 March 2020, available at: https://www.osti .gov/servlets/purl/1780585.

33 Michael Johnson, 'The Emerging Commercial Marketplace in Low-Earth Orbit,' 27 February 2019, available at: https//www.nasa.gov/mission_pages/sta-tion/research/news /b4h-3rd/ev-emerging-commercial-market-in-leo.

34 'Artificial Intelligence and Offensive Cyber Weapons,' December 2019, IISS, ORG, vol. 25, available at: https://www.iiss.org/publications/strategic-com-ments/2019/artificial-intelligence-and-offensive-cyber-weapons.

35 M. Taddeo, D. McNeish, A. Blanchard et al., 'Ethical Principles for Artificial Intelligence in National Defence,' *Philosophy & Technology*, vol. 34 (2021), pp. 1707–29, https://doi.org/10.1007/s13347-021-00482-3.

36 https://www.merriam-webster.com/dictionary/asymmetric%20warfare.

37 Frank Hoffman, *Conflict in the 21st Century: The Rise of Hybrid Wars* (Arlington: Potomac Institute for Policy Studies, 2007).

38 Brian P. Fleming, 'Hybrid Threat Concept: Contemporary War, Military Planning and the Advent of Unrestricted Operational Art (PDF),' 19 May 2019, United States Army Command and General Staff College. Archived (PDF) from the original on 20 March 2017 quoted in Hybrid warfare – Wikipedia. https://indianstrategicknowledgeonline.com/web/2753.pdf.

39 Ashok Kumar Singh, *What Is Grey Zone Warfare?* available at: https://idsa.in/askanexpert/what-is-grey-zone-warfare, accessed on 28 December 2022.

40 Passini et al., 'We vs. Them: Terrorism in an Intergroup Perspective.'

41 Müller, 'Ethics of Artificial Intelligence and Robotics.'

42 United Nations Declaration on Measures to Eliminate International Terrorism annex to UN General Assembly resolution 49/60, 'Measures to Eliminate International Terrorism,' of 9 December 1994, http://hrlibrary.umn.edu/resolutions/49/60GA1994.html.

43 https://www.un.org/securitycouncil/ctc/, accessed on 26 August 2022. Also see United against Terror, *The Hindu* (editorial), 31 October 2022.

44 Resolution 1566 Cumulatively Requires: (a) An Intention to Cause Death or Serious Bodily Injury or Hostage Taking, (b) An Offense Under One of the 19 Existing 'Counterterrorism' Conventions, and (c) A Purpose (or 'Specific Intent') to Provoke a State of Terror in the Public or a Group of Persons, or to Intimidate a Population, or to Compel a Government or International Organization to do or to Abstain from Doing Any Act. Excerpts from article by Ben Saul, The Legal Black Hole in United Nations Counterterrorism, 2 June 2022, available at: https://theglobalobservatory.org/2021/06/the-legal-black-hole-in-united-nations-counterterrorism/.

45 https://www.reuters.com/article/us-david-rohde-drone-wars-idUSTRE80P11I20120126, accessed on 11 September 2022.

46 Seth J. Frantzman, 'Turkey Accused of Killing Kurdish Female Commander in Syria,' *The Jerusalem Post*, 28 July 2022, https://www.jpost.com/middle-east/article-713373, accessed on 11 September 2022.

47 Ben Saul, 'The Unlawful U.S. Killing of Ayman al-Zawahri,' 17 August 2022, https://www.lawfareblog.com/unlawful-us-killing-ayman-al-zawahri accessed on 27 August 2022. Ben Saul tweets at @profbensaul.

48 Michael Walzer, *Just and Unjust Wars: A Moral Argument with Historical Illustrations* (New York: Basic Books, 1977), pp. 216–22.

49 Walzer, *Just and Unjust Wars*, p. 217.

50 Brian Orend, *The Morality of War*, 2nd ed (Ontario: Broadview Press, 2013), p. 197. However, on the issue of reprisals Orend feels that it has 'dubious probability of success.'

51 https://www.marketsandmarkets.com/Market-Reports/unmanned-aerial-vehicles-uav-market-662.html, accessed on 25 August 2022.

52 Colin P. Clarke, 'Approaching a "New Normal": What the Drone Attack in Venezuela Portends,' *The RAND Blog*, 13 August 2018, available at: https://www.rand.org/blog/2018/08/approaching-a-new-normal-what-the-drone-attack-in-venezuela.html, accessed on 11 August 2022.

53 Clarke, 'Approaching a "New Normal": What the Drone Attack in Venezuela Portends.'

54 http://dlibra.bg.ajd.czest.pl:8080/dlibra/docmetadata?id=4404&from=publication, accessed on 11 September 2022.

55 http://dlibra.bg.ajd.czest.pl:8080/dlibra/docmetadata?id=4404&from=publication, accessed on 11 September 2022.

56 https://www.un.org/en/about-us/un-chartull-text, accessed on 10 September 2022.
57 Ben Saul, Lawfare Blog, 'The Unlawful US Killing of Ayman al-Zawahri,' 17 August 2022, available at: https://www.lawfareblog.com/unlawful-us-killing -ayman-al-zawahri, accessed on 24 August 2022.
58 Mitt Regan, 'Do Targeted Strikes Work? The Lessons of Two Decades of Drone Warfare,' Modern Institute of Warfare at West Point, 6 February 2022, https:// mwi.usma.edu/do-targeted-strikes-work-the-lessons-of-two-decades-of-drone -warfare/, accessed on 22 August 2022.
59 Carl von Clausewitz, *On War*, tr. and ed. by Michael Howard and Peter Paret (Princeton: Princeton University Press, 1989), p. 593.
60 Sherry Wasilow and Joelle Thorpe, 'Artificial Intelligence, Robotics, Ethics, and the Military: A Canadian Perspective,' *AI Magazine*, 40,2019, pp. 37–48, https://doi.org/10.1609/aimag.v40i1.2848.
61 'A Security Perspective: Security Concerns and Possible Arms Control Approaches,' in United Nations Office for Disarmament Affairs (UNODA) (2017), Perspectives on Lethal Autonomous Weapons Systems, UNODA Occasional Papers No. 30, November 2017, New York: United Nations, https:// www.un.org/disarmament/publications/occasionalpapers/unoda-occasional -papers-no-30-november-2017/, pp. 19–33.
62 'Lethal Autonomous Weapons Pledge,' available at: https://futureoflife.org/open -letter/lethal-autonomous-weapons-pledge/.
63 chrome-extension://efaidnbmnnnibpcajpcglclefindmkaj/https://dema.az.gov/ sites/default/files/Publications/AR-Terrorism%20Definitions-BORUNDA.pdf, accessed on 5 September 2022.
64 Giancarlo Elia Valori, 'Kissinger and the Current Situation Considering the Development of Artificial Intelligence and the Ukraine Crisis,' 26 November 2022, available at: https://moderndiplomacy.eu/2022/11/26/kissinger-and-the -current-situation-considering-the-development-of-artificial-intelligence-and -the-ukrainian-crisis/, accessed on 12 December 2022.
65 Quoted in Noam Chomsky, *At War with Asia* (New York: Vintage Books, 1970), p. 310.
66 Quoted from Yuval Noah Harari, *Homo Deus: A Brief History of Tomorrow* (New York: Harper, 2016), p. 462.

9
ARTIFICIAL INTELLIGENCE IN MILITARY OPERATIONS

Technology, ethics and the Indian perspective

R.S. Panwar

Introduction

AI has become a field of intense interest and high expectations within the defence technology community. AI technologies hold great promise for facilitating military decisions, minimising human casualties and enhancing the combat potential of forces, and in the process dramatically changing, if not revolutionising, the design of military systems. This is especially true in a wartime environment, when data availability is high, decision periods are short, and decision effectiveness is an absolute necessity.

The rise in the use of increasingly autonomous unmanned aerial vehicles (UAVs) in military settings has been accompanied by a heated debate as to whether there should be an outright ban on Lethal Autonomous Weapon Systems (LAWS), sometimes referred to as 'killer robots.' Such AI-enabled robots, which could be in the air, on the ground, or under water, would theoretically be capable of executing missions on their own. The debate concerns whether artificially intelligent machines should be allowed to execute such military missions, especially in scenarios where human lives are at stake. This chapter focuses on the development and fielding of LAWS against the backdrop of rapid advances in the field of AI, with special emphasis on the legal and ethical issues associated with their deployment. It also reviews the status of AI technology in India, assesses the current capability of the Indian Army to adapt to this technology, and suggests steps that need to be taken on priority to ensure that we do not get left behind other advanced armies in the race to usher in a new AI-triggered Revolution in Military Affairs (RMA).

DOI: 10.4324/9781003421849-10

AI: Current status of technology

AI is a maturing technology. A general definition of AI is the capability of a computer system to perform tasks that normally require human intelligence, such as visual perception, speech recognition and decision-making. Functionally, AI-enabled machines should have the capability to learn, reason, judge, predict, infer and initiate action. In layman's terms, AI implies trying to emulate the brain. There are three main ingredients that are necessary for simulating intelligence: The brain, the body and the mind. The brain consists of the software algorithms that work on available data, the body is the hardware, and the mind is the computing power that runs the algorithms. Technological breakthroughs and convergence in these areas are enabling the AI field to rapidly mature.

In 2016, in a significant development, Google DeepMind's AlphaGo program defeated South Korean Master Lee Se-dol in the popular board game Go, and the terms AI, machine learning and deep learning were used to describe how DeepMind won. The easiest way to think of their inter-relationship is to visualise them as concentric circles, with AI the largest, then machine learning, and finally deep learning – which is driving today's AI explosion – fitting inside both.[1] AI is any technique that enables computers to mimic human intelligence. Machine learning is a subset of AI, which focuses on the development of computer programs that can change when exposed to new data, by searching through data to look for patterns and adjusting program actions accordingly. Deep learning is a further subset of machine learning that is composed of algorithms which permit software to train itself by exposing multi-layered neural networks (which are designed on concepts borrowed from a study of the neurological structure of the brain) to vast amounts of data.

The most significant AI technologies making rapid progress today are natural language processing and generation, speech recognition, text analytics, machine learning and deep learning platforms, decision management, biometrics and robotic process automation. Some of the major players in this space are: Google, now famous for its artificial neural network-based AlphaGo program; Facebook, which has recently announced several new algorithms; IBM, known for Watson, which is a cognitive system that leverages machine learning to derive insights from data; Microsoft, which helps developers to build Android, iOS and Windows apps using powerful intelligence algorithms; Toyota, which has a major focus on automotive autonomy (driverless cars); and Baidu Research, the Chinese firm that brings together global research talent to work on AI technologies.[2]

Today, while AI is most commonly cited for image recognition, natural language processing and voice recognition, this is just an early manifestation of its full potential. The next step will be the ability to reason, and in fact reach a level where an AI system is functionally indistinguishable from a human. With such a capability, AI-based systems would potentially have an infinite number of applications.[3]

In a 1951 paper, Alan Turing proposed the Turing Test to test for AI. It envisages two contestants consisting of a human and a machine, with a judge, suitably screened from them, tasked with deciding which of the two is talking to him. While there have been two well-known computer programs claiming to have cleared the Turing Test, the reality is that no AI system has been able to pass it since it was introduced. Turing himself thought that by the year 2000 computer systems would be able to pass the test with flying colours. While there is much disagreement as to when a computer will actually pass the Turing Test, one thing all AI scientists generally agree on is that it is very likely to happen in our lifetime.[4]

There is a growing fear that machines with AI will get so smart that they will take over and end civilisation. This belief is probably rooted in the fact that most of society does not have an adequate understanding of this technology. AI is less feared in engineering circles because there is a slightly more hands-on understanding of the technology. There is perhaps a potential for AI to be abused in the future, but that is a possibility with any technology. Apprehensions about AI leading to end-of-civilisation scenarios are perhaps largely based on fear of the unknown, and are largely unfounded.

AI in military operations

Is AI a harbinger of a New RMA? Robotic systems are now widely present in the modern battlefield. Increasing levels of autonomy are being seen in systems which are already fielded or are under development, ranging from systems capable of autonomously performing their own search, detect, evaluation, track, engage and kill assessment functions, fire-and-forget munitions, loitering torpedoes, and intelligent anti-submarine or anti-tank mines, among numerous other examples. In view of these developments, many now consider AI and robotics technologies as having the potential to trigger a new RMA, especially as Lethal Autonomous Weapon Systems (LAWS) continue to achieve increasing levels of sophistication and capability.

It is actually difficult to precisely define what is a LAWS. In the acronym 'LAWS,' there is a fair amount of ambiguity in the usage of the term 'autonomous,' and there is a lack of consensus on how a 'fully autonomous' weapon system should be characterised. In this context, two definitions merit mention. A 2012 US Department of Defense (DoD) directive defines an autonomous weapon system as one that, 'once activated, can select and engage targets without further intervention by a human operator.' More significantly, it defines a semi-autonomous weapon system as one that, once activated, is intended to engage individual targets or specific target groups that have been selected by a human operator. By this yardstick, a weapon system, once programmed by a human to destroy a target group (which could well be interpreted to be an entire army) and thereafter seeks and destroys individual targets autonomously, would still be classified as semi-autonomous.[5]

The Human Rights Watch (HRW) definition differs. As per HRW, fully autonomous weapons are those that, once initiated, will be able to operate without Meaningful Human Control (MHC). They will be able to select and engage targets on their own, rather than requiring a human to make targeting and kill decisions for each individual attack. However, in the absence of consensus on how MHC is to be specified, it concedes that there is a lack of clarity on the definition of LAWS.[6]

There is a view that rather than focus on autonomous systems alone, there is a need to leverage the power of AI for increasing the combat power of the current force. This approach is referred to as Narrow or Weak AI. Narrow AI could lead to many benefits, such as using image recognition from video feeds to identify imminent threats, anticipating supply bottlenecks, automating administrative functions, etc. Such applications would permit force restructuring, with a smaller staff comprising of data scientists replacing large organisations. Narrow AI thus has the potential to help the Defence Forces improve their teeth-to-tail ratio.[7]

Another focus area on the evolutionary route to the development of autonomous weapons is what can be termed as human-machine teaming, wherein machines and humans work together in a symbiotic relationship. Like the mythical centaur, this approach envisages harnessing inhuman speed and power to human judgement, combining machine precision and reliability with human robustness and flexibility, as well as enabling computers and humans to help each other to think, termed as cognitive teaming. Some functions will necessarily have to be completely automated, like missile defence lasers or cybersecurity, and in all such cases where there is no time for human intervention. But, at least in the medium term, most military AI applications are likely to be teamwork: Computers will fly the missiles, aim the lasers, jam the signals, read the sensors, and pull all the data together over a network, putting it into an intuitive interface, which humans, using their experience, can use to make well-informed decisions.[8]

LAWS: Legal and ethical issues

LAWS powered by AI are currently the subject of much debate based on ethical and legal concerns, with human rights proponents recommending that the development of such weapons should be banned, as they would not be in line with international humanitarian law (IHL) under the Geneva Convention. The legal debate over LAWS revolves around three fundamental issues. The first is the *Principle of Distinction*. This principle requires parties to an armed conflict to distinguish civilian populations and assets from military assets, and to target only the latter (Article 51[4][b] of Additional Protocol I). The second one is the *Principle of Proportionality*. The law of proportionality requires parties to a conflict to determine the civilian cost of achieving a particular military target and prohibits an attack if the civilian harm exceeds the military advantage (Articles 51[5][b] and 57[2][iii] of Additional Protocol I). The third one is the issue of *Legal Review*. The rule on legal review provides that signatories to the Convention are obliged to determine whether or not new weapons as well as means and methods of warfare are in adherence to the Convention or any other international law (Article 36 of Additional Protocol I).

It has also been argued that fully autonomous weapon systems do not pass muster under the Marten's Clause, which requires that 'in cases not covered by the law in force, the human person remains under the protection of the principles of humanity and the dictates of the public conscience' (Preamble to Additional Protocol I).[9] Then there is the 'Campaign to Stop Killer Robots.' Under this banner, the HRW has argued that fully autonomous weapon systems would be prima facie illegal as they would never be able to adhere to the above provisions of IHL, since such adherence requires a subjective judgement, which machines can never achieve. Hence, their development should be banned at this stage itself.[10]

There exists an equally vocal body of opinion which states that the development and deployment of LAWS would not be illegal, and in fact, would lead to the saving of human lives. Some of their views are given here. LAWS do not need to have self-preservation as a foremost drive, and hence can be used in a self-sacrificing manner, saving human lives in the process. They can be designed without emotions that normally cloud human judgement during battle leading to unnecessary loss of lives. When working as a team with human soldiers, autonomous systems have the potential capability of objectively monitoring ethical behaviour on the battlefield by all parties. Then, the eventual development of robotic sensors superior to human capabilities would enable robotic systems to pierce the fog of war, leading to better informed 'kill' decisions. Autonomous weapons would have a wide range of uses in scenarios where civilian loss would be minimal or non-existent, such as naval warfare. The question of legality depends on how these weapons are used, not their development or existence.[11] Finally, it is too early to argue over the legal issues surrounding autonomous weapons because the technology itself has not been completely developed yet.

It is better to take a middle path between these two extreme positions. The middle path involves discussion on the degree of autonomy and meaningful human control (MHC). LAWS have been broadly classified into three categories. Human-in-the-Loop LAWS can select targets, while humans take the 'kill' decision. Human-on-the-Loop weapons can select as well as take kill decisions autonomously, while a human may override the decision by exerting oversight. Then human-out-of-the-loop LAWS are those that may select and engage targets without any human interaction. Entwined within this categorisation is the concept of MHC, i.e., the degree of human control which would pass muster under IHL. Despite extensive discussions at many levels, there is no consensus so far on what is meant by full autonomy nor on how MHC should be conceptualised.[12]

Triggered by the initiatives of HRW and other NGOs, an informal group of experts from a large number of countries debated the issue of LAWS for several years at the United Nations Office of Disarmament Affairs (UNODA) forum, Convention on Certain Conventional Weapons (CCW). In December 2016, several countries agreed to formalise these deliberations, and as a result, a Group of Governmental Experts (GGE) was established, the first of which was held from 13–17 November 2017, chaired by Ambassador Amandeep Gill of India. Approximately 90 countries along with many other agencies participated in the meeting. Some of the conclusions arrived at during the meeting were as follows. States must ensure accountability for lethal action by any weapon system used by them in armed conflict, acknowledging the dual nature of the technologies involved. The Group's efforts should not hamper civilian research and development in these technologies, and there is a need to keep potential military applications using these technologies under review.[13] Thereafter, the GGE has been holding meetings regularly, with the latest one concluding its deliberations in May 2023.

AI in military operations: International perspective

As of now, near-autonomous defensive systems have been deployed by several countries to intercept incoming attacks. Offensive weapon systems, in contrast, would be those which may be deployed anywhere and actively seek out targets. However, the difference between offensive and defensive weapons is not watertight. The most well-known autonomous defensive weaponry are missile defence systems, such as the Iron Dome of Israel and the Phalanx Close-In Weapon System used by the US Navy. Fire-and-forget systems, such as the Brimstone missile system of the United Kingdom and the Harpy Air Defence Suppression System of Israel, are also near-autonomous. South Korea uses the SGR-A1, a sentry robot with an automatic mode, in the Demilitarised Zone with North Korea. One example of an offensive autonomous system likely to be deployed in the near future is Norway's Joint Strike Missile, which can hunt, recognise and detect a target ship or land-based object without human intervention.[14]

The USA has put AI at the centre of its quest to maintain its military dominance. In November 2014, the then US Secretary of Defense Chuck Hagel announced a new Defense Innovation Initiative, also termed the Third Offset

Strategy. Secretary Hagel modelled his approach on the First Offset Strategy of the 1950s, in which the US countered the Soviet Union's conventional numerical superiority through the build-up of America's nuclear deterrent, and on the Second Offset Strategy of the 1970s, in which it shepherded the development of precision-guided munitions, stealth, and intelligence, surveillance and reconnaissance (ISR) systems to counter the numerical superiority and improving technical capability of Warsaw Pact forces. As a part of its Third Offset Strategy, the Pentagon reportedly dedicated $18 billion for its Future Years Defense Program. A substantial portion of this amount has been allocated for robotics, autonomous systems, human-machine collaboration and cyber and electronic warfare.[15]

China is also focusing in a big way on AI-enabled autonomous systems. Chinese military leaders and strategists believe that the nature of warfare is fundamentally changing due to unmanned platforms. High-level support for research and development (R&D) in robotics and unmanned systems has led to a myriad of institutes within China's defence industry and universities conducting robotics research. China's leaders have labelled AI research as a national priority, and there appears to be a lot of coordination between civilian and military research in this field.[16]

Indian perspective on the use of AI-enabled weapons

Perhaps as a result of being preoccupied with the huge challenges being faced on operational and logistic fronts, including issues related to modernisation, the AI/robotics/LAWS paradigm is yet to become a key driving force in the doctrinal thinking and perspective planning of the Indian Army (IA). The above discussion dictates that this needs to change. The following paragraphs shed some light on the relevance of AI and LAWS in our context and what we need to do in order to keep pace with twenty-first century warfare.

The Indian military landscape is comprised of a wide variety of scenarios where autonomous systems (AS), and more specifically LAWS, can be deployed with advantage. With the progressive development of AI technologies, various scenarios in increasing degrees of complexity can be visualised and they are discussed below.[17]

One is Anti-Improvised Explosive Device (IED) Operations. Autonomous systems designed to disarm IEDs are already in use in some form, although there is scope for further improvement. Such autonomous systems are non-lethal and defensive in nature. An AI-enabled swarm of surveillance drones (as opposed to manually piloted aerial vehicles or UAVs or Unmanned Undersea Vehicles (USVs) could greatly boost India's surveillance capabilities. Such a system would be non-lethal, but could support both offensive and defensive operations. There is scope for deployment of Robot Sentries, duly tailored to India's requirements, along the international border (IB)/ line of control (LC), on the lines of SGR-A1. Such a deployment would be categorised as lethal and defensive in character. India is currently in the process of procuring armed UAVs. Future armed UAVs/USVs with increasing degrees of autonomy in navigation, search, detection, evaluation, tracking, engagement and killing functions may be visualised. Such systems would be classified as lethal and offensive.

Offensive or killer robots deployed in land-based conventional offensive operations would require a much higher technological sophistication to become a feasible proposition. If robot soldiers are to be successfully deployed in counter-insurgency (COIN) operations, a very high AI technology threshold would need to be breached. In addition to a more sophisticated 'perceptual' ability to distinguish an adversary from amongst a friendly population, qualities such as empathy and ethical values similar to humans would need to be built into such systems. As per one school of thought, such capability can never be achieved, while others project reaching such a technological 'singularity' within this century.

India's response in international fora has been to hedge against the future and, until such weapons are developed, attempt to retain the balance of conventional power that it currently enjoys in the subcontinent. At the Informal Meeting of Experts on LAWS held in Geneva in April 2016, India reiterated this strategy. India's permanent representative at the UN, Ambassador D.B. Venkatesh Varma, stated that the UN CCW on LAWS should be strengthened in a manner that does not widen the technology gap amongst states, while at the same time endorsing the need to adhere to IHL while developing and deploying LAWS.[18]

International deliberations on legal and ethical issues related to LAWS are unlikely to slow the pace of their development and deployment by various countries. China is already well on its way to becoming a technology leader in this field, and Pakistan is expected to leverage its strategic relationship with China to obtain these technologies. India, therefore, needs to take urgent steps to ensure that it remains well ahead in this race. It can do this by leveraging the strengths of players from both the public and private sectors. The challenge for the Indian political leadership is to put together a cooperative framework where civilian academia and industry can collaborate with bodies like the Defence Research and Development Organisation (DRDO) to develop autonomous systems. Also, steps should be taken to ensure that the United States becomes India's strategic ally in autonomous technologies.[19]

The DRDO stated way back in 2013 that they were developing robotic soldiers and that these would be ready for deployment around 2023. Given DRDO's credibility based on past performance, these statements must be taken as an expression of intent rather than as the final word on delivery timelines. DRDO's main facility working in this area is the Centre for Artificial

Intelligence and Robotics (CAIR), whose vision, mission and objectives all refer to the development of intelligent systems, AI and robotics technologies. CAIR has achieved some headway in making some prototype systems, such as Muntra UGV, Daksh remotely operated vehicle, wall climbing and flapping wing robots, etc. It is now in the process of developing a Multi Agent Robotics Framework (MARF) for catering to a myriad of military applications. However, in order to keep in step with progress in the international arena, these efforts alone may not suffice.[20]

The Indian Defence Forces, and the Indian Army in particular, are still a long way off from operationalising even older generation technologies pertaining to the Network Centric Warfare (NCW), Information Operations (IO) and C4I2SR systems.[21] As regards next-generation technologies such as AI and robotics, presently there appears to be a void even in terms of concepts, doctrines and perspective plans. Occasional interactions with CAIR and other agencies do take place, mostly at the behest of the DRDO. Despite good intentions, DRDO is not likely to be successful in developing lethal and non-lethal autonomous systems without the necessary pull from the IA. It is also worth noting that worldwide, R&D in these technologies is being driven by the private commercial sector rather than the defence industry. Unfortunately, Indian equivalents of Baidu, Amazon, Google, and Microsoft, etc. are yet to rise to the occasion, despite the strengths of India's IT industry. Clearly, much more needs to be done.

Given the very high level of sophistication involved in AI/robotics technologies, together with the fact that the Indian public as well as private sector defence industry is not too mature, her project management interface with R&D agencies cannot afford to be based on purely operational knowledge. Therefore, while the Military Operations and Perspective Planning Directorates, in conjunction with HQ Army Training Command, would necessarily be central to formulation of concepts and doctrines, it is imperative to institute, in addition, a lead agency which, while being well-versed with operational requirements, has a clear grasp of these sophisticated technologies. Currently, Military College of Electronic and Mechanical Engineering (MCEME) is the designated Centre of Excellence for Robotics. Since AI is a sub-discipline of Computer Science, Military College of Telecommunications Engineering (MCTE) appears to be best placed to play the role of a lead agency for the development of AI-based autonomous systems, provided the Corps of Signals develops AI as an area of super-specialisation. It would be prudent, at this juncture, to brainstorm this issue at the apex level and take urgent follow-up action.

Conclusion

Given India's extended borders with her adversaries on two fronts and the volatile COIN scenarios in Jammu and Kashmir and the North-East, it is well appreciated that having sufficient boots on the ground is an absolute must. At the same time, it is imperative that the Indian Army keeps pace with the changing nature of warfare in the twenty-first century, driven by rapid advances in technology on many fronts. AI and robotics technologies, after decades of false starts, today appear to be at an inflection point and are rapidly being incorporated into a range of products and services in the commercial environment. It is only a matter of time before they manifest themselves in defence systems in ways significant enough to usher in a new RMA. Notwithstanding the worldwide concern on development of LAWS from legal and ethical points of view, it is increasingly clear that, no matter what conventions are adopted by the UN, R&D by major players in this area is likely to proceed unhindered.

Given India's own security landscape, the adoption of AI-based systems with increasing degrees of autonomy in various operational scenarios is expected to yield tremendous benefits in the coming years. Perhaps there is a need to adopt a radically different approach for facilitating the development of AI-based autonomous systems, utilising the best available expertise within and outside the country. As with any transformation, this is no easy task. Only a determined effort, with specialists on board and due impetus being given from the apex level, is likely to yield the desired results.

Notes

1 Michael Copeland, *What's the Difference Between Artificial Intelligence, Machine Learning, and Deep Learning?*, 29 July 2016, available at: https://blogs .nvidia.com/blog/whats-difference-artificial-intelligence-machine-learning-deep -learning-ai/, accessed on 18 Jan 2024.
2 Margaret Rouse, *Deep Learning (Deep Neural Networking)*, available at: http:// searchbusiness-analytics.techtarget.com, accessed on 6 June 2017.
3 Christina Cardoza, *What Is the Current State of AI?* 30 Sep 2016, available at: https://sdtimes.com/ai/current-state-ai/, accessed on 18 Jan 2024.
4 Bernado Goncalves, Can machines think? The controversy that led to the Turing Test, 11 Jan 2022, AI and Society, Vol 38, available at https://www.lan-stad.site/pro-editor/YYisVZWXyIwi, accessed on 18 Jan 2024
5 Ashton B. Carter, *Autonomy in Weapon Systems*, US Depart of Defence Directive 3000.09, 21 November 2012, available at: https://www.esd.whs.mil/portals/54/ documents/dd/issuances/dodd/300009p.pdf, accessed on 18 Jan 2024.
6 Mary Wareham, *Presentation on Campaign to Stop Killer Robots*, PIR Centre Conference on Emerging Technologies, Moscow, 29 September 2016, available at: https://www.stopkillerrobots.org/wp-content/uploads/2013/03/KRC _Moscow_29Sept2016.pdf, accessed on 18 Jan 2024.
7 Benjamin Jensen and Ryan Kendall, 'Waze for War: How the Army can Integrate Artificial Intelligence,' 2 September 2016, available at: https://warontherocks .com/2016/09/waze-for-war-how-the-army-can-integrate-artificial-intelligence/, accessed on 18 Jan 2024.
8 Sydney J Freedberg Jr, *Centaur Army: Bob Work, Robotics, & The Third Offset Strategy*, 9 November 2015, available at: https://breakingdefense.com/2015/11/ centaur-army-bob-work-robotics-the-third-offset-strategy/ ., accessed on 18 Jan 2024.
9 Protocol Additional to the Geneva Conventions of 12 August 1949, and relating to the Protection of Victims of International Armed Conflicts, 8 June 1977,

(hereafter Additional Protocol I), available at https://www.ohchr.org/en/instruments-mechanisms/instruments/protocol-additional-geneva-conventions-12-august-1949-and-0 , accessed on 18 Jan 2024.

10 International Human Rights Clinic, *Losing Humanity: The Case against Killer Robots,* Human Rights Watch, November 2012, available at: https://www.hrw.org/sites/default/files/reports/arms1112_ForUpload.pdf, accessed on 18 Jan 2024.

11 Ronald Arkin, *Counterpoint*, Communications of the ACM, December 2015, available at: https://cacm.acm.org/magazines/2015/12/194632-the-case-for-banning-killer-robots/abstract, accessed on 18 Jan 2024. .

12 Noel Sharkey, 'Towards a Principle for the Human Supervisory Control of Robot Weapons,' *Politica & Società*, no. 2 (2014), pp. 305–24; Kevin Neslage, 'Does "Meaningful Human Control" Have Potential for the Regulation of Autonomous Weapon Systems?' *University of Miami National Security and Armed Conflict Review* (2019), pp. 151–76.

13 United Nations, Report of the 2017 Group of Government Experts on Lethal Autonomous Weapon Systems (LAWS), Geneva, 22 Dec 2017, available at https://documents-dds-ny.un.org/doc/UNDOC/GEN/G17/367/06/PDF/G1736706.pdf?OpenElement, accessed on 18 Jan 2024.

14 R. Shashank Reddy, *India and the Challenge of Autonomous Weapons*, Washington, DC: Carnegie Endowment for International Peace, June 2016, available at https://carnegieendowment.org/files/CEIP_CP275_Reddy_final.pdf, accessed on 18 Jan 2024

15 Chuck Hagel, *The Defence Information Initiative*, Memorandum Secretary of Defence, 15 November 2014, available at: https://defenseinnovationmarketplace.dtic.mil/wp-content/uploads/2018/04/DefenseInnovationInitiative.pdf, accessed on 18 Jan 2024.

16 Jonathan Ray et al., *China's Industrial and Military Robotics Development*, Centre for Intelligence Research and Analysis, October 2016, available at: https://www.uscc.gov/sites/default/files/Research/DGI_China%27s%20Industrial%20and%20Military%20Robotics%20Development.pdf, accessed on 18 Jan 2024.

17 R.S. Panwar et al., *International Perspectives: Autonomy and Counter-Autonomy in Military Operations*, Panel Discussion, Carnegie Endowment for International Peace, Washington, 31 October 2016, available at https://carnegieendowment.org/2016/10/31/rise-of-artificial-intelligence-implications-for-military-operations-and-privacy-event-5392 , accessed on 18 Jan 2024.

18 Panwar et al., *International Perspectives: Autonomy and Counter-Autonomy in Military Operations.*

19 Bedavyasa Mohanty, *Command and Control: India's Place in the Lethal Autonomous Weapons Regime*, ORF Issue Brief, May 2016, available at https://www.orfonline.org/wp-content/uploads/2016/05/ORF_Issue_Brief_143_Mohanty.pdf, accessed on 18 Jan 2024.

20 Reddy, *India and the Challenge of Autonomous Weapons.*

21 R.S. Panwar, *NCW: Concepts and Challenges*, The Army War College Journal (Winter 2015), available at: https://www.academia.edu/43982158/Network_Centric_Warfare_A_Command_and_Control_Concept, accessed on 18 Jan 2024.

10
FIGHTING INSURGENTS WITH ROBOTS AND DRONES

A case study of the Indian armed forces

Kaushik Roy

Introduction

India's linguistic, religious and ethnic diversities along with the traumatic history of the partition of the subcontinent in 1947 are breeding grounds for numerous insurgencies. Pakistan, and to a lesser extent, China, provides moral and material (finance, weapons, training) support to the multitude of insurgents operating in India. Most of the insurgencies are concentrated in Kashmir and North-East India. There is an ongoing insurgency in Central India led by the Maoists (Naxals). However, not the armed forces but the civilian police and the paramilitary units are combating it. But in Kashmir and North-East India, the Indian Army plays the most conspicuous part in the counterinsurgency (COIN) tasks. The infantry is the principal branch dealing with the insurgents. Frequently, the infantry is backed up by paramilitary forces and armed police. In the last two decades, there has been some talk in the politico-military circles about using the Indian Air Force (IAF) and Indian Navy for COIN operations.

This chapter focuses on the role of Artificial Narrow Intelligence (ANI) in India's armed forces for combating armed internal rebellions. In this chapter, artificial intelligence (AI) will stand for ANI. At the beginning of the second decade of the twenty-first century, the Indian government and its armed forces woke up to the challenges and opportunities posed by AI. The non-state actors fighting the Indian state are eagerly adopting AI-related technologies. This chapter analyses the impact of AI-embedded weapons on the doctrine, tactics, command and ethical-legal aspects related to COIN operations. For the sake of brevity and lack of space, I will not deal with

DOI: 10.4324/9781003421849-11

the AI-related cyber challenges posed by the non-state actors and the Indian armed forces' cyber response to this threat.

Strategic scenario

From the very beginning, Pakistan could not come to terms with the reality that Muslim-dominated Kashmir would be part of India. Pakistan had fought three conventional wars (1947–48, 1965, 1971) and one quasi-conventional war (Kargil 1999) to occupy Kashmir but had eventually failed. Hence, Pakistan's strategy is to weaken India through a 'thousand cuts' by encouraging *jihadi* militancy in different parts of the country. Inter-Services Intelligence (ISI, an organ within the Pakistan Army) is the principal mechanism through which Islamabad recruits, trains and finances the various militant organisations that target India. Besides providing material and moral support to the insurgent groups operating in North-East India, China provides diplomatic and military support to Pakistan in the latter's endeavour to weaken India by pursuing the strategy of a 'thousand cuts.' Beijing's strategic aim is that India, embroiled in multiple armed internal rebellions, should thus remain engaged within South Asia and not be in a position to counter the People's Republic of China's (PRC) rising influence in South-East Asia and West Asia. In addition to China, Myanmar and Bangladesh occasionally provide arms and sanctuaries to the Mizo and Naga militants of North-East India.[1]

In desperation, India conceived of a strategy of launching a limited conventional strike by the integrated battle groups (IBGs) under the nuclear umbrella, lest Pakistan continue to aid the insurgents in Kashmir. This is known as India's Cold Start doctrine. *Indian Army Doctrine 2004* (ARTRAC 2004) is popularly known as Cold Start. It refers to quick mobilisation and then multiple offensives across a wide front. The Cold Start doctrine emerged in reaction to Operation Parakram (December 2001–June 2002) in which the strike corps mobilised very slowly, thus allowing Pakistan to strengthen its own defence. In fact, mobilisation of the three strike corps (each about 25,000 strong) during Operation Parakram took one month. In contrast, the Cold Start doctrine is designed to launch rapid limited and calibrated conventional strikes across the line of control (LOC) to prevent the use of nuclear weapons by Pakistan. The IBGs comprising T-90 and T-72 tanks, self-propelled guns (SPGs) supported by tactical air force (helicopter gunships providing ground support and air cover to be given by the fighter jets) would make rapid thrusts into Pakistani territory within 96 hours of ordering them. The IBGs would go up to a depth of 80 km within the Pakistani territory, and then New Delhi would start pressuring Islamabad to come to the negotiation table. Thus, Cold Start was a doctrine of rapid limited war by India.[2]

Pakistan's civilian and military elite raised the ante by lowering the nuclear threshold. The official Pakistani position was that even in case of a limited conventional strike by India across the India-Pakistan border, Islamabad would resort to a countervalue nuclear response. China is actively aiding Pakistan to create an operational nuclear force. Both Pakistan and China, unlike India, did not accept the policy of No First Use of nukes.[3] In December 2021, Pakistan tested Babur-1B cruise missile with a range of 900 km, which could carry both conventional and nuclear warheads. India is in the process of buying the S-400 missile defence system (Ballistic Missile Defence/BMD) from Russia. This BMD will be deployed in Punjab and in the region around Delhi. In retaliation, Pakistan has deployed Baber-3 (Sea Launched Cruise Missile) which has Multiple Independent Re-entry Vehicles (MIRV) capabilities.[4]

Thus, Pakistan's much smaller nuclear arsenal, in comparison to India, has more or less neutralised India's nuclear option and conventional superiority. According to one estimate, in 2009, Pakistan had 60 nuclear warheads. Both India and Pakistan are densely populated. In the case of a nuclear exchange between India and Pakistan, one estimate is that 12 million people would die. Hence, India lacks the option of launching a decisive conventional strike across the border under the nuclear umbrella to teach Islamabad a lesson for encouraging insurgencies and terrorist strikes in India. This allows Pakistan to operate with impunity, the terrorist and insurgent camps across the LOC in Pakistan Occupied Kashmir.[5]

Pakistan's internal scenario also supports the promotion of *jihadi* terrorism and insurgency. In 2010, Pakistan's population was 185 million and it is still growing. In 2050, it is estimated that Pakistan will have about 300 million people. Further, Pakistan's population has a youth bulge. Some 60% of the population is below the age of 25. Many of these male youths are educated in the *madrassas*, and are underqualified for employment in the service and manufacturing sectors, which are the principal growth areas of Pakistan's economy. Most of them strongly support the idea of Pakistan being an Islamic state. A substantial chunk of Pakistan's population is concentrated in the urban centres which are emerging as principal hubs of Islamic extremism and recruitment bases for the radical groups.[6]

David Kilcullen asserted in one of his monographs that in the twenty-first century the guerrillas would come out of the mountains and attack the big cities of the nation states. Instead of the countryside, the mega urban centres located along the coastline would become the principal playgrounds of the insurgents. This is because a significant chunk of the world's population (and the most important and powerful segment of the populace of the countries) resides in the megacities. The big cities are the administrative centres, transportation and communication hubs, and economic pivots of the countries. Since the high value targets reside in such centres, it makes sense for the urban guerrillas to target the megalopolises.[7] Kilcullen's assertion holds water to a great extent as regards the South Asian scenario.

In the new millennium, *Lashkar-e-Taiba* (Army of Allah), which operates from Pakistan, has emerged as the most potent terrorist threat for India. This organisation specialises in conducting urban terrorism. In July 2006, this outfit launched suicide bombers in the crowded local trains of Mumbai killing 160 Indians. On 26 November 2008, 10 Pakistani terrorists armed with AK-47 rifles and hand grenades belonging to this organisation from Karachi crossed the Arabian Sea in boats and reached Mumbai. For three days and nights, these terrorists conducted mayhem in Mumbai resulting in the death of 163 people.[8]

India's Maritime Military Strategy published in 2007 and *Indian Maritime Doctrine* which came out in 2009, considered COIN duties (termed constabulary functions) as third in importance after military and diplomatic functions.[9] After the 2008 Mumbai incident, the Indian Navy with the Coast Guard has been assigned the task of protecting the coastline of India against the infiltration and exfiltration of the insurgents. During the Mumbai train blasts in July 2006, the terrorists used RDX explosives provided by the ISI.[10] COIN operations are continuing in North-East India, and Jammu and Kashmir. In 2018, the Indian Army eliminated 15 infiltration bids which resulted in the death of 35 insurgents. The figures for 2017 were 33 infiltration bids across the LOC, resulting in the death of 59 insurgents. In total, during 2018, 254 armed rebels were eliminated by the Indian security forces (army and paramilitary units) in Jammu and Kashmir. In the same year, the security forces killed 16 insurgents and arrested 604 terrorists in North-East India.[11]

COIN being a manpower-intensive task is proving to be more costly to the Indian Army compared to the conventional wars it has waged.[12] Further, COIN is proving practically to be the principal task of the Indian Army. According to one estimate, in the 1990s, the Indian Army deployed 300,000 security personnel (army and paramilitary) in Kashmir.[13]

Virtue and India's COIN doctrine

The three branches of the Indian armed forces started publishing their official doctrines in the public domain after 2000. Successive Indian government's assertions in the public domain refer to the internal rebels being misguided youths, led astray by hostile foreign governments. Analysis of the *Ministry of Defence Annual Reports (MODAR)*, articles in the Service journals, and memoirs of military leaders enable us to understand the various strategic and doctrinal shifts at different moments of time.

Maintenance of law and order is the duty of the state government. Generally, when an armed rebellion breaks out, then the state government uses civil police and provincial armed police (PAP) and then asks for the central government's paramilitary forces. When they fail, the central government, in consultation with the state government, deploys army units. The paramilitary forces have until now have been able to tackle the insurgencies in Punjab, Tripura, and against the Maoists in Central India. In Punjab, after Operation Bluestar (5 June 1984), the army guarded the borders of Punjab with Pakistan while the armed police and the paramilitary forces conducted COIN in the countryside. However, in terms of training, weapons and combat motivation, the PAP and paramilitary forces by themselves could not tackle the *jihadis* in Kashmir and the insurgents of North-East India. In these regions, the army functions as the principal COIN player. In extreme cases, the central government dismisses the state government and then appoints a governor. Under the governor's rule, the army, paramilitary forces and the concerned state's PAP conduct joint operations.[14] The PAP and the paramilitary forces never use the term COIN. In other words, when the insurgency crosses a certain threshold, then the military is deployed by its civilian masters.

From the 1980s, the Indian Army has been following the strategy of 'iron fist in a velvet glove.' General J.J. Singh, Chief of Army Staff (COAS, 2005–2007), notes in his autobiography: 'The "iron fist" denotes a ruthless and "no-nonsense" approach while tackling the insurgents or terrorists, and the "velvet glove" demonstrates the compassion and humane face of the security forces while dealing with innocent citizens.'[15] This got its official stamp in the first ever COIN doctrine published in the new millennium (*Doctrine for Sub Conventional Operations* 2006). In the 1990s, about 44% of the Indian Army's infantry battalions were engaged in COIN duties. So much so, many COAS started arguing in public forums that internal policing was harming the training of the Indian infantry for high intensity conventional war.[16] Many COAS in their memoirs have asserted that India is preparing for waging a three-front war: Pakistan, China and COIN.[17]

India's warfighting doctrine is shaped by the ideas of *dharmayuddha* tradition just as the norms of Just War influence the theory and praxis of combat by the Western powers. In accordance with the *dharmayuddha* code of conduct, force is to be used as a last resort when other techniques like *sama* (conciliation), *dana* (welfare measures) and *bheda* (divide and rule policy) fail. The decision to use force lies with the rulers (politicians). Once the decision is taken, organised violence is to be implemented by the professional practitioners who operate the legally constituted organ of the state (Kshatriyas manning the *danda*, i.e. armed forces). These elements in *dharmayuddha* are somewhat equivalent to *jus ad bellum* (just cause, right authority, right intention, use of violence as a last resort) of Just War theory. Force is to be used in a proportionate and graduated manner. One here is reminded of *jus in bello* (distinction and use of proportionality in the use of organised violence). The objective of using force is not to annihilate the persons involved in *kopa* (internal rebellions) but to win them over. In order to achieve this purpose, along with a small amount of force, various welfare measures are to be instituted for people residing in the disturbed region.[18]

In the new millennium, the Indian armed forces conceptualise doctrine in the following manner: 'The terms "doctrine" and "strategy" are interrelated but not interchangeable…. Whilst Doctrines provide precepts for development and employment of military power, Strategy is a plan of action for developing and deploying military forces so as to achieve National Security Objectives.'[19] The Indian armed forces' joint doctrine provides a foundation upon which the three Services operate in synergy. Thus, joint doctrine underpins the development of Service-specific strategies which complement the former.[20]

The Indian armed forces use a plethora of acronyms like Low Intensity Conflict (LIC), Sub-conventional Warfare, Asymmetric Warfare, etc. for categorising COIN operations. India's military officers differentiate between insurgency and terrorism. Lieutenant-Colonel Vivek Chadha of the Indian Army insightfully opines:

> Terrorism which enjoys little popular support, will employ sensational and visible violence to terrorise and break the will of the people and in turn make a government accede to its demand. Insurgency, however, will enjoy considerable popular support and will therefore need to be handled with far more caution and understanding, lest it progresses to the stage of a revolution or civil war.[21]

In India, either the insurgency, which had begun in the countryside, gradually spread its tentacles in the cities where the insurgents carried out terrorist acts; or it is the other way round. Terrorism, which had initially started in the big cities, gradually spread to the surrounding countryside where insurgent militias started operating.

The Indian armed forces make no distinction between COIN and counterterrorism (CT). CT is regarded as a component of COIN. The Indian COIN doctrine has evolved within the format of *dharmayuddha* and hence displays a high level of ethics. There is absolutely zero tolerance for ethical failings. It is emphasised that though the insurgents will deliberately flout the International Humanitarian Law (IHL) of warfare, the COIN forces should strictly follow the laws of warfare. The insurgents do not respect the warrior's code of honour but there should be no transgression on the part of the security forces. This is considered necessary for gaining public support and retaining legitimacy. The temptation for vengeance, revenge, reprisal or the gratuitous use of force against the armed non-state actors is not allowed.[22] The Indian Army Doctrine issued in 2004 quotes from the *Bhagavad Gita*:

'Be humane, cultured and compassionate' as part of the code of conduct of the warrior.[23] Some of the cardinal elements of India's COIN doctrine are to display political flexibility by engaging in negotiations and compromise. The army personnel deployed in the disturbed region aim to win the 'hearts and minds' of the people by providing civic services (building roads, schools, hospitals, etc.).[24] By following and implementing these principles, claim the Indian armed forces, the insurgents will be totally delegitimised.

The Indian armed forces accept the primacy of politics in military matters. For the uniformed personnel, the kinetic element is just one component of COIN. Military force can never, in the eyes of the Indian armed forces, eliminate insurgencies. Application of a moderate amount of military force in a graduated manner in combination with welfare measures could only prepare the ground for a political solution to the problem. The Indian Navy's maritime doctrine originally issued in 2009, and then reissued in 2015, notes:

> Counter Insurgency (CI) Operations encompass all measures instituted by the Government against insurgency, including combat operations by security forces, economic development, political and administrative steps and psychological operations. These are aimed at reducing insurgent capability for waging violence, while simultaneously shaping a conducive environment amongst the populace, to diminish the support for the insurgency.[25]

The IAF doctrine issued in 2012 discusses the various tactics that should be used in sub-conventional warfare:

> Aerial reconnaissance can play a key role in tracking terrorist/insurgent activities and identifying operating bases, training camps and supply nodes.... Electronic support measures... can be used to locate terrorist/insurgent lines of communication and operating bases, as well as hiding places of key leaders, even across international borders.... Military air power can also be used to transport police and other civil security elements in response to crisis. Mobility greatly enhances the speed and reach of humanitarian relief and civil support operations, all of which support government legitimacy and reduce adversary influence.[26]

The above-mentioned IAF doctrine is an operational doctrine and hence takes into account the recent advances in the field of technology. In contrast, *Joint Doctrine* 2017 is a fundamental doctrine which deals with the basic principles of conflict. It is to be noted that the IAF doctrine as regards sub-conventional warfare emphasises the non-kinetic rather than the kinetic component of COIN. However, if necessary, notes the 2012 IAF doctrine, force should be applied against the insurgents in a calibrated manner very quickly. In accordance with this doctrinal principle, on 26 February 2019, the IAF carried out the Balakot Strike against the insurgent training camps in Khyber Pakhtunwa.[27]

COIN tactics and India's AWS arsenal

Officers of the Indian armed forces are working to chart out a comprehensive scheme for using AI in a variety of ways for fighting insurgents and terrorists. The Indian military officers are eager to utilise the image recognition algorithms while conducting COIN along the India-Pakistan border. Deep learning algorithms can, with extreme accuracy, read human lips, synthesise speech, and to a great extent, simulate facial expressions. Deep learning is a branch of machine learning (ML) that involves algorithms which can analyse data through multiple layers of complex processing. Each layer's output becomes the input to the next layer in order to carry out pattern analysis and classification. ML means a computer has the ability to learn without being explicitly programmed, i.e., unsupervised learning. Civilian and military analysts are eager to use these technologies to stop the movement of insurgents between India-Pakistan and India-China borders. The Defence Research and Development Organisation (DRDO) set up the Centre for Artificial Intelligence and Robotics (CAIR) in 1986. In the new millennium, the CAIR is carrying out research in artificial neural networks and computer vision among other things.[28]

The length of India's land border is 15,200 km. Incorporation and integration of AI-enabled deep learning algorithms, image recognition mechanism and speech recognition (the capacity of the computers to recognise and interpret spoken languages) software with the border management scheme will increase the speed and efficiency of border patrols to prevent the movement of the insurgents across the India-Pakistan and India-China borders.[29]

In 2002, one IAF officer named A.K. Tiwary wrote about the various possible ways in which the Unmanned Aerial Vehicle/drone (UAVs) could carry out COIN tasks. UAVs fitted with cameras can supplement satellite images. As the UAVs, compared to the satellites, fly closer to the surface of the earth, the scale and resolution of the images taken by the former are better. Tiwary notes that though computer software may be used for processing and interpreting the images, human interpretation is necessary, especially if one is looking for small fleeting targets like a few insurgents in a large area. Tiwary is realistic in his evaluation. He writes that UAVs could at best provide near real-time intelligence rather than real-time intelligence as the targets after observation would move away or conceal themselves better.[30] India is operating UAVs equipped with high-resolution cameras and sensors for Intelligence collection, Surveillance and Reconnaissance (ISR) of the borders as well as sensitive areas of the countryside and the cities. New Delhi plans to equip the surveillance and monitoring drones with software and algorithms that are capable of reading identity cards and possessing facial recognition capability.[31] This will aid the security forces to track down the insurgents and terrorists before they can strike.

In 2013, India's Ministry of Defence published a future roadmap for the Indian armed forces. This document noted that with the passage of time, not only should UAVs (also termed by the Indian armed forces as Remotely Piloted Aircraft System [RPAS]) be miniaturised but they would also become more intelligent. Swarming drones which are capable of working together, and teaming of UAVs with manned platforms will become the norm in the near future. Swarm intelligence stands for the collective behaviour of decentralised self-organised systems. The principal characteristics of a swarm intelligence system are its capacity to function in a coordinated manner. Swarming occurs when several units conduct a convergent attack on a target from several axes. Similarly, ground robots/Unmanned Ground Vehicles (UGVs) singly or in bunches will be deployed in extracting mines and Improvised Explosive Devices (IEDs), and in urban search and rescue missions.[32] The IED is the favourite device of the *jihadi* insurgents and the Maobadis.

The era of urban warfare is coming. According to one estimate, by 2050, two-thirds of humans will reside in cities.[33] There are distinct advantages to using robots in built-up urban areas. Small robots can see whatever satellites and airborne sensors are unable to view. Further, the use of lethal robots will keep soldiers out of harm's way in the confined terrain of megacities.[34] For urban warfare, the CAIR is working on a project named Multi Agent Robotics Framework, which is a collection of intelligent mobile robots that can function as a combined team. The Indian Army already possesses wheeled robots, four-legged and six-legged robots, and wall-climbing robots that can operate in a variety of terrains.[35] CAIR is working to make these robots mission-capable. The camera in such a robot will capture the image of the intruder and then check with its database. If there is no match with the existing database, the picture will be sent by internet or satellite link to the main base for storage for future use. If the snapshot of the intruder matches the database containing information about insurgents, then a message will be sent to the local security base or to the human security teams operating in close vicinity.[36] A particular type of ground robot deployed by the Indian Army could travel 20–30 km away from the base and use its metal detector to search for and locate bombs.[37] Each of the bomb disposal ground robots used by the Indian Army could remove explosives weighing up to 20 kg. The operator sitting in the control room can watch video images received from the wireless cameras mounted at the side of the mobile robot.[38]

The Indian Army has at its disposal the multipurpose Daksh UGV for COIN and related CT tasks. It is used to detect intruders, harmful gases and bombs with the metal detector, temperature sensor and gas sensor carried by this 'smart' machine. The built-in microphone in the camera carried by this UGV allows the operator at the base to hear people talking in the vicinity of this robot. This robot is operated both manually and in an automatic mode. An ultrasonic sensor is used to drive the robot automatically. The aim is to increase the range of communication through which this robot is controlled manually. A DTMF decoder is placed in this robot, which allows the handler to control this machine from his/her cell phone. The cameras in this robot allow the operator to navigate it in difficult terrain. In order to increase the operating range and endurance of this UGV, the DRDO is working on adding solar panels in addition to the batteries.[39]

The DRDO is planning to manufacture swarming micro-drones which could be used profitably in anti-terrorist operations. These drones are to carry out both lethal (shooting) and non-lethal tasks (surveillance, reconnaissance, tracking, etc.). The Indian Army, IAF and the Indian Navy are interested in this projected scheme. In fact, the officers want such drones to communicate with each other while conducting an operation, and also to exchange data with the human combat team operating on the ground and sea. The future tasks are to develop the range, endurance, speed, and communication software that are to be embedded in such drones.[40]

In fact, the non-state actors in South Asia already possess drones. Either they have manufactured it by buying the components from the open market or acquired it from China and Pakistan or both. Plugging in internet dongles with Global System Mobile Communication SIM cards can be used to control the drones via the internet through a personal computer or smartphone. Compact internal combustion engines used for small drones are easily available in the open market. Between 2015 and 2017, Indians bought 40,000 drones. The cost of these drones varies from Indian Rs (INR) 2,000 to INR 50,000 (1 US $ = INR 82).[41] These are bought over the counter without any regulatory control. In November 2019, the Maoists used drones in Bastar to acquire intelligence about the Border Security Forces' (a paramilitary organisation) camps. From 2020, drones have been appearing regularly over the army camps and airfields in Kashmir. In fact, in late December 2020, the Chief of the Air Staff of India publicly acknowledged that drones in the hands of non-state actors make them more dangerous.[42]

The Indian policymakers assume that Pakistan is supplying drones to the *jihadis* operating across the LOC. During 2018, Pakistan obtained four Cai HONG MALE UAVs from China. In addition, China agreed to sell 48 Wing Long II drones to Pakistan. In March 2021, Pakistan displayed Shapur II, which is a Medium Altitude Long Endurance (MALE) Drone. Pakistan borrowed technology for manufacturing this drone from Turkey. In fact,

Shapur II is similar to the Turkish Bayraktar TB2. Pakistan has entered into agreements with the Turkish Aerospace Industry to develop MALE combat drones. All these developments have made India nervous.[43]

On 17 July 2021, India's Home Minister Amit Shah declared:

> Drones have become a serious issue of security concern. Defence Research and Development Organisation (DRDO) is working on developing anti-drone "*Swadeshi*" technology to get over this danger. All R&D projects have been sanctioned by the government to develop anti-drone technology.[44]

What Shah meant by *swadeshi* (indigenous/homegrown) is technology not imported from foreign countries but developed within India as part of Prime Minister Narendra Modi's *Atmanirbhar Bharat* (self-sufficient India, not dependent on foreign-imported technologies) policy.

DRDO's Hyderabad-based laboratory named Centre for High Energy Systems and Sciences (CHESS) is working on several anti-drone systems. One is Directed Energy Weapons (DEWs), which are laser-based or microwave-based weapons that can disable hostile drones. Electro-optical laser pulses and radars are to be used to track enemy drones and then jam the radio frequency between the UAV and the operator or destroy the UAV using laser technology. In 2018, the DRDO developed a KW anti-UAV laser system, which is carried on a truck and has a range of 1 km. DRDO plans to develop powerful lasers with longer ranges and reduce the weight of the system in order to make it more mobile. The Indian government is thinking of deploying this anti-drone system in airports to protect them from possible drone strikes launched by the insurgents.[45]

However, progress for making high-quality indigenous drones is quite slow in India. The head of the DRDO has veto power over the issue of what military items India can import. Stephen Cohen and Sunil Dasgupta rightly assert that DRDO's dual role as both supplier and evaluator of weapon systems has resulted in reduced accountability for its failures to provide state-of-the-art military systems at the right time and without incurring prohibitive cost.[46]

Commanding the robots and drones

> The relationship of man and robot, or creator and creation, has therefore always been seen as a potentially very problematic one.
>
> Armin Krishnan[47]

Historian Martin Van Creveld writes that command for him is equivalent to the management of the military organisation.[48] Generally, control and communications are regarded as separate from command. Hence the acronym

C3 is used. Then, the acronym C3I (command, control, communications and intelligence) came into use. From the last decade of the twentieth century, as the military analysts started discussing Network Centric War (NCW), the acronym C4ISR (command, control, communications, computers, intelligence, surveillance and reconnaissance) became common in the military literature.[49] To me, it is mere hair-splitting. Control is an integral component of command. Control requires communication. From the 1980s, computers (a tool) have replaced radio as the principal mode of communication. Intelligence is another component of command. Surveillance and reconnaissance are vital to the acquisition of intelligence. Then again, training is also a crucial component of command but is generally left out of discussion.

In the present era, command is now based on computers. The automated and digitised battlefield of the future is reshaping the art of generalship. There is a general consensus among the military officers of India and the world that in the age of information war, victory will go to the side which has control over the flow of information. As regards acquisition and processing of large amounts of data from various sensors, computers are better and quicker than the human brain. So, there is talk of using AI-embedded software for interpreting and analysing the data.

Steven I. Davis, a United States naval officer, notes that human oversight over machine-supported decision-making is necessary and could be maintained if ANI (which could perform limited tasks that have been preprogrammed into these machines) is utilised in different and separate components of the military planning process. What Davis is implying is that ANI rather than Artificial General Intelligence (AGI) should be relied upon. AGI embedded in a decision-making apparatus could not only copy human thought but might also invent novel methods of critical analysis. Instead of replicating and automating simple decisions made by the human commanders, AGI would serve as an adjunct and provide different manners of solutions to the problems facing the military command. AGI is not only costly, but being extremely complex, is highly unpredictable in behaviour, especially during chaotic situations. It is highly probable that in the near future a squadron of swarming drones controlled by AGI might turn against its own side.[50] Though Davis is speaking of the operational level of war, his insights can be applied also at the tactical level.

For Manish Chowdhury, an officer of the Indian Navy, ANI is to be used to augment and supplement (not replace) human judgement in the tactical sphere. He writes:

> Victory in battle is largely achieved through tactical moves by the commanders based on perceptions, experiences and procedures. ML [Machine learning] embedded Decision Support Systems with training

data comprising force level, statistics of past decisions and situational intelligence can assist in efficient Multi Criteria Decision Making.[51]

Such systems will supplement the judgement of the commander in the digital age and reduce reaction time.

Another aim is to reduce the Clausewitzian 'fog of war' by enhancing battlefield situational awareness of the commanders through the use of Internet of Things (IoT). Group Captain Ashish Gupta writes:

> Situational awareness (SA) encompasses a wide range of activities on the battlefield to gain information about the enemy's intent, his capability and actual position.... IOT can play a vital role in raising situational awareness by collecting, analysing and delivering the synthesised information in real-time for expeditious decision-making for appropriate military action.[52]

Overall, for the present, the uniformed officers of India are for using AI to reduce the cognitive load of the commanders.

Since the insurgents operate in small bands, for COIN tasks, the Indian Army relies on its Battlefield Management System (BMS) for communicating with the small scattered bands of soldiers hunting and fighting the armed rebels. The BMS has three principal components: C3 network, cloud-based database and manned and unmanned systems. The BMS communicates with the brigade headquarters, battalion headquarters and the subunits under its control (companies, platoons and sections). The brigade headquarters control the MALE UAVs, the battalion headquarters have at their disposal tactical UAVs, and the squad possesses the small and micro-UAVs. The BMS could communicate and assign mission orders to these UAVs. The high-altitude long-endurance UAVs are under the control of the corps headquarters, which operate the Command Information Decision Support System and do not come into the picture during COIN duties. The BMS receives information from the Battlefield Surveillance System, which in turn can access data in the cloud database. The cloud database contains the information provided by the intelligence agencies and is always updated by the findings of the UAVs on the battlefield.[53]

The IAF considers COIN as part of all domain warfare. For conducting COIN duties, the IAF is working on a machine-man interactive command system. This command system will be able to integrate UAVs (geared for ISR duties and provided by the IAF), manned aerial vehicles (for firepower support and these crafts could either come from the IAF or the Indian Army) and ground teams (from the Indian Army).[54] Integration of UAVs with manned aircraft and prevention of accidents require sensor equipment to detect flying objects in the near vicinity, and also obstacles on the ground, which would allow the operators to take timely corrective actions. The Traffic Collision Avoidance System (TCAS) is an airborne system, originally developed for manned aircraft. It needs to be modified for UAVs to identify collision threats and to take action autonomously. UAVs need light-weight infrared sensors, optical sensors and laser radars (LIDAR) for detecting non-cooperative hostile traffic. In addition, radio telephony (voice) contact between the UAV operator and the air traffic control management is a necessity.[55]

Training is the lifeblood of war and an essential aspect of command. Proper control of the Autonomous Weapon System (AWS) in near combat and combat scenarios requires intense training between military personnel and intelligent machines. From the 2010s, the Indian Army and IAF regularly conducted joint exercises in the Thar Desert of Rajasthan. These joint exercises involved extensive use of UAVs. The three services possess UAVs, but interoperability is an issue due to differences in their command and control systems.[56] What is required is a common joint training framework. Most of these UAVs are geared for ISR and communication roles. There have been attempts to incorporate micro UAVs for directing artillery fire on enemy targets.[57] Colonel Deepak Kumar Gupta of the Indian Army is for raising specialised mission-orientated combat teams equipped with swarm intelligence-based systems. These units, writes Gupta, need to conduct regular intensive physical exercises (not merely computer simulations) to hone their skills for operating across varied terrains (urban centres, rural areas, airports, and military installations). Then only these units would be able to counter both vapour and cloud swarming drone tactics that might be used by the insurgent groups sponsored by hostile foreign powers.[58] Tuneer Mukherjee, an Indian civilian analyst, rightly emphasises:

> There will be a need to develop a new training curriculum to produce technologically literate and AI-calibrated personnel, and an equally proficient testing and validation regime to test these new technologies.... Most notably, operational paradigms and tactics need to be revamped.[59]

In case an AWS takes a wrong decision and engages in fratricide of friendly troops and kills civilians, damages non-combatant property, who will be held responsible? This brings us to the issue of ethics and legality as regards the deployment and use of Lethal Autonomous Weapon Systems (LAWS).

Ethics, COIN and killer robots

In modern times, India, like other civilised nations, adheres to the IHL and tries to follow the guidelines of International Human Rights Law (IHRL). This section evaluates how far the ethics inherent in the principles of *dharmayuddha* are operable in the context of using AI-embedded weapons in COIN scenarios within India.

The strategic elite and academicians in and outside India are aware of the challenges posed by AI-embedded weapon systems on the ethics and legality of warfare, and are debating the issue. Many argue that killer robots will make warfare immoral, illegal, unethical and inhuman. For instance, Robert H. Latiff, a retired officer of the United States Air Force, writes: 'New technologies, like autonomous systems – those that perform their functions without any human control or intervention – chip away at our humanity and must be more carefully considered before being employed in weaponry.'[60] From the opposite pole, another strand of argument claims that the robo-soldiers will make anti-insurgency operations more humane. One proponent of this point of view is Jai Galliott, who opines that unmanned systems, by improving precision, are making use of force more effective, discriminate and proportionate in accordance with the tenets of the Just War theory. Hence, the use of AWS will reduce collateral damage.[61]

The imperatives behind developing fully autonomous intelligent war machines are strong. Though the technology for such a kind of weapon does not exist yet, it will come into existence in the near future. Paul Scharre, a former US Army Ranger, notes that those countries which are arguing for a ban on AWS are really militarily weak polities attempting to tie down the hands of the militarily stronger powers. In the long run, Scharre continues, deployment of a fully autonomous intelligent military machine will be favoured by a comparatively backward and weaker military power because keeping a human in the loop in contested environments requires protected communications, which is far more challenging than building a weapon which can target and hunt on its own.[62] After all, drone jammers which block Global Positioning Systems, cellular communications and WIFI signals could be used in jamming the communication system between the drones and their operators. Worse, the hackers could take control of the drones and UGVs. One way to overcome this problem, write two Indian analysts, is to manufacture fully autonomous intelligent drones and UGVs.[63] Paul Springer says that it is the bounden duty of those states which are developing or might develop fully autonomous intelligent killing machines to insert safeguards in these devices. Some sort of self-destruct system, or a remote means to override or shut down a LAWS which is showing symptoms of malfunctioning (disobedience), is necessary.[64] The problem is that the self-destructive software may also be hacked by hostile hackers or it could malfunction. What then?

A minority view in India is that of U.C. Jha (a retired IAF officer). Jha writes that it is not at all illogical to ban LAWS before they actually come into existence, as blinding laser weapons were banned before they were developed and deployed.[65] In fact, his assumption is that it would be more difficult to ban LAWS once they have been manufactured and are ready for operational deployment by the big powers. In contrast, Lieutenant-General R.S. Panwar (from the Signal Corps of the Indian Army) claims that despite the global concern about the development of LAWS from legal and ethical points of view, and the measures adopted or which will be adopted by the United Nations, the major powers (India's concern is China, and with Beijing's aid, Pakistan) will develop and deploy AWS.[66] This line of interpretation is more or less accepted by all the civilian and military analysts of India. The bulk of the strategic and military analysts, at present, concern themselves with the level of autonomy to be granted to the AWS.

Armin Krishnan takes a more practical middle position. He writes that a regulation that explicitly prohibits the development and use of self-learning military robots would close the door on the future possibility of evil machines attacking mankind. If super-intelligent computers are developed, then they should have only weak links with the physical world. At least on higher levels of decision-making a human should always remain in the command loop. However, like Scharre, Krishnan holds the belief that those states which feel threatened would feel forced to deploy autonomous machines for defensive purposes or to launch a retaliatory strike. These fully autonomous machines need to be outlawed.[67] So, Krishnan is making a case for semi-autonomous LAWS. On a similar note, IAF's retired Wing Commander, Ajey Lele, claims that military robotics should not be allowed to develop to such a level that killer robots could cause unnecessary and unintended harm. He continues that autonomy cannot be absolute. There might be either a low or high level of autonomy.[68] In fact, the focus should be on developing a man-machine interface.[69]

Nevertheless, there is a problem with semi-autonomous lethal machines also. Nik Hynek and Anzhelika Solovyeva write that most of the AI-embedded weapons that are in operation or are being developed have a switchable mode of control which allows a balance between human control and system autonomy depending on the battlefield situation. The very possibility of such a mixed mode will be the principal stumbling block along the way towards some sort of a regulatory or prohibitory treaty. The reason is that it will virtually be impossible to determine whether any particular decision to kill is made with the autonomous mode being on or off, at the moment when the target is eliminated.[70]

Bashir Ali Abbas, a civilian researcher associated with the National Maritime Foundation of India, notes that a blanket ban on LAWS is an impossibility. Since such weapons are bound to be used in the near future, what we need is to develop new rules for certain types of AWS. For instance, the human operator must always exercise effective control over the AWS. So, meaningful human control (MHC) will ensure that the responsibility to follow IHL falls on the 'human in the loop.'[71]

MHC, which involves use of an appropriate level of human judgement concerns not merely the issue of activating the weapon system of the LAWS but also the very targeting process. Merel Ekelhof stresses that very often the AWS is given full autonomy as regards target selection. The introduction of autonomous technologies presents a fundamental challenge to the human decision-making process. To shorten the Observe, Orient, Decide, and Act (OODA) loop, autonomous intelligence processing technologies are introduced to reduce the cognitive burden on human analysts and to augment actionable intelligence, which will enhance the decision-making process by increasing tempo. It is the automated technologies that decide, out of a huge amount of data, which specific datum will be shown to the operators and whatever data are to be ignored. Thus, the black box of autonomous technologies plays a crucial role in generating situational awareness and preparing the final target list for the human handlers.[72]

Nishant Gupta, an IAF officer, perceptively notes that AI must be fair, accountable and transparent. Most of the current AI systems and their algorithms work in a manner that only input data and the results are known to the developers. What is happening in between is unknown. This is known as the 'Black Box Phenomenon.' Gupta asserts: 'But, understanding of the decision making process would be significant from the legal and ethical perspectives.'[73] The ability to have an understanding of the AI's decision-making process will allow the operator of the AWS to assess and judge the intelligent machine's decision before the handler approves or disapproves of it.

In order to trust the AI's decisions, we need to understand such systems better. A solution provided by AI to a problem is not adequate. We humans must have a clear comprehension of why a particular decision is taken by AI in a specific circumstance. Herman Cappelen and Josh Dever assert that for this purpose, we need to provide content to AI's program. This will help us to interpret and understand the outputs of the AI system. They write that AI content is a problem at the level of environmental and sociological detail, and is not a problem at the level of programming and computational detail. An internal computational structure is unnecessary for the explainability of AI's decisions and recommendations. So, externalist approaches have much to offer for explaining AI content. Cappelen and Dever conclude that the externalist tradition in the philosophy of language and mind could aid humans in interpreting the solutions provided by AI.[74] The military policymakers and generalist civilian analysts need to pay heed to the interpretation offered by these two philosophers. Ronald Arkin stresses the option of direct encoding of prescriptive ethical codes within the AWS which can govern its actions in a manner consistent with the Laws of War and Rules of Engagement.[75] Till now, there has been no talk among the Indian civilian and military leaders and scientists about this possibility, probably because they think that the manufacture of such software is beyond the realm of possibility.

Possible future scenarios

We can surmise that India will use LAWS with human controllers in greater numbers more frequently and extensively against the insurgents in the coming days. Some of the probable scenarios may be checked out. In 1966, the Indian government used the IAF Hunter jets against the fighters of the Mizo National Front. There was a furore within the domestic society because the democratic government had used combat aircraft against the citizens of India. Further, the government was criticised for using indiscriminate force, which went against one of the established COIN principles (use of minimum force) of the country.[76] Use of UAVs in place of manned combat aircraft against the guerrillas in the near future will not generate such a high level of anger among the people of India.

Moreover, thanks to advanced technology, the lethal force that will be used will be discriminatory in nature, thus keeping within the ambit of India's *dharmayuddha* tradition. The UAVs and the UGVs will not engage in rape, molestation and torture of the people as AWS lacks human emotions (biases and prejudices). AWS will not suffer from tiredness, battle fatigue and frustrations. As per the norms of India's COIN doctrine, the security forces conduct an attritional campaign which involves the gradual use of a limited amount of force backed up by the initiation of a host of welfare measures. This is because the objective of COIN, as per India's sub-conventional operations doctrine, is not to eliminate the enemy but to make them understand that they should submit and cooperate with the civilian administration of the government. This in turn necessitates a lot of boots on the ground, and the soldiers and officers are required to display utmost restraint even in the face of provocations.[77]

Most of the Indian Army personnel find such policing tasks taxing for their mental and physical health. Further, the senior military officers complain that COIN duties (cordon and search, road opening duties, armed policing as per aid to civil tasks, establishing picquets, etc.) detract the infantry battalions from training for high-intensity NCW. Use of AWS for such dull, dirty, dangerous and repetitive tasks will prove to be beneficial for the country and its army. The Indian Army has for quite some time been suffering from a serious shortage in the officer cadre. Not only are mid-level officers taking premature leave, but there has also been a serious drop in recruitment at the junior level. For COIN tasks, young and middle-level officers commanding sections, platoons and battalions are extremely important. Two factors are responsible for the officer shortage in the Indian Army. One is the growing liberalisation of the Indian economy which offers jobs with greater economic largesse compared to the pay and perquisites offered to the officer cadre. Second is the arduous COIN campaign, which involves deployments in dangerous inaccessible zones and associated physical and mental problems. These two factors are, to a great extent, responsible for university-educated urban middle-class youths not considering the army as their preferred career.[78] One way the Indian Army could overcome the officer shortage is by using more intelligent warring machines (UGVs plus UAVs with human commanders at the back) in place of infantry units for tackling the insurgents.

Further, the Indian Army is thinking of using loitering munitions fitted with precision-guided munitions in sub-conventional warfare. A loitering munition is a modified UAV which can engage out-of-sight ground targets with explosive warheads. These UAVs are fitted with high-resolution electro-optical and infrared cameras which allow the targeter to locate, surveil and track the target. This sort of UAV is able to hover over the target area for an extended period before striking, thus enabling the operator to decide what to target and when. The robust sensor-to-shooter communications will also allow a human-in-the-loop command system to continue. Loitering munitions are perfect tools for targeted assassinations of insurgent commanders and also for crowd control by taking out the leaders. In addition, in urban centres, the loitering drones with non-lethal weapons could temporarily neutralise/inactivate the hostile crowd. This will prevent non-combat fatalities as collateral damage in urban operations always causes a political hue and cry.[79]

Conclusion

Technology is neutral. Both India and its non-state opponents are using AI-related technologies, and their use by both sides will increase in scope and frequency in the near future. The issue is which side will use it more effectively. Complete autonomy for LAWS remains a chimera in the foreseeable future. At most, in the near future, we will see semi-autonomous intelligent weapon systems deployed for operations in combination with human soldiers and human-manned weapon platforms. The Indian armed forces need to update their joint and service-specific doctrines for conceptual blending of AWS with other conventional weapon systems manned by humans. Along with doctrine, training with and against AWS needs to be elaborated on and implemented. A seamless amalgamation and deployment of AI-related technologies and non-AI technologies are necessary for effective force planning, force structuring and force deployment. For the time being, the officers of the Indian armed forces are content with using AI as a junior partner with the human commander in tactical and operational decision-making processes. However, in the future, there might be a demand that to reduce the OODA loop and to increase the tempo, more autonomy should be given to AWS. Such a scenario will run counter to the Indian government's prescription of using AI as an adjunct of humans in the managerial/command spheres. The 'face of war' is indeed changing. Probably in the near future, we will see robo-soldiers, UGVs, and drones fighting the insurgents and terrorists in the urban megacenters and countryside of India. More challenging is to establish a balance between human and machine judgement while conducting COIN tasks. Clausewitzian 'fog' must not be replaced with 'AI generated fog.'

Notes

1 K.P. Bajpai, *Roots of Terrorism* (New Delhi: Penguin, 2002), pp. 95–102.
2 A. Ahmed, *India's Doctrine Puzzle: Limiting War in South Asia* (London: Routledge, 2014), pp. 1–2, 13, 54–5, 58–9; Walter C. Ladwig III, 'A Cold Start for Hot Wars? The Indian Army's New Limited War Doctrine,' *International Security*, vol. 32, no. 3 (2007/8), pp. 158–90.
3 Jing-dong Yuan, 'Foe or Friend? The Chinese Assessment of a Rising India after Pokhran-II,' and T.V. Paul, 'The Causes and Consequences of China-Pakistani Nuclear/Missile Collaboration,' in Lowell Dittmer (ed.), *South Asia's Nuclear Security Dilemma: India, Pakistan, and China* (New Delhi: Pentagon Press, 2005), pp. 150–74, 175–88.
4 Sher Bano, 'India launches 3rd Arihant-Class Nuclear-Powered Submarine,' p. 6, and Ahyousha Khan, 'Babur Cruise Missile: Pakistan Strengthening its Deterrence,' pp. 13–14, in *SVI Foresight*, vol. 8, no. 1 (2022), Strategic Vision Institute Islamabad, thesvi.org/wp-content/uploads/2022/02/SVI-Foresight-Vol -8-No1-Jan 2022, accessed on 21 June 2022.
5 S. Ganguly and S.P. Kapur, *India, Pakistan, and the Bomb: Debating Nuclear Stability in South Asia* (New York: Columbia University Press, 2010), pp. 74–122.
6 T.V. Paul, *The Warrior State: Pakistan in the Contemporary World* (London: Random House, 2014), p. 190.
7 D. Kilcullen, *Out of the Mountains: The Coming Age of the Urban Guerrilla* (2013, reprint, New Delhi: Oxford University Press, 2014).
8 S. Wolpert, *India and Pakistan: Continued Conflict or Cooperation?* (Berkeley: University of California Press, 2010), p. 3.
9 Ashok Sawhney, 'The Navy in India's Socio-Economic Growth and Development,' in Rajesh Basrur, Ajaya Kumar Das and Manjeet S. Pardesi (eds.), *India's Military Modernization: Challenges and Prospects* (New Delhi: Oxford University Press, 2014), p. 33.
10 Brigadier S.M. Limaye, 'Urban Terrorism-A New Face of Asymmetric Warfare,' *Army War College, Mhow*, vol. 43 (July 2014), p. 63.
11 *Annual Report*, 2018–19, Ministry of Defence (hereafter *ARMOD*), Government of India, p. 20, https://mod.gov.in, accessed on 20 November 2022.
12 Colonel H. Singh, *Doda: An Insurgency in the Wilderness* (New Delhi: Lancer, 1999), p. 245; J. Gill, 'Military Operations in the Kargil Conflict,' in Peter R. Lavoy (ed.), *Asymmetric Warfare in South Asia: The Causes and Consequences of the Kargil Conflict* (2009, reprint, Cambridge: Cambridge University Press, 2010), p. 122.
13 Peter R. Lavoy, 'Introduction: The Importance of the Kargil Conflict,' in Lavoy (ed.), *Asymmetric Warfare in South Asia*, p. 18.
14 Rajesh Basrur, Ajaya Kumar Das, and Manjeet S. Pardesi, 'Introduction,' in Basrur et al. (eds.), *India's Military Modernization*, pp. 13–14; Lieutenant-General K.S.

Brar, *Operation Blue Star: The True Story* (1993, reprint, New Delhi: UBSPD Pvt. Ltd., 2011).

15 Quoted from General J.J. Singh, *A Soldier's General: An Autobiography* (New Delhi: HarperCollins, 2012) p. 211.

16 R. Rajagopalan, 'Innovations in Counterinsurgency: The Indian Army's Rashtriya Rifles,' *Contemporary South Asia*, vol. 13, no. 1 (2004), pp. 25–6.

17 See for instance the memoir of General Shankar Roychowdhury (COAS 1994–1997), S. Roychowdhury, *Officially at Peace: Reflections on the Army and its Role in Troubled Times* (New Delhi: Penguin, 2002).

18 Kaushik Roy, *Hinduism and the Ethics of Warfare in South Asia: From Antiquity to the Present* (New York and Cambridge: Cambridge University Press, 2012) pp. 13–39; 106–60; B. Schneider, 'India's Drones: Assessing the rationale for Unmanned Aerial Vehicle Acquisition,' *The Cornell International Affairs Review*, vol. 12 (Fall 2018), p. 34.

19 Quoted from *Joint Doctrine Indian Armed Forces* (New Delhi: Directorate of Doctrine, Headquarters Integrated Defence Staff, Ministry of Defence, 2017) p. 58.

20 *Joint Doctrine* 2017, p. 4.

21 Quoted from V. Chadha, *Low Intensity Conflicts in India: An Analysis* (New Delhi: SAGE, 2005), p. 403.

22 Lieutenant-General S.R.R. Aiyengar, 'Ethical Issues in Asymmetrical Warfare,' *Army War College, Mhow*, vol. 43 (July 2014), pp. 7–11.

23 *Indian Army Doctrine* 2004, Shimla, Headquarters Army Training Command, Part 1, Preface, https:// files.ethz.ch/isn/157030/India/202004, accessed on 12 December 2021.

24 D.B. Shekatkar, 'India's Counterinsurgency Campaign in Nagaland,' in Sumit Ganguly and David P. Fidler (eds.), *India and Counterinsurgency: Lessons Learned* (Oxon: Routledge, 2009), pp. 9–27.

25 Quoted from *Indian Maritime Doctrine Indian Navy*, Naval Strategic Publication 1.1 (2009, reprint, New Delhi: Integrated Headquarters Ministry of Defence, 2015), p. 18.

26 Quoted from *Basic Doctrine of the Indian Air Force* (New Delhi: Indian Air Force Air Headquarters, 2012), p. 106.

27 A. Subramaniam, *The Indian Air Force, Sub-Conventional Operations and Balakot: A Practioner's Perspective* (New Delhi: Observer Research Foundation Issue Brief no. 294, 2019), pp. 1–9.

28 S.S. Vempati, *India and the Artificial Intelligence Revolution* (Washington, DC: Carnegie Endowment for International Peace, 2016), pp. 9, 20–21.

29 P.K. Mallick, 'Artificial Intelligence in Armed Forces: An Analysis,' *CLAWS Journal* (Winter 2018), pp. 70–6.

30 Air Commodore A.K. Tiwary, *Air Power and Counterinsurgency: A Review* (New Delhi: Lancer, 2002), pp. 171–3.

31 R. Swaminathan, *Drones & India: Exploring Policy and Regulatory Challenges Posed by Civilian Unmanned Aerial Vehicles* (New Delhi: Observer Research Foundation, 2015), pp. 6–7.

32 Technology Perspective and Capability Roadmap (TPCR), April 2013, Headquarters Integrated Defence Staff, Ministry of Defence, pp. 8, 19, https:// www.mod.gov.in, accessed on 20 December 2021.

33 L. Royakkers and Rinie van Est, *Just Ordinary Robots: Automation from Love to War* (London: CRC Press, 2016), p. 34.

34 A. Krishnan, *Killer Robots: Legality and Ethicality of Autonomous Weapons* (Surrey: Ashgate, 2009), p. 30.

35 K. Roy, *Advances in ICT and the Likely Nature of Warfare* (New Delhi: KW Publishers in association with CLAWS, 2019), pp. 80–1.

36 N.S. Usha et al., 'Military Reconnaissance Robot,' *International Journal of Advanced Engineering Research and Science*, vol. 4, no. 2 (2017), pp. 49–55.

37 R. Arun, P. Gokulsrinath, A. Ravi, A. Surendran, and A. Koliraj, 'Future Soldiers for Future Indian Army: By Artificial Intelligence and Technology,' *International Journal of Innovative Research in Science, Engineering and Technology*, vol. 5, no. 5 (2016), p. 6959.

38 P.K. Pal et al., 'A Mobile Robot that Removed and Disposed Ammunition Boxes,' *Current Science*, vol. 92, no. 12 (2007), pp. 1673–77.

39 T. Kaur and D. Kumar, 'Design of Cell Phone Operated Multipurpose Security Robot for Military Applications Using Solar Panel,' *International Journal of Scientific Engineering and Technology Research*, vol. 3, no. 16 (2014), pp. 3472–5.

40 A. Pant, *Aerial Drones in Future Wars: A Conceptual Perspective* (New Delhi: KW Publishers in association with Manohar Parikkar Institute for Defence Studies and Analyses, 2020), pp. 7–10.

41 A. Shrivastava, 'Mass Attack by Drones: Facing the Challenge,' *AIR POWER Journal*, vol. 13, no. 2 (2018), pp. 56–7, 63–4.

42 K. Taneja, 'Asymmetric Warfare, Technology and Non-State Actors: What India Must Do?,' in Manoj Joshi and Pushan Das (eds.), *The Future of War in South Asia: Innovation, Technology and Organisation* (New Delhi: Observer Research Foundation, 2021), pp. 63–70.

43 Ahyousha Khan, 'Advancement of UAV and the Future of Warfare in South Asia,' *SVI Foresight*, vol. 8, no. 1 (2022), pp. 15–16, Strategic Vision Institute Islamabad, available at: thesvi.org/wp-content/uploads/2022/02/SVI-Foresight-Vol-8-No-1-Jan 2022, accessed on 21 June 2022.

44 Quoted from Brigadier K. Bhardwaj, 'Emergence of Drone Warfare and Implications for India,' *Journal of the United Service Institution of India*, vol. CLI, no. 625 (2021), p. 351.

45 R.K. Narang, 'India's Drone Regulations-1.0: Progress, Policy Gaps and Future Trajectory,' *AIR POWER Journal*, vol. 13, no. 4 (2018), p. 127; A. Sud, 'UAVs and Counter UAVs Technologies in the World and the Indigenous Capability,' *CLAWS Journal* (Summer 2020), pp. 82–3.

46 S.P. Cohen, and S. Dasgupta, *Arming Without Aiming: India's Military Modernization* (New Delhi: Viking, 2010), pp. 32–3.

47 Krishnan, *Killer Robots*, p. 13.

48 M.V. Creveld, *Command in War* (Cambridge: Harvard University Press, 1985), p. 1.

49 See for instance, David Webb, 'Missile Defence-The First Steps Towards War in Space,' in Edward Halpin, Philippa Trevorrow, David Webb, and Steve Wright (eds.), *Cyberwar, Netwar and the Revolution in Military Affairs* (Hampshire: Palgrave Macmillan, 2006), pp. 83–5.

50 S.I. Davis, 'Artificial Intelligence at the Operational Level of War,' *Defense & Security Analysis*, vol. 38, no. 1 (2022), pp. 74–90.

51 Quoted from M. Chowdhury, 'Emerging Dynamics of Warfare—Role of Artificial Intelligence (AI) and Robotics and How India can Exploit It,' *Journal of the United Service Institution of India*, vol. CLI, no. 623 (2021), p. 13.

52 Quoted from A. Gupta, '"Internet of Things": A New Paradigm for Military Actions,' *AIR POWER Journal*, vol. 10, no. 2 (2015), p. 45.

53 D. Dash, *Autonomy and Artificial Intelligence: The Future Ingredient of Area Denial Strategy in Land Warfare* (New Delhi: KW Publishers in association with CLAWS, 2018), pp. 24–7.

54 *Doctrine of the Indian Air Force, IAP 2000-22* (New Delhi: Indian Air Force Air Headquarters, 2022), pp. 8, 24, 27–8, 65, 74–6.

55 R.K. Narang, 'Technological and Regulatory Challenges in Integration of UAVs in Non-Segregated Air Space,' *AIR POWER Journal*, vol. 11, no. 3 (2016), p. 117.

56 P.K. Mehra, 'The Indian Air Force of Tomorrow: Challenges,' in Basrur et al. (eds.), *India's Military Modernization*, p. 70.

57 K. Bommakanti, 'AI in the Indian Armed Services: An Assessment,' *CLAWS Journal*, vol. 14, no. 1 (2021), p. 108.

58 D.K. Gupta, *Military Potential of Swarm Intelligence*, Issue Brief, no. 139 (New Delhi: Centre for Land Warfare Studies, 2018), pp. 1–5.

59 Quoted from T. Mukherjee, *Securing the Maritime Commons: The Role of Artificial Intelligence in Naval Operations*, no. 159 (New Delhi: Observer Research Foundation, 2018), p. 18.

60 Quoted from R.H. Latiff, *Future War: Preparing for the New Global Battlefield* (New York: Alfred A. Knopf, 2017), p. 9.

61 J. Galliott, *Military Robots: Mapping the Moral Landscape* (Surrey: Ashgate, 2015), p. 109.

62 P. Scharre, *Army of None: Autonomous Weapons and the Future of War* (2018, reprint, New York: W.W. Norton & Co., 2019), p. 350.

63 S.U.M. Chary and B.P.C. Sekhar, 'Design and Optimization of Drone Jammer Antenna using Structural Analysis,' *Journal of Interdisciplinary Cycle Research*, vol. 12, no. 11 (2020), pp. 425–30.

64 P.J. Springer, *Outsourcing War to Machines: The Military Robotics Revolution* (Santa Barbara: Praeger, 2018), p. 218.

65 Wing Commander U.C. Jha, *Killer Robots: Lethal Autonomous Weapon Systems Legal, Ethical and Moral Challenges* (New Delhi: Vij Books Pvt. Ltd., 2016), p. 188.

66 Lieutenant-General R.S. Panwar, 'Artificial Intelligence in Military Operations: An Overview-Part II,' *Future Wars*, 29 September 2017, p. 6, available at: https://futurewars.rspanwar.net/, accessed on 2 December 2022.

67 Krishnan, *Killer Robots*, 164–5.

68 A. Lele, 'Missile and Rocket-Defence Weapon System,' In Expert Meeting, Autonomous Weapon Systems: Implications of Increasing Autonomy in the Critical Functions of Weapons, Versoix, Switzerland, 15–16 March 2016, p. 31, available at: icrcndresourcecentre.org/wp-content/uploads/2017/11/4283-0002, accessed on 12 November 2022.

69 A. Lele, *Strategic Technologies for the Military: Breaking New Frontiers* (New Delhi: SAGE, 2009), p. 67.

70 N. Hynek and A. Solovyeva, *Militarizing Artificial Intelligence: Theory, Technology, and Regulation* (London and New York: Routledge, 2022), p. 176.

71 B.A. Abbas, 'Lethal Autonomous Weapons Systems under Existing Norms of International Humanitarian Law,' *Journal of Defence Studies*, vol. 14, no. 3 (2020), pp. 51, 55.

72 M.A.C. Ekelhof, 'Lifting the Fog of Targeting: "Autonomous Weapons" and Human Control through the Lens of Military Targeting,' *Naval War College Review*, vol. 71, no. 3 (2018), pp. 61–87.

73 Quoted from N. Gupta, 'Artificial Intelligence: Emerging Opportunities and Associated Challenges,' *AIR POWER Journal*, vol. 15, no. 4 (2020), p. 23.

74 H. Cappelen, and J. Dever, *Making AI Intelligible: Philosophical Foundations* (Oxford: Oxford University Press, 2021).

75 R. Arkin, *Governing Lethal Behaviour in Autonomous Robots* (London: CRC Press, 2009), p. 38.

76 V. Chadha, 'India's Counterinsurgency Campaign in Mizoram,' in Ganguly and Fidler (eds.), *India and Counterinsurgency*, pp. 37–8.

77 W.C. Ladwig III, 'Insights from the Northeast: Counterinsurgency in Nagaland and Mizoram,' in Ganguly and Fidler (eds.), *India and Counterinsurgency*, pp. 50–1; *Doctrine for Sub Conventional Operations*, Integrated Headquarters of Ministry of Defence (Army). Shimla, Headquarters Army Training Command, 2006, pp. 15–16, available at: https://indianstratgeicknowledgeonline.com, accessed on 26 November 2022.

78 D. Kumar, 'The Officer Crisis in the Indian Military,' *South Asia Journal*, vol. 33, no. 3 (2010), pp. 442–67.

79 Colonel R. Prabhu, 'Loitering Munitions: Bridging Sensor to Shooter Voids in Artillery Fires by Precision,' *CLAWS Journal* (Winter 2020), pp. 190–200.

11

ARTIFICIAL INTELLIGENCE AND THE FUTURE OF ARMED CONFLICTS

Contextualising Africa in the global matrix

Gabriel Udoh

Introduction

AI is one of the fastest-growing fields in modern times. Just as its use has been extended into fields like agriculture, health, education, Earth observation, marketing and several others, the importance of AI has spread rapidly into the military sector. The use of AI has far-reaching effects in the planning and execution of armed conflicts. These effects could cause severe breaches in the laws and ethics of armed conflicts and, for instance, promote the unnecessary loss of lives and property. While IHL requires that civilians and those not actively participating in armed conflicts, civilian properties, and their source of livelihood be spared from attacks, AI-generated Earth observation imagery can help identify dense civilian-populated areas, homes, vegetation, property, and other sources of livelihood, like water bodies. Unethical uses can also trigger unnecessary loss of lives and property.

Amidst concerns from different actors, military planning in the near future may depend wholly on the AI-embedded autonomous weapons capacity of states. Also, Africa seems to be either backward or totally unperturbed by the developments in the military use of AI, especially in areas like using satellite imagery to coordinate strikes. There are major concerns regarding the ethics and legality of the use of these systems. The aim of this chapter is to analyse general developments in the use of AI for military planning, autonomous weapon systems, the extent of involvement by other actors in comparison to Africa, and ethical and political reasons that presuppose the slow pace of military AI developments in Nigeria, the 'hub' of Africa. It considers how far automated and autonomous systems have helped in military

DOI: 10.4324/9781003421849-12

circles, and how much more can be achieved in the war against terrorism and internal insurrections.

The evolution of technology has taken a rather fast pace in the last century. This speedy growth has been witnessed by almost all sectors involved in human development: Agriculture, education, health, Earth observation and several others. The introduction of AI in military and defence circles has raised, as expected, questions as to how far humanity is willing to give up certain controls without jeopardising its own safety.[1] There is also the opinion that AI and the use of automatically generated data and machinery can enhance precision and save civilian lives during attacks. In what has become very competitive, governments have settled strategies towards effectively engaging AI and higher levels of autonomy in different sectors. The USA rolled out its policy through the Department of Defense in 2012.[2] China had, in 2017, laid out its strategy to lead the world in AI by 2030.[3]

In military planning, the importance of involving AI and autonomous technologies cannot be over-emphasised. World powers like the USA and Russia have made this quite clear. The Russian President, Vladimir Putin, stated that 'the one who becomes the leader in this sphere will be the ruler of the world.'[4] The Department-of-Defense-powered US National Defense Strategy indicated the plan to utilise AI as a major tool with which it will ensure 'the US will be able to fight and win the wars of the future.' To this end, several states have already included AI and data in military strategy, planning and executing attacks during armed conflicts. There are automated reconnaissance and surveillance systems, cyber tools, logistics and military vehicles (including Unmanned Aerial Vehicles, Unmanned Ground Vehicles and Unmanned Marine Vehicles), defence systems, ordnance disposal systems, offensive systems and even lethal autonomous weapon systems. G. Reitmeier lists weapons already on their way to autonomy including the Israeli HARPY drone, the KUB-BLA Kamikaze drone of Russia, the Chinese Blowfish AB Helicopter drone, etc.[5] As has been earlier stated, the use of AI in military strategy goes way beyond weapon systems. AI-generated imagery from satellites and unmanned surveillance vehicles has been used to plan and execute accurate attacks.[6]

Proponents of AI-powered systems and data in armed conflicts have argued that AI could increase effectiveness and precision in strikes, given the danger of unplanned civilian casualties. Others have raised ethical issues including the 'inhumanity' of allowing machines to make decisions concerning human life. Ongoing discussions by the group of Governmental Experts of the Committee on Conventional Weapons spanning from 2014 have received petitions from NGOs and position papers from governments of independent states as to whether or not a treaty is necessary towards placing a ban on the use of AI in weapon systems.[7] This chapter notes that the argument could also be extended to the unethical use of AI-generated data (as in the case of Earth observation imagery) to plan military strikes.

These concerns notwithstanding, military strategists have consistently developed plans that involve an active improvement and involvement of AI in military planning and systems as vital elements in armed conflicts. As Paul Scharre rightly puts it:

> Militaries could desire greater autonomy to take advantage of computers' superior speed or so that robots can continue engagements when their communications to human controllers are jammed. Or militaries might build autonomous weapons simply because of a fear that others might do so.[8]

The fear of competition is not unfounded. There have been reports of these developments involving the use of weapons outrightly banned by governments and organisations that are considered to have little respect for IHL and ethics.[9]

Documentation indicates that the use of AI in military strategy and planning has taken centre stage in countries like the USA, Russia, China and India.[10] This follows years of development and inclusion of AI in weapon systems, reconnaissance, logistics, etc. To further the purpose of IHL regarding the distinction of civilian and military objectives in times of armed conflicts, there are investments in satellite imagery, ranging from large satellites from NASA to small military-grade satellite systems to help in identifying places and objects that should be excluded from strikes. The downside to this has also been considered: The possibility that unethical and illegal uses of AI-generated Earth observation data can lead to the loss of civilian lives and property.

Considering the fast pace of these developments in other states and continents, Africa appears to be falling behind. With the increase in security threats against the sovereignty of some African states and other internal insurrections have become a major source of concern. Many intrastate conflicts are occurring in various parts of Africa. There are ethical and political explanations for this seeming slow growth of AI use in military planning. Some of these include limitations imposed by a deep cultural stigma and concerns about insecurity and potential threats to privacy posed by domestic drone surveillance, corruption and several others.[11] However, there are developments in Africa towards eliminating the growing rate of conflicts and conflict zones within the continent. These include increases in defence budgets, the introduction of small military satellites into orbit, and the purchase and building of automated and autonomous military systems.[12]

This chapter poses the following research questions: One, what are the current major developments in the use of AI in military systems in Africa and the rest of the world? Second, how far can the various uses of AI in military planning and strategy enhance the safety of civilian lives and property during armed conflicts? Third, what are the major concerns about the use of AI in military planning and strategy, especially in Africa (specifically Nigeria)? Lastly, how can Africa (specifically Nigeria) benefit from the advantages of AI in armed conflicts, given that it plays host to several conflict zones?

This interdisciplinary research involves qualitative methods. Basically, this chapter will be a literature review of existing research on the topic. This will involve searching academic databases and other sources for relevant studies and articles, and analysing the findings and arguments presented in these works. In addition to this, the chapter will also incorporate expert opinions and insights from individuals working in the field of AI and military strategy. These sources will be identified through a systematic search of libraries, online platforms and professional networks.

This research will investigate developments in the use of AI in armed conflict, military planning and strategy. The ethical and legal implications of these will also be discussed, in line with provisions of the IHL and international law generally. Finally, this research will highlight these developments in Africa (with particular attention to Nigeria), considering the continent's conflict situation, and how far the inclusion of AI in military planning can help in the prevention of unnecessary deaths and suffering in its fight against insecurity.

AI, autonomy and autonomous systems

This chapter seeks to clarify the meaning of certain interdisciplinary concepts like AI and autonomy, as used in the context. According to M. Coeckelbergh, 'the hype surrounding AI has given rise to all kinds of speculations about the future of AI and indeed, of what it is to be human.'[13] AI is (arguably) intelligence demonstrated by non-natural entities, or machines. B. Raphael defines AI as an idea and the process: 'of making machines do things that would require human intelligence if done by man.'[14] It would require human intelligence to drive a car from point A to B without hitting the sidewalks between both locations. As such, any mechanical system that has the capacity to carry out this task possesses some level of autonomy, and can be described as operating on AI. This simplified description of AI is built on the presumption that 'intelligence' is focused on other human-like capabilities like emotions, consciousness, adaptability, etc. According to M.U. Scherer, since 'humans are the only entities that are universally recognised (at least among humans) as possessing intelligence, it is hardly surprising that definitions of intelligence tend to be tied to human characteristics.'[15] Other researchers have attributed AI to the pre-instruction of systems with codes.[16] This argument seems to suggest the constant involvement of human elements in these activities, albeit indirectly. One very striking definition of AI is found in Section 238 of the National Defense Authorization Act. This section considers AI as 'any artificial system that performs tasks under varying and unpredictable circumstances without significant human oversight, or that can learn from experience, and improve performance when exposed to data sets.' This idea poses a few vital points worthy of note; unpredictability, the limitation of significant human oversight, a capacity to learn from experiences, and the introduction of datasets, which exposes some ethical challenges and opportunities not found in many other definitions.

Autonomy denotes independence. It must be noted here that there are diverse perceptions of autonomy given its context in different disciplines. R. Crootof suggests that due to the absence of a universal definition, negotiating a new treaty to regulate or ban autonomous weapon systems often ends up unproductive.[17] Paul Scharre defines autonomy in line with the minor differences between autonomous and automated systems. Here, while the internal cognitive structures of an automated system can always be back-traced to a human in the loop, autonomous systems are 'goal-oriented,' 'self-directed' systems, 'sophisticated enough that their internal cognitive processes are less intelligible to the user, who understands the task the system is supposed to perform but not necessarily how the system will perform that task.'[18] The Proba-1 satellite system, though commanded essentially from the ESA's ground station in Redu, Belgium, has onboard autonomy and ground-segment automation systems that enable it to control satellite passes 'with no operator attendance.'[19] The US Defense Science Board defines autonomy as 'a capability (or set of capabilities) that enables a particular action of a system to be automatic or, within programed boundaries, "self-governing".'[20] It can be concluded from these descriptions that while autonomy connotes independence, its use in machine systems still indicates a minimal amount of human interference; a lower level of interaction than what is available in automated and automatic systems.

Autonomous systems are 'systems that are able to make decisions based on a set of rules and/or limitations. It is able to determine what information is important in making a decision.'[21] This chapter admits that there is an undeniable human interaction in the functionality of autonomous systems, notwithstanding its complexity, albeit in its initial programming stages. To trace responsibility for actions and inactions of autonomous systems, faults in autonomously generated data, etc., this ideology becomes useful.

Military uses of AI

AI has military potential. This is evidenced in the multiplicity of its use and functionalities, including the capability to carry out complex financial transactions, flag potential terrorists using facial recognition, identify activities and environmental patterns through satellite imagery and remote sensing, and even perform document reviews.[22] This makes it attractive also for military purposes. Autonomy and automation stand out as one of the most significant trends that affects military operations today. Unmanned systems are used in military supply logistics and transport, health diagnostic technologies for armed personnel in the field, data management systems, surveillance systems including both drone and Earth observation satellite surveillance systems, unmanned military vehicles (Unmanned Aerial Vehicles, Unmanned Ground Vehicles, Unmanned Maritime Systems, Unmanned Ground Systems and Vehicle Automation Systems), swarm technologies, human augmentation systems, logistic information systems, power and energy management systems,[23] with enabling technologies such as navigation, mission sensing, communications and piloting, machine intelligence for planning, learning and data analysis, mobile manipulation, energy storage and management, propulsion/mobility mechanisms, human-machine interfaces and multi-agent coordination.[24] This research is concerned more about the uses of AI in LAWS and in the collation and use of Earth observation data for military planning and strategy.

Autonomous weapon systems

What is considered as autonomous weapon systems (AWS) today had, until recently, only existed in science fiction and is already in use by several countries including Israel, Russia, the USA and South Korea.[25] While there is no universally accepted definition of AWS, Crootof describes it as weapon systems that 'based on conclusions derived from gathered information and pre-programed constraints, is capable of independently selecting and engaging targets.'[26] In this view, it is not important whether other functionalities of the system are manually controlled by humans, as long as it maintains a high level of independence in its 'kill-cycle': target selection and engagement. The US Department of Defense (DoD) defines autonomous weapons systems as

> any system that, once activated, can select and engage targets without further intervention by a human operator. This includes human supervised AWS that are designed to allow human operators to override operations of the weapons system but can select and engage targets without further human input after activation.[27]

This definition, again, indicates the presence of a human in the operational loop of AWS, as opposed to the perception of complete autonomy. Reitmeier[28] distinguishes autonomy in weapons systems into three distinct levels:

1. A person 'in the loop' if they retain a high degree of oversight and control over the system (automatic systems).
2. A person 'on the loop' if the system operates autonomously, but the person continues to monitor and control the process (semi-autonomous systems).
3. Finally, a person 'off the loop' if the system operates completely autonomously without any human intervention (fully autonomous system).

The third group of AWS raises the argument of its very existence. Do *"fully"* autonomous systems exist? Researchers have doubted the capabilities of machines to be 'fully' autonomous since, as indicated in all the descriptions, there is always a human somewhere in the loop. The definition as to where the loop actually begins and ends is yet unclear. Where the loop is only limited to certain functionalities (like the kill-cycle), then systems like the Israeli HARPY fall within this category.[29] This chapter identifies the existence of defensive and offensive military systems that have incorporated high levels of autonomy. This ranges from autonomous combat drones such as the British Taurus Stealth combat drone, the Chinese Blowfish helicopter drone, the Israeli HARPY drone, the Russian T-14 Armata battle tank, the Israeli Iron Dome Missile Defence System, the American Sea-Hunter and several others. Generally, AI in military systems has been heralded as a major enabler of new generations of weaponry. These developments will likely determine policies, sizes of defence budgets, and even the nature of future armed conflicts.[30]

AI-generated Earth observation data

Discussions surrounding the use of AI in Earth observation and surveillance for military strategy mostly arise within the context of distinction in strikes. IHL requires that a clear distinction be made during attacks; civilians, civilian objectives, and even persons hors de combat should not be targeted. It is difficult to properly identify areas with potential high civilian population density, except with proper surveillance data, be it from UAVs or Earth observation satellite systems. The National Air and Space Intelligence Centre has warned that foreign entities 'are integrating space and counterspace technologies into warfighting strategies.'[31] The Gulf Conflict opened the doors for military space capabilities. The US INMARSAT was used to simultaneously transmit still electronic images from soldiers in the Persian Gulf directly to the Pentagon to ease the dynamic planning and strategising of the war. R. Morgan called this a vivid illustration of the 'significant and pivotal role played by military and civilian satellite communications systems in international crises.'[32] Today, small and less expensive military satellites are built in constellations. NASIC has reported the existence of several satellites in outer space, 666 owned by 38 countries, being used for intelligence, surveillance, reconnaissance and remote sensing.[33] In June 2022, Germany's SARah-1 began a 10-year mission to provide reconnaissance imagery to the Bundeswehr.[34]

According to R. Merrifield:

> AI can make a satellite aware of its surroundings and decide autonomously when and how to carry out operational tasks, such as gathering images, analysing and processing them, and then selecting only the essential data for downloading to the Earth station.[35]

This could help in target identification, monitoring and tracking, for buildings, ships and even cars. As these capabilities continue to be enhanced, there are advantages for military strategists to plan towards defence in line with international law and IHL.

Current developments in AI-military uses

This research posits that a vast majority of the developments of AI in military technologies are in the realm of weapons systems, military surveillance, and planning. The evolution of automated and autonomous systems in armed conflicts dates back to the Gatling gun used in the American Civil War (1861-1865). Though this itself was not an autonomous weapon, it paved the way for the subsequent development and introduction of autonomy into weapons systems. For the era, a weapon that could fire 300 rounds per minute was an advantage against an enemy who needed to recoil and refire manually. Others followed, including the German G7e/T4 Falke ('Falcon') torpedo. Then came the precision-guided munitions (PGMs), mostly considered autonomous due to the inability of the human operator to recall them once launched. There are also loitering munitions and homing missiles which are autonomous in target identification and engagement.[36]

As Scharre predicted, 'the trend of creeping automation that began with the Gatling gun will continue.'[37] The incorporation and expansion of autonomous capabilities has been extended into numerous military support systems. Artificially intelligent support systems that are automated at various levels are relevant for all dimensions of the operational areas (land, air, sea, and space, as well as the cyber and information space). One could speak of assisted perception and action in the increasingly complex techno sphere in which military operations are carried out.[38]

This rapid spread of AI technology into areas that were considered a human-exclusive preserve has been remarkable. Countries like Germany, the USA, China, the UK, Russia, France, India, and North Korea are ahead in the military AI race. The South Asia Monitor of 16 January 2021 reported that 'the Indian Army is investing heavily into artificial intelligence, autonomous weapons systems, quantum technologies, robotics, cloud computing, and algorithm warfare, in order to achieve convergence between the Army's warfighting philosophies and the military attributes of these technologies.'[39] This follows the prediction that India's unique security situation, insurgency, and the presence of two hostile neighbours will drive its need for autonomous weapons systems.[40]

The US, one of the biggest players in military AI development, has insisted that it would not support any calls for a ban on AWSs. As of 2011, the government had already built the X-47B; the first unmanned aircraft to perform an autonomous take-off and landing on an aircraft carrier. With a defence budget as high as 754 billion dollars (for the fiscal year 2021), the US reportedly spends more on weapons systems development than any other G7 country.[41] According to Scharre:

Many applications of military robotics and autonomy are non-controversial, such as uninhabited logistics convoys, tanker aircraft, or reconnaissance drones. Autonomy is also increasing in weapons systems, though with next generation missiles and combat aircraft pushing the boundaries of autonomy. A handful of experimental programs show how the US military is thinking about the role of autonomy in weapons. Collectively, they are laying the foundations for the military of the future.[42]

Already, the US boasts of military systems with autonomous capacities ranging from anti-ship and anti-land missiles like the Tomahawk anti-ship missile (TASM) and the Tomahawk land attack missile (TLAM), to the Long Range Anti-ship Missile (LRASM) which, according to its developers, employs precision routing guidance, a multi-modal sensor suite, weapon data link, and an enhanced digital anti-jam GPS to detect and destroy specific targets within a group of numerous ships at sea. DARPA is also building technologies that enable swarm drones to hunt in coordinated packs.[43]

The Chinese also lean on the assurance that autonomy and AI-enabled weapons systems promise, more speed, reach, precision, and lethality of future operations.[44] China has consistently worked towards becoming an AI superpower. This has created a pseudo-arms race between itself and other competitors like the US, Russia, and Israel. Currently, the People's Liberation Army boasts of the Blowfish A3 helicopter, which can autonomously perform more complex combat missions; including fixed-point timing, detection, fixed-range reconnaissance, and targeted precision strikes. These copters have already been marketed to governments in the Middle East, including Pakistan and Saudi Arabia.[45] A recent report by the Centre for Security and Emerging Technology has made several findings on China's broader long-term strategy and long-term planning, involving the use of AI in its defence planning thus:

> While we can only estimate a floor for Chinese military AI spending. It is likely that the PLA spends more than 1.6 billion dollars yearly on AI related equipment. Whereas some PLA officers have expressed serious reservations about developing lethal autonomous weapon systems (LAWS), laboratories affiliated to the Chinese military are actively pursuing AI-based target recognition and fire control research, which may be used in LAWS.[46]

The report also mentions that the 209 billion-dollar Chinese defence budget for 2021 is only second to that of the US, and has most of it apportioned for the acquisition of equipment,[47] most of which are believed to be in pursuit of the 'intelligentization' agenda of the Chinese PLA.[48]

Russia has proven to be one of the strongest opponents to the idea of a ban on LAWS.[49] The Kremlin currently owns the Orion UAV; a drone reported to be capable of engaging other drones in combat. The Uran 9 UAV has already been marketed publicly as well. The Russians had already indicated their intention to develop robotic swarms and other technologies that would pull their soldiers out of the front lines.[50] Turkey has built the STM Kargu for asymmetric warfare. This system can be used both in manual and autonomous modes. A UN Security Council Panel of Experts on Libya reported that the Kargu carried out the first autonomous drone attack in history.[51] Also, it has been suspected of several strikes during the 2020 Nagorno-Karabakh War,[52] even though Azerbaijani authorities were yet to verify this information at the time of this research.

Israel has consistently automated its weapon systems and has indicated its strong stance against an international regulation to ban or even regulate AWS. Its long war with Palestine has been supported by an automated border control sentry technology placed along the Gaza borders.[53] These automated machines loaded with 7.62 calibre machine guns are trained to identify human movements along its 1.5 km automated kill-zone and to open fire at unpermitted entries.[54] Israel's Iron Dome Missile Defence system is heralded as one of the most impressive in the world, with the capacity to autonomously defend Israeli territories against incoming missile attacks. China, South Korea, Turkey and Chile have all purchased and some have even reengineered their own version of the Israeli HARPY drone to a personalised variant.[55]

Several countries have shown similar capacities and intentions to develop future military strategies in line with emerging technologies, autonomous systems, and data from Earth observation systems. Also, non-state actors have automated certain attack systems, like UAVs and UGVs, in their artillery. In 2015, the Shitte militia in Iraq fielded an armed ground robotic system.[56] Future wars will be fought autonomously. This futuristic development has the potential to spare the lives of people, especially soldiers who get killed on war fronts.

The use of AI-generated Earth observation data has also shown promise in the planning and execution of armed attacks in future wars. Notably, there is a growing range of satellites, both commercial and military. M. Borowitz discusses the rapid increase of small satellites in Earth's orbit. The author opines that due to the minimisation of major satellite components, 'small satellites are now capable of providing high quality data and services across different mission areas including remote sensing; communications; position, navigation and timing; and on-orbit rendezvous and proximity operations.'[57] Of the reported 666 satellites for intelligence, surveillance, reconnaissance and remote sensing as of 2018, 353 belonged to entities and the government of the US, China had 122, Russia had 23 and the rest of the world had the remaining 168.[58] The US Army had already experimented with the capability of satellite systems to generate battlefield imagery, leveraging the use of both commercial and military satellites to rapidly transmit targeting information and help commanders make real-time decisions.[59] The Project Convergence which started in 2020 at Arizona, USA reported that 'combined with new capabilities on data fusion and artificial intelligence, the Army reported cutting down the sensor to shooter timeline from 20 minutes to 20 seconds.'[60] There are instances where the US defence has used geospatial intelligence to find and eliminate terrorists and their organisations around the world.[61]

Africa

Introducing AI into war planning and military strategies in the African countries poses several ethical questions and opportunities. In most instances, there is always the concern as to how it can help in saving more lives or in unnecessarily killing people and destroying property during armed conflicts. The major concerns are the capabilities of AI-controlled weaponry to distinguish between military objectives and civilian objectives; being a basic principle of IHL. Also, there is concern as to how satellite information can help avoid mistaken targets and unnecessary loss of lives and property, not forgetting how its unethical use can cause untold consequences and breach major rules of engagement in armed conflicts. There are sub-topics like the issues of proliferation of not-so-expensive autonomous systems, wrongful use of open-access satellite data, misidentification of targets, and unlawful targeting of prohibited areas, etc.

Many African countries do not have technological capabilities as advanced as their counterparts in other parts of the world. Also, the contribution of these countries to the discussions surrounding AWS remains low. However, the involvement of Africa in discussions related to AWS is very important. Thompson Chengeta believes that this is important, more so since some of the weapons and the regulations thereof will end up affecting African countries even if they did not participate in their development or formulation.[62] The need to boost up military capacities in the African region is enhanced by the alarming number of armed conflicts and terrorist activities within. Libya is still struggling with years of instability. South Sudan, Central African Republic, Ethiopia and Cameroon are conflict hotbeds to watch out for in 2022.[63] In 2023, some of these conflicts took more brutal dimensions in some places like Sudan and Ethiopia, while some others - Niger and Guinea Bissau - witnessed military coups. Nigeria and some of its neighbours have been battling with Boko Haram and Islamic State in the West African Province (ISWAP) insurgency for about two decades. Al-Shabaab, Somali pirates and connections of Al-Qaeda are valid threats in different parts of Africa.[64]

The defence industry's response has not matched the same fast pace as the activities of others. Nigeria and South Africa appear to be military leaders, with advances in autonomous technologies, though not comparable to those of other more powerful countries outside Africa. This chapter argues that though at a comparatively slower pace than their counterparts in the Global North, recent activities in Africa show advancements in upgrading its autonomous/automated security systems architecture, to measure up with the very intense and growing threats of terrorist activities. This has been necessitated, in some states, by the influx of Islamist militant groups and the establishment of new cells within the region.

The BBC reports that the Turkish-made Bayraktar TB2 drone has been ordered in large numbers and received by the governments of Togo, for combating *jihadist* threats from the neighbouring Burkina Faso.[65] Other states have also acquired the same drone. Niger Republic got a consignment in May 2022.[66] Ethiopia, Morocco and Tunisia have also bought the very useful drone to boost their military surveillance and air strike capabilities.[67] Burkina Faso has received the Bayraktar[68] while the Ghanaian Navy had, also in May 2022, received two Delta Squad surveillance drones to increase security patrols in maritime boundaries.[69] South Africa has also improved its drone fleet from the locally designed Indiza hand-launched border patrol and peacekeeping drone to more advanced systems.[70] As in 2021, the country's Paramount Advanced Technologies (PAT) exhibited the N-Raven; a long-range UAV with the ability to carry out precision strikes on targets both static and those in motion, and capable of flying for two hours at 180 km/h.[71] According to the report, the N-Raven 'addresses a myriad of mission requirements, including "future warfighter engagements" where intelligent swarm technologies combined with multiple munition loitering and attack operations have been proven to ensure mission survivability.'[72]

This research is centred around Nigeria, given its relatively strong military capacities and economy within the continent.[73] Nigeria struggles with several security threats, ranging from petty crimes, through kidnappings and assassinations, to serious terrorist activities, especially in the northern region. Nigeria's weapon systems are mostly imported from other countries like China, the USA and Russia.[74] Currently, the Defence Industries Corporation of Nigeria (DICON), established in 1964, serves as Nigeria's official weapons' developer.[75] DICON is still known to produce weapons as basic as ceremonial swords, hand grenades, sub-machine guns, the Nigerian rifle 1 model 7.62 mm and the OBJ 007 (a Nigerian version of the Russian AK-47). There have also been sightings of locally-made light patrol vehicles and Igirigi, an armoured vehicle manufactured by DICON.[76]

Commentators admit that Nigeria's military drone industry has received a major boost in the last decade. The Tsuigami, Nigeria's first UAV, was produced by Nigerian Airforce aero-engineers in collaboration with UAVision of Portugal, to carry out surveillance and reconnaissance, search and rescue, convoy protection, maritime and land military patrol, etc.[77]

Nigeria has also made some heavy investments in the areas of space science and technology. Nigeria has launched its NigeriaSat-2, NigeriaSat-X and NigComSat-1R satellites.[78] NigeriaSat-1 was built by Surrey Satellite Technology Limited (SSTL) of the United Kingdom. It was first launched into Low Earth Orbit (LEO) from Plesetsk, Moscow, on 27 September 2003, carried by the Kosmos Rocket and loaded with Disaster Monitoring Constellation (DMC) micro-satellites – UK DMC and BILSAT.[79] It must be noted that the aforementioned satellites were built for purposes such as disaster management and communication. NigComSat-1, Africa's first communication satellite, was launched in 2007 but suddenly failed in 2008 due to power-failure related faults and was replaced by the NigComSat-1R. There are no defence-specific satellite systems owned by the Nigerian government so far.

This research finds that there is a need to further enhance its automated military capacities. In 2018, the BBC reported that the Nigerian Army had confirmed the death of 23 of its men in Northern Nigeria, having been attacked by UAVs from Boko Haram fighters.[80] Other non-state armed groups like Hamas, Houthi, the Taliban, Al Qaeda and ISWAP had recorded drone surveillance and attacks as early as 2014, and some as far back as two decades ago.[81] These are all indications that the Nigerian military machine and planning still have lots of room for improvement. A policy brief by the Friedrich Ebert Stiftung Nigeria on Defence and Budget Spending reports that:

> Nigerians hold the view that our national security architecture is not
> as effective as it ought to be, coming on the heels of very tangible and

discernable realities: insurgency, pervasive militancy, kidnapping, piracy, herders/farmers' clashes, armed robbery, ritual killings, cultism, rape and huge numbers of internally displaced persons (IDPs).[82]

Aside from these numerous counts of insecurity, it is scary to think that some violent non-state actors have been benefitting from the use of weaponised UAVs for about two decades now.[83] The enemy is adjusting to major technological war-fighting changes. Nigeria's military planners need to be proactive.

It is argued that the major factor slowing down the inclusion of AI in Nigeria's military planning is the lack of funding. Military expenditure in Nigeria rose from $697 million to $469.6 billion in the decade between 2010 and 2020. Nigeria also serves as the largest contributor in military funding to ECOWAS.[84] First, there is the problem of corruption and lack of accountability. The Friedrich Ebert Stiftung Nigeria rightly reports that 'Nigeria is not getting value for the money it continues to invest in the defense and security sector.'[85] There are grey areas between approved funds for military expenditure and the actual spending of such funds; towards the purchase of military equipment and other purposes for which they were originally meant. A BBC report in 2015 indicates Nigerian soldiers decried the lack of adequate weaponry and fled from the Boko Haram terror group equipped with superior weapons.[86] Dismissed, subsequently, for mutiny, one of the ex-soldiers described that besides the lack of weaponry, their armoured vehicles were 'in such poor mechanical conditions ... and would not move.'[87] Others confirmed that the vehicles would often run out of fuel during exchanges with the terrorists.[88] According to Ibrahim, one of the dismissed soldiers, 'because of the corruption of the nation, it makes everything run down.'[89] Automating military systems as an antidote towards waging wars against insecurity and terrorism will require more accountability in the handling of funds for these purposes. Investments must be made and, as it were, corruption stands in its way.

Another politically motivated perspective on the slow pace of the development in Nigeria's military growth is the suspected involvement of some top government officials in the plot to enhance insecurity for certain political purposes and other motivations. A study conducted in 2021 includes statements by soldiers claiming that helicopters were used to supply food and weaponry to the terrorists' camps and how the military's tactical locations were leaked and overrun by the terrorists. According to the research, this prolongs the war against insecurity, thereby increasing budgetary allocations for defence and dropping into private pockets.[90]

The International Committee of the Red Cross and several other organisations, states and many authors, etc., have reaffirmed the need to implement a preemptive ban on autonomous weapons systems.[91] This position is mostly backed by the argument that these systems undermine the conceptualisation of humanity; both as a philosophy and as a foundational principle of IHL. Considering this, added to the fact that Nigeria appears to be uncomfortable with technologies that are not totally being controlled by human hands, this research finds that this 'fear' is a contributing factor to the slow pace of developments of technologies generally (including military technologies). According to the *Guardian* newspaper, writing on this 'fear' reports thus:

> An array of factors, from political repression to currency controls and rampant inflation, have fueled the stunning rise of crypto currencies in Nigeria. In February, the government took fright and banned cryptocurrency transactions through licensed banks. In late July, it announced a pilot scheme for a new government-controlled digital currency - hoping to reduce incentives for those wanting to use unregulated crypto.[92]

The headline of a newspaper, 'Out of Control and Rising: Why Bitcoin has Nigeria's Government in Panic,' summarises the point being made here. Automating systems can be viewed in its most elementary form as reducing human controls in those systems. The negative view has been extended to the use of weapon systems that are automated. Chengeta, discussing the notion of humanity in the use of AWS, encourages the actors to consider the value of human dignity in all discussions pertaining to autonomous weapons systems.[93]

There are several other factors mitigating against the advancement of Nigeria's military development generally, and the use of AI in military systems specifically. Lack of education and awareness, accountability problems, theft of equipment, which once again, serves as a pointer to the high spate of insecurity in the country, all fall in this group. To conclude this chapter, some solutions. The primary objective is to instil the understanding that the use of AI gives an added advantage towards the improvement of military systems and their optimisation would benefit Africa when facing different enemies in future conflicts.

Conclusion

This chapter set out to discuss why Nigeria needs to consider enhancing the pace of its incorporation of AI into all phases of its military planning. The main objectives of this research included discussing how far AI has been used in military planning and technologies both in Africa and the rest of the world, how advantageous AI is in military planning, and how important it is to involve AI when planning for future wars. This chapter also set out to discuss the major setbacks faced by Nigeria in the implementation of an AI-laced military plan and made suggestions on how these can be overcome in the near future. This final section of this chapter concludes, among other issues, that it is important that Africa considers a speedier implementation of AI in its military planning for future wars, especially given its exposure to numerous insurrections.

Although there is a fear that AI is capable of completely eliminating human control of weapons systems, one can also agree that one of the ethical high points of AI in armed conflicts is that it brings the advantage of replacing human soldiers with non-human systems, thereby reducing human casualties in armed conflicts. This has been, from inception, the primary aim of IHL and its rules of armed conflicts. In April 2021, it was reported that about 33 soldiers had been rounded up and killed by members of the Boko Haram terror group.[94] According to the report, having fought for several hours, the remaining soldiers called for back-up until a military jet came to the rescue.[95] The point being made here is that these lives could have been preserved if the location had been manned by autonomous systems, or surveilled by UAVs, controlled by soldiers in safer areas far away from the conflict zone. The assailants would have been confronted with autonomous systems or even fully weaponised robots. The responses would have been more efficient; lives would have been saved. In the worst case scenario, the machines would be destroyed, not human lives. In the same year, Flight Lieutenant Abayomi Dairo barely escaped with his life when his military jet was shot out of the sky by terrorists.[96] The use of UAVs for surveillance could forestall these situations and preserve the lives of human pilots. The capacity of certain autonomous robots and surveillance systems to access rough terrains and areas either too dangerous or impossible for human access would provide support in areas of military transportation and logistics, supplies and weaponry, search and rescue missions, reconnaissance and many others.

It is suggested here that automated weapons or AWS are capable of higher levels of precision, speed and efficiency than human soldiers; an attribute that will effectively reduce or eliminate the risk of civilian harm or casualty. The USA submitted in its 2018 report to the Group of Governmental Experts on lethal autonomous weapons systems - convened through the Conference on Conventional Weapons, that AI is useful in providing increased awareness to commanders on the presence of civilians and/or civilian objectives by automating data analysis.[97] It could also, generally speaking, give a foresight of the likely effects of different weapons, thereby helping reduce civilian casualties during strikes. There are indications that most times, especially during terror attacks, security forces find it difficult to strike without incurring unnecessary civilian damage since they get hurdled up with the target. This difficulty would be eased with the use of more specific weapons and targeting systems, supported by the speed and precision of autonomous weaponry.

Notes

1 B. Stauffer, 'Stopping Killer Robots: Country Positions on Banning Fully Autonomous Weapons and Retaining Human Control,' 10 August 2020, available at: https://www.hrw.org/report/2020/08/10/stopping-killer-robots/country -positions-banning-fully-autonomous-weapons-and-retaining-human-control accessed on 22 October 2022.
2 Department of Defense, 'Autonomy in Weapon Systems,' 8 May 2017, available at: https://www.esd.whs.mil/portals/54/documents/dd/issuances/dodd/300009p .pdf, accessed on 22 October 2022.
3 Marina Reshetnikova, 'Future China: AI Leader in 2030?' (2021), available at: https://www.researchgate.net/publication/357737833_Future_China_AI _Leader_in_2030, accessed on 23 October 2022.
4 CNBC, 'Putin: Leader in Artificial Intelligence Will "Rule the World",' 4 September 2017, available at: https://www.cnbc.com/2017/09/04/putin-leader -in-artificial-intelligence-will-rule-world.html, accessed on 23 October 2022.
5 G. Reitmeier, *Licence to Kill: Artificial Intelligence in Weapon Systems and New Challenges for Arms Control* (Potsdam: Friedrich Naumann Foundation for Freedom, 2020), p. 6.
6 V. Vilisov, 'Logging a War: How Digitalization has Changed the Perception of Modern Warfare and the Documentation of War Crimes,' 27 March, available at: https://theins.ru/en/society/251631, 23 October 2022.
7 The Group of Governmental Experts at the Committee on Conventional Weapons has been meeting frequently since 2014, in an attempt to regulate the use of autonomous weapons systems. This committee has received inputs from states and concerned organisations like the International Committee of the Red Cross (ICRC), Human Rights Watch (HRW), etc.
8 P. Scharre, *Army of None: Autonomous Weapons and the Future of War* (New York: W.W. Norton & Company, 2018), p. 7.
9 B. Duggan, 'Amnesty Says Sudan Used Deadly Chemical Weapons in Darfur Conflict,' *CNN*, 29 September 2016, available at: https://edition.cnn.com/2016 /09/29/africa/sudan-chemical-weapon-darfur/index.html, accessed on December 2022.
10 M. Horowitz et al., *Strategic Competition in an Era of Artificial Intelligence* (Washington, DC: Centre for a New American Security, 2018).
11 K. Sandvik, 'African Drone Stories,' *Behemoth: A Journal of Civilization*, vol. 8, no. 2 (2015), p. 74.
12 For instance, Defense News reports that Morocco has just concluded an air defense base and purchased military equipment to stock up the facility. Read the full article here https://www.defensenews.com/global/mideast-africa/2022 /01/04/satellite-images-show-morocco-has-built-an-air-defense-base-near-its -capital.
13 M. Coeckelbergh, *AI Ethics* (Cambridge: The MIT Press, 2020), p. 11.
14 B. Raphael, *The Thinking Computer* (San Francisco: W.H. Freeman, 1976).
15 M.U. Scherer, 'Regulating Artificial Intelligence Systems: Risks, Challenges, Competencies, and Strategies,' *Harvard Journal of Law & Technology*, vol. 29, no. 2 (2016), pp. 354–400; see especially p. 359.
16 Coeckelbergh, *AI Ethics*, p. 64.
17 R. Crootof, 'The Killer Robots are Here: Legal and Policy Implications,' *Cardozo Law Review*, vol. 36, (2015), pp. 1837, 1843.
18 Scharre, *Army of None*, p. 30.
19 S. Blair, 'Autonomy in Action: Ten Years of Proba-1,' *ESA Bulletin*, vol. 148 (2011), pp. 38–44.

20 Defense Science Board, DoD, 'The Role of Autonomy in DOD Systems, 2018 (Task Force Report),' p. 1, available at: https://irp.fas.org/agency/dod/dsb/autonomy.pdf.

21 Defense Department (DOD), *DOD Integrated Roadmap FY2011-2036* (Washington, DC: Government Printing Office, 2011), p. 43.

22 M.U. Scherer, 'Regulating Artificial Intelligence Systems: Risks, Challenges, Competencies, and Strategies,' p. 354.

23 K. Ivanova, G.E. Gallasch and J. Jordans, 'Automated and Autonomous Systems for Combat Service Support: Scoping Study and Technology Prioritisation, Land Division, Defence Science and Technology Group, DST-Group-TN-1573,' 2016, p. 1, available at: https://www.dst.defence.gov.au/sites/default/files/publications/documents/DST-Group-TN-1573.pdf.

24 Ivanova et al., 'Automated and Autonomous Systems for Combat Service Support,' p. 2.

25 Crootof, 'The Killer Robots are Here: Legal and Policy Implications,' p. 1840.

26 Crootof, 'The Killer Robots are Here: Legal and Policy Implications,' p. 1837.

27 US Department of Defense (DOD), Directive 3000.09, 8 May 2012, Autonomy in Weapon Systems (incorporating change 8 May 2017), available at: https://www.whs.mil/directives/issuances/dodd/10/8/2017, accessed on 13 July 2022.

28 Reitmier, *Licence to Kill*, p. 5.

29 Scharre, *Army of None*, p. 46.

30 'Congressional Research Service, Defense Primer: US Policy on Lethal Autonomous Weapon Systems,' 2020 available at: https://crsreports.congress.gov, accessed on 21 June 2022.

31 'National Air and Space Intelligent Centre, Competing in Space,' 2018, p. 1, available at: https://www.nasic.af.mil/Portals/19/documents/Space_Glossy_FINAL--15Jan_Single_Page.pdf?ver=2019-01-23-150035-697, accessed on 19 December 2022.

32 R. Morgan, 'Military Use of Commercial Communication Satellites: A New Look at the Outer Space Treaty and Peaceful Purposes,' *Journal of Air Law and Commerce*, vol. 60, no. 1 (1994), p. 60.

33 National Air and Space Intelligent Centre, 2018, p. 4.

34 SARah (Satellite-Based Radar Reconnaissance System), a Next-Gen Constellation of the German Bundeswehr, (2022), available at: https://www.eoportal.org/satellite-missions/sarah#spacecraft, accessed on 19 July 2022.

35 R. Merrifield and R. Merrifield, 'Fleets of Autonomous Satellites to Coordinate Tasks Among Themselves,' 27 March 2019. available at: https://ec.europa.eu/research-and-innovation/en/horizon-magazine/fleets-autonomous-satellites-coordinate-tasks-among-themselves, accessed on 17 July 2022.

36 Scharre, *Army of None*, p. 35.

37 Scharre, *Army of None*, p. 36.

38 J. Beyerer and P. Martini (eds.), 'Rise of Intelligent Systems in Military Weapon Systems Position Paper (2020),' available at: https://www.fraunhofer.de/content/dam/zv/de/forschungsthemen/schutz-sicherheit/rise-of-intelligent-systems-in-military-weapon-systems-position-paper-fraunhofer-vvs.pdf, accessed on 20 May 2022.

39 S. Singh, 'Indian Army Employing Autonomous Weapon Systems,' 16 January 2021, available at: https://www.southasiamonitor.org/india/indian-army-employing-autonomous-weapon-systems, accessed on 2 September 2022.

40 Shashank R. Reddy, 'India and the Challenge of Autonomous Weapons,' 2016, available at: https://carnegieendowment.org/files/CEIP_CP275_Reddy_final.pdf. Page 9, accessed on 3 September 2022.

41 'Budget Basics: National Defense,' 1 June 2022, available at: https://www.pgpf
.org/budget-basics/budget-explainer-national-defense, accessed on 10 October
2022.

42 Quoted from Scharre, *Army of None*, p. 59.

43 Scharre, *Army of None*, p. 61.

44 E. Kania, 'AI Weapons' in Chinese Military Innovation, Brookings Institution,
p. 2, 2020, available at: https://www.brookings.edu/wp-content/uploads/2020
/04/FP_20200427_ai_weapons_kania.pdf, accessed on 20 October 2022.

45 K. Grenier, 'DefenseOne: SECDEF - China Is Exporting Killer Robots to the
Mideast,' 11 February 2021, available at: https://www.nscai.gov/2019/11/05/
secdef-china-exporting-killer-robots-to-mideast, accessed on 20 October 2022.

46 R. Fedasuik et al., 'Harnessing Lightning: How the Chinese Military Is Adopting
Artificial Intelligence. CSET Report of October,' 2021, p. iv, available at: https://
cset.georgetown.edu/wp-content/uploads/CSET-Harnessed-Lightning.pdf.

47 R. Fedasuik et al., 'Harnessing Lightning,' p. 3.

48 M. Pomerleau, 'China Moves Toward New "Intelligentized" Approach to
Warfare, Says Pentagon,' 18 August 2022, available at: https://www.defense-
news.com/battlefield-tech/2020/09/01/china-moves-toward-new-intelligentized
-approach-to-warfare-says-pentagon/, access on 23 July 2022.

49 Russia's Commentary Submitted at the CCW Can Be Found at https://docu-
ments.unoda.org/wp-content/uploads/2020/09/Ru-Commentaries-on-GGE-on
-LAWS-guiding-principles1.pdf.

50 S. Bendett, 'Strength in Numbers: Russia and the Future of Drone Swarms,' 20
April 2021, available at https://mwi.usma.edu/strength-in-numbers-russia-and
-the-future-of-drone-swarms/, accessed on 22 July 2022.

51 H. Nasu, 'The Kargu-2 Autonomous Attack Drone: Legal & Ethical Dimensions,'
10 June 2021, available at: https://lieber.westpoint.edu/kargu-2-autonomous
-attack-drone-legal-ethical/, accessed on 2 October 2022.

52 'Mr. Reed Is a Published Author, & Mr. Rife Is a US Naval Institute Member and
Senior Historian at SJR Research,' 20 January, 2022. New Wrinkles to Drone
Warfare, available at: https://www.usni.org/magazines/proceedings/2022/janu-
ary/new-wrinkles-drone-warfare, accessed on 21 May 2022.

53 A. Bensaid, '"Israel's Autonomous "robo-snipers" and Suicide Drones Raise
Ethical Dilemma",' 9 April, 2021, available at: https://www.trtworld.com/
magazine/israel-s-autonomous-robo-snipers-and-suicide-drones-raise-ethical
-dilemma-44557, accessed on 13 September 2022.

54 Anon, 'The Israeli Arsenal Deployed Against Gaza During Operation Cast
Lead,' *Journal of Palestine Studies*, vol. 38, no. 3 (2009), pp. 175–91, https://doi
.org/10.1525/jps.2009.xxxviii.3.175.

55 Scharre, *Army of None*, p. 46.

56 Scharre, *Army of None*, p. 102.

57 M. Borowitz, 'The Military Use of Small Satellites in Orbit, European Space
Governance Initiative/Center for Security Studies,' 4 March 2022, p. 1, available
at: https://www.ifri.org/sites/default/files/atoms/files/m._borowitz_military_use
_small_satellites_in_orbit_03.2022.pdf.

58 'National Air and Space Intelligent Centre,' 2018, p. 4.

59 Borowitz, 'The Military Use of Small Satellites in Orbit,' p. 4.

60 N. Strout, 'Satellites Played a Starring Role at Project Convergence,' 17 August
2022,available at: https://www.defensenews.com/digital-show-dailies/ausa
/2020/10/12/us-army-uses-satellites-to-affect-the-state-of-the-battlefield/,
accessed on 20 October 2022.

61 A. Olaide, H. Onuoha, and M. Ibrahim, 'Optimization of Nigerian Satellites and
Geo-Spatial Intelligence on National Security,' *IOSR Journal of Environmental*

Science, Toxicology and Food Technology, vol. 7 (2013), pp. 14–20. See especially p. 20.

62 T. Chengeta, 'Dignity, Ubuntu, Humanity and Autonomous Weapon Systems (AWS) Debate: An African Perspective,' *Revista De Direito Internacional,* vol. 13, no. 2 (2016), pp. 460–502.

63 'ISSAfrica.org (2021),' African Conflicts to Watch in 2022, available at: https://issafrica.org/pscreport/psc-insights/african-conflicts-to-watch-in-2022, accessed on 21 August 2022.

64 'Dr Alex Vines OBE Managing Director, Terrorism in Africa,' 15 September 2021, available at: https://www.chathamhouse.org/2021/09/terrorism-africa, accessed on 21 August 2022.

65 'Turkey's Bayraktar TB2 Drone: Why African States Are Buying Them,' 25 August 2022. available at: https://www.bbc.com/news/world-africa-62485325, accessed on 20 January 2023.

66 'Turkey's Bayraktar TB2 Drone.'

67 'Turkey's Bayraktar TB2 Drone.'

68 P. Kenyette, 'Burkina Faso Buys 5 Bayraktar TB2 Drones from Turkey,' 10 September, 2022, available at: https://www.military.africa/2022/09/burkina -faso-buys-5-bayraktar-tb2-drones-from-turkey/, accessed on 20 January 2023.

69 L. Ekene, 'Ghana Navy Gains Drone Surveillance Capability, 19 January 2023, available at: https://www.military.africa/2023/01/ghana-navy-gains-drone-sur-veillance-capability/, accessed on 20 January 2023.

70 S. Lesedi, 'South African Army Seeks to Acquire Reconnaissance Drones,' 12 January, 2023, available at: https://www.military.africa/2023/01/south-afri-can-army-seeks-to-acquire-reconnaissance-drones/, accessed on 20 January 2023.

71 S. Lesedi, 'IDEX 2021: Paramount Unveils N-Raven Swarming UAV,' 23 February 2021, available at: https://www.military.africa/2021/02/idex-2021 -paramount-unveils-n-raven-swarming-uav/, accessed on 20 January 2023.

72 Lesedi, 'IDEX 2021.'

73 Scharre, *Army of None,* p. 102.

74 Military Weapons in Nigeria, 2016, available at: https://oec.world/en/profile/bilateral-product/military-weapons/reporter/nga?redirect=true, accessed on 20 August 2022.

75 Official information about DICON is available at its homepage https://dicon.gov .ng/.

76 A. Frahan, 'Nigerian Defense Budget 2022 Increased for Additional Equipment and Capabilities,' 5 January 2022, available at: https://www.armyrecognition .com/analysis_focus_army_defence_military_industry_army/nigerian_defense _budget_2022_increased_for_additional_equipment_and_capabilities.html, accessed on 20 July 2022.

77 Premium Times Newspaper (2018, February 16). Video: Nigeria Air Force drone Tsaigumi in military exercise, available at: https://www.premiumtimesng.com/news/more-news/258930-video-nigeria-air-force-drone-tsaigumi-military-exer-cise.html, accessed on 21 July 2022.

78 M. Iderawumi, 'Replacing Nigerian Satellite Gone Past Design Life: The Journey So Far,' 9 April 2021, available at: https://africanews.space/replacing-nigerian -satellite-gone-past-design-life-the-journey-so-far/, accessed on 20 July 2022.

79 A. Ganiy, Current Trends in Nigeria's Space Development Programme to Facilitate Geospatial Information (GI) Sharing and Implementation of the NGDI, 2008, pp. 4–7, available at https://2001-2009.state.gov/documents/organization /110820.pdf, accessed on 20 July 2022.

80 Boko Haram Dey Use Drones and Foreign Fighters to Attack US - Nigerian Army - BBC News Pidgin, 29 November 2018, available at: https://www.bbc .com/pidgin/tori-46382348, accessed on 20 July 2022.
81 H. Haugstvedt, 'A Flying Threat Coming to Sahel and East Africa? A Brief Review,' *Journal of Strategic Security*, vol. 14, no. 1 (2020), pp. 92–105. https:// www.jstor.org/stable/26999979.
82 O. Obaze et al., 'Defense Budget and Spending: Implications and Options for Post Covid-19 Nigeria, Friedrich Ebert Stiftung Nigeria Report,' 18 August, 2020, p. 2, available at: https://library.fes.de/pdf-files/bueros/nigeria/18220.pdf.
83 Haugstvedt, 'A Flying Threat Coming to Sahel and East Africa? A Brief Review,' p. 1.
84 Obaze, 'Defense Budget and Spending: Implications and Options for Post Covid-19 Nigeria,' p. 8.
85 Quoted from Obaze, 'Defense Budget and Spending: Implications and Options for Post Covid-19 Nigeria,' p. 4.
86 W. Ross, 'The Soldiers Without Enough Weapons to Fight Jihadists,' 22 January, 2015, available at: https://www.bbc.com/news/magazine-30930767, accessed on 12 June 2022.
87 Ross, 'The Soldiers Without Enough Weapons to Fight Jihadists.'
88 Ross, 'The Soldiers Without Enough Weapons to Fight Jihadists.'
89 Quoted from Ross, 'The Soldiers Without Enough Weapons to Fight Jihadists.'
90 T. Oriola, 'Nigerian Soldiers on the War Against Boko Haram,' *African Affairs*, vol. 120, no. 479 (2021), pp. 147–75. See especially p. 169.
91 ICRC Position on Autonomous Weapon Systems, May 2021, available at https:// www.icrc.org/en/download/file/166330/icrc_position_on_aws_and_back-ground_paper.pdf, accessed on 21 July 2022.
92 E. Akinwotu, 'Out of Control and Rising: Why Bitcoin Has Nigeria's Government in a Panic,' 31 July 2021, available at: https://www.theguardian.com/technology /2021/jul/31/out-of-control-and-rising-why-bitcoin-has-nigerias-government-in -a-panic, accessed on 21 August 2022.
93 Chengeta, 'Dignity, Ubuntu, Humanity and Autonomous Weapon Systems (AWS) Debate: An African Perspective,' p. 477.
94 Reuters, 'Sources: More Than 30 Nigerian Soldiers Killed in Militant Attack,' 26 April 2021, available at: https://www.voanews.com/a/africa_sources-more -30-nigerian-soldiers-killed-militant-attack/6205065.html, accessed on 21 June 2022.
95 'Sources: More Than 30 Nigerian Soldiers Killed in Militant Attack.'
96 'Dairo: An Officer's Exemplary Heroism,' n.d., available at: https://www.this-daylive.com/index.php/2021/07/25/dairo-an-officers-exemplary-heroism/, accessed on 21 June 2022.
97 United States of America, Human-Machine Interaction in the Development, Deployment and Use of Emerging Technologies in the Area of Lethal Autonomous Weapons Systems, 2018. A report submitted to the Group of Governmental Experts of the High Contracting Parties to the Convention on Prohibitions or Restrictions on the Use of Certain Conventional Weapons Which May Be Deemed to Be Excessively Injurious or to Have Indiscriminate Effects on 28 August, 2018 in Geneva, available at: https://reachingcriticalwill.org/images /documents/Disarmament-fora/ccw/2018/gge/documents/GGE.2-WP4.pdf, accessed on 23 January 2023.

INDEX

Printed in the United States
by Baker & Taylor Publisher Services